Technology and the Air Force
A Retrospective Assessment

Edited by

Jacob Neufeld, George M. Watson, Jr.,
and David Chenoweth

Air Force History and Museums Program
United States Air Force
Washington, D.C. 1997

Preface

The history of the United States Air Force is inextricably bound up in the history of aerospace technology. Major revolutions have influenced the evolution of Air Force capabilities and systems, most notably those of atomic weaponry, the turbojet revolution, supersonic flight, avionics, aerial refueling, spaceflight, precision weaponry, electronic flying controls, composite materials, and stealth. It is worthwhile to take a retrospective look at some of the aerospace challenges and opportunities the Air Force faced and how it took advantage — or failed to take advantage — of them.

With this in mind, the Air Force History and Museums Program organized a symposium on October 23 and 24, 1995, in which leading historians, technologists, and military decisionmakers met at Andrews Air Force Base, Maryland, to present case studies on a series of technological challenges, opportunities, and problems. This symposium, co-sponsored by the Air Force Historical Foundation, covered relevant technological histories ranging from the turbojet revolution of the 1930s to the stealth revolution of the 1990s. This volume presents the texts of the papers in the order they were given. Many people within the Air Force History and Museums Program helped put this program together. I especially wish to acknowledge the contributions of Jacob Neufeld of the Air Force History Support Office for his efforts in bringing both the symposium and this publication to fruition.

RICHARD P. HALLION
Air Force Historian

Contents

Contents

Illustrations and Photographs

Technology and the Air Force

Technology and the Air Force
A Retrospective Assessment

James O. Young is the historian at the Air Force Flight Test Center, Edwards AFB. He earned the B.A., M.A., and Ph.D. degrees in history from the University of Southern California. His dissertation, *Black Writers of the Thirties,* was published by Louisiana State University. Dr. Young has taught history and American Studies at several colleges. In addition to writing official histories, he has produced numerous film documentaries and video briefings, several of which have been broadcast on cable television, at the Air Force Museum, and are and are included in university curricula. Dr. Young has written many articles and book reviews and served as a technical advisor for film documentarians, most recently for the Smithsonian's "Frontiers of Flight" and cable television's "X Planes" series. His latest writings include *The USAF Test Pilot School, 19441989, Supersonic Symposium: The Men of Mach* 1, and three chapters in *The Hypersonic Revolution.* He is an honorary member of the Jet Pioneers' Association.

Riding England's Coattails:
The Army Air Forces and the Turbojet Revolution

James O. Young

In 1928, twenty-one-year-old Royal Air Force flight cadet Frank Whittle speculated that it would be possible to attain very high speeds — speeds in excess of 500 mph — if one could achieve stratospheric flight. He also perceived that the piston-engined, propeller-driven airplane would never do the job. To achieve the speed and altitude he envisioned, some alternative form of propulsion system uniquely suited to those conditions was essential. His deductions were prophetic.[1]

During the 1930s, the prop-driven, piston-engined airplane underwent a dramatic metamorphosis. Streamlined, all-metal, light-weight, monocoque fuselages, retractable landing gear, and a host of other airframe innovations reduced aircraft weight and drag to previously unimagined levels. And the engines? The Wright Brothers had powered their first airplane with an engine providing about 12 horsepower — or one horsepower per 15 pounds of engine weight. In the early years of World War II, engine designers would squeeze more than 2,000 horsepower out of the churning pistons of their ever more complex, turbosupercharged combat designs (by the end of the war, the Wasp Major would deliver up to 3,500 horsepower), and they had achieved a power-to-weight ratio of better than 1:1. To fully exploit this power, there had been major improvements in fuels and propeller design as well. During the 1930s, for example, the Army Air Corps adopted 100-octane fuel, and prop designers had developed aerodynamically efficient, variable-pitch propellers which could be adjusted, in flight, for optimum performance at different speeds and altitudes.[2]

In their quest for ever greater speeds during the 1930s, designers came up with aircraft that appeared to be little more than engines with empennage and wings. Indeed, the world speed record leaped upward throughout the decade following Whittle's original speculations. Perhaps no aircraft better epitomized this trend than Willie Messerschmitt's Me 209V–1 that, in April 1939, pushed the record to 469.22 mph. (Although unofficially surpassed during the coming war, this mark remained the official record for the next three decades). The Me 209 defined the practical limits of prop-driven aircraft. Its engine, the 12-cylinder, liquid-cooled Daimler-Benz DB 601ARJ, provided 1,800 horsepower — and could be boosted up to 2,300 horsepower for short bursts — but it had a service life of only 30 minutes.[3] And, like many of its kind, the Me 209 was extremely difficult to fly; its pilot, Fritz Wendel, later recalling that it "was a brute. Its flying characteristics still make me shudder. . . . In retrospect, I am inclined to think that its main fuel was a highly volatile mixture of sweat from my brow and the goose pimples from the back of my neck!"[4]

Aeroengine pioneer Ernest Simpson once described the reciprocating engine as "an invention of the devil." Although marvelous examples of mechanical ingenuity and precision engineering, they were infernally complicated and temperamental. Maintenance was "difficult, frequent, and often painful." Added to this was the fact that, by the late 1930s, designers found themselves caught in a vicious circle. Higher speeds required ever-larger engines, which consumed greater amounts of fuel and resulted in larger and heavier airframes, whose size and weight served to negate the increased performance of the engines. And the engines, whether air- or liquid-cooled, posed monumental problems. In air-cooled engines, for example, the peak power output of an individual cylinder was something less than 175 horsepower, and thus, to boost power, designers were forced to add more and more pistons to a single crankshaft. The ever-increasing mechanical complexity of such linkages became an engineering and maintenance nightmare. Moreover, each additional row of cylinders had a detrimental impact on thermal efficiency. Instead of converting the engine's heat into useful mechanical work (i.e., power to drive the propeller), much of it — along with the airplane's aerodynamic efficiency, as well — had to be wasted in the cooling of these behemoths. Propellers also created seemingly insurmountable problems. As their blade tips approached supersonic speeds, for example, they encountered "compressibility burble" — shock waves that caused an unacceptable increase in drag — and, as the air thinned out with increasing altitude, props lost their "bite."[5]

The field of aeronautics was approaching a crossroads by the mid-1930s. Aerodynamicists, who had made such great strides since the mid-1920s, were pointing in a new direction. Indeed, at the Fifth Volta Congress of High Speed Flight, which met at Campidoglio, Italy, in 1935, the world's leading aerodynamicists began to seriously consider the theoretical possibility of flight beyond the speed of sound.[6] It was readily apparent to those assembled that the piston engine-prop combination could never meet that challenge. It was also becoming apparent to many that, in the not too distant future, the reciprocating engine would reach a plateau beyond which only minutely small improvements in performance could be expected in return for enormous expenditures in terms of time, money, and engineering effort.[7]

Though he certainly had not considered the possibility of supersonic flight, Frank Whittle had forecast many of these developments in 1928; and while undergoing flight instructor's training the following year, he saw the solution, not in any refinements to the existing technology, but in a radically new

4

The second reconstructed version of Whittle's bench test engine in 1938.

approach. He had already rejected rocket propulsion and a gas turbine-driven prop as impractical. Next, he had examined the possibility of a ducted-fan system — a jet propulsion system in which a conventional piston engine powered a low-pressure blower. The blower and engine would both be located in the duct and fuel would be burned in the flow stream aft of the engine to generate thrust. He had concluded, however, that this system would be far too heavy and would, in fact, offer no real advantage over the piston engine-prop combination.[8] Then, in late 1929, as he later recalled, "the penny dropped":

> It suddenly occurred to me to substitute a turbine for the piston engine [in the ducted fan system]. This change meant that the compressor would have to have a much higher pressure ratio than the one I had visualized for the piston-engined scheme. In short, I was back to the gas turbine, but this time of a type which produced a propelling jet instead of driving a propeller. Once the idea had taken shape, it seemed rather odd that I had taken so long to arrive at a concept which had become very obvious and of extraordinary simplicity.[9]

Thus, after less than two years of self-directed study and speculation, he had deduced that, for very high speeds and altitudes, employing a gas turbine to produce jet propulsion was the most feasible and, ultimately, obvious answer. As originally conceived in his patent application of 1930, air entered the engine inlet and was initially compressed by a 2-stage axial compressor and then further compressed by a single-stage, one-sided centrifugal compressor; after passing through a diffuser which transformed its kinetic energy into pressure, the highly compressed air entered a ring of combustors into which fuel was injected and then ignited; the hot, expanding gases were then expelled at high velocity through a two-stage axial-flow turbine, which drove the compressor stages by means of a shaft, and then exited through a ring of nozzles to produce forward thrust. With all of its moving parts on a single rotating shaft, Whittle believed, it would be much simpler and far lighter than piston engines.[10]

Like so many revolutionary breakthroughs, Whittle's idea was elegant in its simplicity, and like so many such ideas, it was scorned by the "experts" as impractical. He had not been the first to speculate about the possibility of employing a gas turbine for aircraft propulsion. The idea had been studied throughout the 1920s, though usually in the context of employing a turbine to drive a propeller. Based on the generally negative findings of these studies, conventional wisdom scoffed at Whittle's proposal: compressor and turbine efficiencies would be insufficient, the temperatures and stresses imposed on a constant-pressure gas turbine would far exceed the capabilities of materials then in existence, the weight of any such engine would far exceed its thrust, and so on. They characterized his proposal as visionary, a very long-term proposition, at best. Whittle, on the other hand, believed that the application of modern aerodynamic theory would permit virtually quantum increases in compressor and turbine efficiencies and that lightweight, heat- and stress-resistant alloys could be developed which would enable him to achieve adequate thrust-to-weight ratios in the near term. Moreover, the combined effects of ram air at high speeds and low temperatures at altitude would augment the work of the compressor, making a jet engine vastly more efficient the faster and higher an aircraft flew. Scoffers there were aplenty, and in what has to rank as one of history's prime examples of official obtuseness, the British Air Ministry denied his request for a modest amount of funding to support development of the concept.[11]

By late 1935, he still had not overcome official disinterest, but after having all but given up, he had finally secured an extremely modest amount (about $10,000) of private funding to begin the design of an engine for bench tests. By March of 1937, his backers had managed to increase the total to about $30,000 and his first bench-test engine, the W.U. (Whittle Unit), was ready for its initial test run. It was an incredibly ambitious undertaking. Whittle set out to build an engine that would produce 1,200 pounds of thrust at 17,500 rpm. At a time when the most efficient supercharger compressors were capable of compressing about 120 pounds of air per minute to a pressure of about twice that of the atmosphere, he strove for one which could handle 1,500 pounds per minute and achieve a remarkable 4:1 pressure ratio. He dispensed with the upstream axial compressor stages and employed a single-stage double-sided centrifugal compressor to achieve the desired 4:1 compression ratio within a relatively small-diameter area. Surrounding the compressor impeller was a scroll-type

6

volute leading into a vertical expanding diffuser pipe containing a honeycomb of divergent channels. At the top of the diffuser the air was turned 90 degrees by a cascade of vanes in an elbow before it entered the single combustion chamber. Once ignited, the expanding gases were to exit through a nozzleless scroll-shaped turbine inlet into a single-stage axial-flow turbine which was supposed to provide just over 3,000 horsepower to drive the compressor (or more than the net power then produced by any piston engine). While he felt confident he could achieve the targeted compressor and turbine efficiencies, Whittle was somewhat daunted when informed by experts that the combustion intensities for which he was striving were at least 20 times greater than had ever before been achieved.[12]

On April 12, 1937, he ran up the W.U. for the first time and it nearly blew apart. For the next two years, he struggled with burned out combustors, erratic fuel pressures, turbine failures and a host of other problems. During that span, he had to completely rebuild the W.U. three times with leftover parts and whatever new components his meager funds would permit. Although he faced almost insurmountable odds, Whittle was determined. Very patiently and ever so slowly, he began to overcome those odds as, with each engine reconstruction, he incorporated significant modifications. As he had intended, for example, he applied theoretical aerodynamics to the design of his turbine and, with the third version of the engine, was able to convincingly demonstrate the advantages of a "free-vortex" design. Each blade was fabricated with a twist in it to compensate for differential radial velocity and pressure across its diameter, producing dramatic improvements in turbine efficiency.[13]

Meanwhile, and although he was unaware of it, hundreds of miles to the east, a brilliant young German physicist was also developing a jet engine of his own design. Based on his study of aerodynamics, Dr. Hans von Ohain had deduced that modern streamlining and structural theory would permit speeds much higher than those possible with the piston engine-prop combination. Thus, like Whittle, he had concluded that a radical new form of propulsion — one uniquely suited for high-speed flight — would be required to exploit the full potential of airframe design. Although he had independently conceived the idea of a gas turbine-driven centrifugal-flow jet propulsion engine much later than Whittle, von Ohain had the good fortune to catch the attention of aircraft manufacturer Ernst Heinkel. In stark contrast to Whittle's impoverished circumstances, von Ohain's efforts to build a bench-test engine were handsomely subsidized by the enthusiastic Heinkel. Employing hydrogen as fuel and providing a thrust of about 550 pounds, von Ohain's engine was actually tested, for the first time, about a month before Whittle's first unit and the success of these tests led to the development of a flight-rated engine and a small single-engined experimental airplane. Powered by von Ohain's 1,100-pound thrust He S-3b on August 27, 1939, the Heinkel He 178 became the first jet-powered aircraft ever to take to its wings.[14]

Technology and the Air Force

Even before this flight, however, official government interest had long since entered into the equation. For, unlike the situation in England, a number of other German engineers — both in industry and government — had also already perceived the virtues of the turbojet solution. Most notable among them were Herbert Wagner and Max Adolph Muller of the Junkers Aircraft Company and Helmut Schelp of the German Air Ministry. By mid-1937, Wagner and Muller had settled on the turbojet as "the shortest path to high aircraft speeds," and by the end of the year, they had an engine under test. Unlike Whittle and von Ohain, their very meticulous studies had indicated that an axial-flow compressor was preferable because it would permit the straightest possible path for the air to flow through the engine and it would offer the advantages of a much smaller diameter and lower drag than a centrifugal-flow design. Schelp had arrived at the same conclusion by mid-1937, and by early 1939, he had engaged all four of the major German engine manufacturers — Daimler-Benz, Junkers Motors, B.M.W., and Bramo — in reaction propulsion programs. By the fall of that year, Junkers was well along in the initial development of a design which would ultimately evolve into the Jumo 004–B, an axial-flow engine producing 1,980 pounds of thrust which would begin to enter mass production in the spring of 1944. And, equally important, by the fall of 1939, Schelp had also already been instrumental in issuing Messerschmitt a contract to design and develop a twin-engine turbojet interceptor which, within five years, would begin to make a name for itself in the skies over western Europe.[15]

Thus, even before a turbojet-powered aircraft had yet flown, the German military had already begun to sponsor a massive effort aimed at the development of jet-powered combat airplanes. Unlike the British (and, later, the Americans), the Germans focused on the development of more efficient axial-flow engines from the outset. They were to suffer, however, from a severe shortage of skilled workers and, even more important, a near-total lack of the high-grade metals and alloys so essential to the development of efficient turbines and combustors. As a result, their engines were frequently inferior both in terms of materials and design. Thus, while designed for a modest service life of 25-35 hours, the Jumo 004B seldom exceeded ten hours of flying time in actual practice. Nevertheless, German efforts would bear fruit in a whole series of turbojet-powered aircraft which would actually enter combat service. The most notable of these was, of course, the sleek Me 262, the twin-engine, sweptwing fighter first conceived back in 1939. Capable of speeds in excess of 540 mph, the Me 262 would be unleashed with devastating effect against American bomber formations over western Europe by the fall of 1944.[16]

Whittle was completely unaware of any of these efforts when, after a successful twenty-minute demonstration of the third reconstruction of his engine to the Air Ministry in late June 1939, he finally won official support and, with that, came the go-ahead to build a flight-rated engine designated the W.1. The ministry also approved the design and construction of a small single-

The Gloster E.28/39 "Squirt" prior to its first flight on May 15, 1941.

engined experimental aircraft, the Gloster E.28/39. With its W.1 unit, which weighed only 623 pounds and provided almost 1,000 pounds thrust, this airplane completed its maiden flight on May 15, 1941. Curiously, and even though approval had already been granted to proceed with the development of an up-rated engine to be known as the W.2B and power the twin-engined Gloster Meteor, an official request to have the event filmed was inexplicably ignored. We have some poor quality motion picture film of this milestone event only because someone violated security regulations and shot it with his own camera.[17]

Among those on hand to witness the early taxi tests of the E.28/39 in April of 1941, however, was an American who was very interested and, indeed, shocked by the enormous potential promised by the new propulsion system. Maj. Gen. Henry H. "Hap" Arnold, Chief of the U.S. Army Air Corps, had been informed of British efforts the previous September, and prompted by alarming intelligence reports of German work in reaction propulsion, he had already launched a high-level inquiry into the subject. On February 25, 1941, he had asked Dr. Vannevar Bush, then chairman of both the National Defense Research Committee and the National Advisory Committee for Aeronautics, to establish a special committee of leading scientists to undertake this effort. Bush, in turn, had asked 82-year-old Dr. William F. Durand, the "dean" of the American engineering community, to head up such an effort under the auspices of the NACA, and by April, the Special Committee on Jet Propulsion commenced its investigation with tentative inquiries into the potential of rocket-assisted takeoff, turbine-driven props and ducted fan engines. But, by that time, Arnold had already witnessed the pure jet Whittle engine in operation on an airplane and was absolutely stunned by how far the British had advanced. And, if the

British had done it, he reasoned, there could be little doubt that the Germans were at least as far along.[18]

The fact that the United States lagged behind Great Britain and Germany and was, indeed, "taken by surprise" has been described as the "most serious inferiority in American aeronautical development which appeared during the Second World War."[19] And it has inevitably raised the question: why? In his pioneering study, *Development of Aircraft Engines* (1950), Robert Schlaifer concluded that it was "simply the result of a historical accident: Whittle, von Ohain, and Wagner were not Americans."[20] In his penetrating and highly interpretive analysis, *The Origins of the Turbojet Revolution* (1980), Edward Constant considered this a "catastrophically inadequate" explanation and argued, instead, that the reason could be found in different national-cultural approaches to science and technology. The British and, particularly, the Germans were steeped in a tradition of theoretical science which encouraged fundamental research into such areas as high-speed aerodynamics and axial-turbo compressor phenomena. They were mentally and psychologically prepared to question the basic assumptions of aeronautical science, and both England and Germany became natural spawning grounds for bold leaps into the unknown — for truly radical innovations such as the turbojet. The United States, on the other hand, "was possessed of a scientific tradition extreme in its empiricism and utilitarianism." The emphasis, Constant persuasively argued, was not on theory but on applied research leading to incremental refinements to existing technology. With a focus almost exclusively on immediately obtainable results, Americans excelled at subsonic aerodynamics, squeezing more and more horsepower out of piston engines, and achieving ever greater efficiencies in propeller design. Thus, while Europeans were exploring the high-speed frontier and even looking over the horizon toward supersonic flight, Americans were focused on the here-and-now as they built the best commercial airline system in the world. Apart from a small group of immigrants, such as the Hungarian-born and German-trained Theodore von Kármán, American scientists and engineers were generally ill-equipped to question the assumptions on which the existing technology was based because their whole techno-cultural orientation was focused on palpable, here-and-now solutions to immediate problems. "The object," he concluded, "was flight, not science, practice, not theory."[21]

The question of why the turbojet was "not invented here" may never be answered to everyone's complete satisfaction. But, apart from national pride, it is not nearly so important as why the United States was so tardy in adopting and developing the new technology even after its revolutionary implications had become so clear to so many within the aeronautical community in this country. General Arnold and other Air Corps commanders may have been taken by surprise (though they should not have been), but an awareness of the potential offered by — indeed, the necessity for — some form of jet propulsion was fairly

10

widespread in this country, especially after the 1935 Volta Congress on high-speed flight. During the late 1930s, for example, Ezra Kotcher served as the senior instructor at the Air Corps Engineering School. While specializing in aerodynamics, he was well enough versed in all fields to be able to teach most of the academic curriculum and was widely regarded as one of the few truly brilliant aeronautical engineers at Wright Field. Looking back on that period, he recalled with a certain amount of sarcasm that "it reached the point that you couldn't throw a whiskey bottle out of a hotel window at a meeting of aeronautical engineers without hitting some fellow who had ideas on jet propulsion."[22] Indeed, in August 1939, just days before the first flight of the He 178, he submitted a report to General Arnold's office (Air Corps Materiel Division Engineering Section Memorandum Report 50-461-351) recommending an extensive transonic research program and suggesting that gas turbine or rocket propulsion systems would have to be developed to support such an effort because of compressibility limitations on prop-driven aircraft at high speeds. His recommendations were apparently ignored by Arnold's staff.[23]

In hindsight, it may seem remarkable that Kotcher's bold recommendations should have been greeted with so little interest. At the time, however, Arnold and his staff were riveted on the immediate problem of building an air force to fight an imminent war, and that meant focusing on the accelerated production of aircraft and related systems already under development. Indeed, by June 1940, Arnold informed his staff that the Army was only interested in airplanes that could be delivered "within the next six months or a year, certainly not more than two years hence" and that all research and development activity would be curtailed in order to ensure timely production of existing designs.[24] Within this context, proposals to develop radical new technologies were relegated to the back burner. This was particularly true with regard to something as exotic as jet propulsion because the assumption in the United States — as it had been in England — was that its development would, at best, be a very long-term proposition.

Military interest in exploring the feasibility of the concept in this country actually dated back to the early 1920s. In 1922, the Air Service Engineering Division at McCook Field asked the Bureau of Standards to investigate the practicality of reaction propulsion. While conducting this study, Edward Buckingham based his calculations on a compressor driven by a reciprocating engine and did not consider any form of gas turbine. In his report, published by the NACA in 1923, he concluded that "propulsion by the reaction of a simple jet cannot compete, in any respect, with airscrew propulsion at such flying speeds as are now prospect." Fuel consumption at those speeds, for example, would be about four times higher. That was true, in 1922, when the airspeeds envisioned were only about 250 mph. But he went even further, concluding that there was "no prospect whatsoever that jet propulsion ... will ever be of practical value, even for military purposes." Unfortunately, his conclusions were based on a number of erroneous assumptions. Because he failed to consider the possibility that aircraft might someday be able to fly at speeds well in excess of 250 mph, he failed to consider the possibility that fuel efficiency might significantly improve at higher speeds. Like his counterparts elsewhere, he also assumed that compressors would necessarily have to be huge and heavy devices similar to those then used for industrial purposes. At the Langley Memorial Aeronautical Laboratory, NACA researchers would accept Buckingham's conclusions as their own, and his erroneous assumptions would cast a pall over serious research into the subject for more than a decade. Thus, even the very few research studies that were conducted by the NACA and the Bureau of Standards during this period merely confirmed Buckingham's conclusions because they were all largely based on those same assumptions.[25]

Indeed, the piston engine-prop combination was such a given that the NACA virtually abandoned the field of propulsion research to industry and the military services and opted, instead, to commit the bulk of its resources to the study of aerodynamics. Under this circumstance, James R. Hansen has noted: "The LMAL had but one comparatively small research division devoted to engine research, but the outlook of its members was 'slaved so strongly to the piston engine because of its low fuel consumption that serious attention to jet propulsion was ruled out.'"[26]

The aeroengine industry shared this assumption and was certainly not about to shift toward any radical new concepts. Like their counterparts elsewhere, Wright Aeronautical and Pratt & Whitney poured enormous resources into progressive refinements to basically unchanging air-cooled designs. Between 1926 and 1939, the procurement system under which they were forced to operate actually discouraged radical innovation. There were virtually no military contracts issued exclusively for experimental research for its own sake. All such costs had to be recouped or amortized in subsequent production contracts. Radical innovations could well require years of trial-and-error development effort before they might prove worthy of mass production; thus, there was little incentive to pursue such a course.[27] The engine manufacturers had a vested interest in the status quo and seemed to be largely unaware of — or unconcerned about — the implications of the pending revolution in high-speed aerodynamics until very late in the game. Wright Aeronautical conducted no studies of its own on gas turbines, and it was only in 1941, after it had somehow obtained intelligence on the success of Whittle's experiments, that the company attempted to obtain a license for the manufacture of his engine in this country.[28] Prior to 1940, some individuals at Pratt & Whitney had briefly examined the potential of gas turbines, and by May 1941, the company was conducting some preliminary tests on components for a compound engine (gas turbine wheel geared to the crankshaft of a piston engine) designed by Andrew Kalitinksy of M.I.T. This was an extremely low priority effort, however, and nothing ever came of it.[29]

The major engine manufacturers' priorities were well established and it was certainly by design that, when the NACA Special Committee on Jet Propulsion was formed in the spring of 1941, General Arnold expressly prohibited their participation. He wanted them to concentrate on the production of conventional engines to meet the crisis at hand, and backed by advice from Vannevar Bush and the chief of the Navy's Bureau of Aeronautics, he also suspected that they would be resistant to any radical new departures.[30] Despite Pratt & Whitney's subsequent claim that it was late in getting into turbojet development only because of Arnold's decision, company officials apparently expressed very little interest in entering the field even after it was invited to participate. Lt. Gen. Donald L. Putt, then a project officer at Wright Field, recalled a conference with Pratt & Whitney personnel during which Brig. Gen. Franklin O. Carroll, Chief of the Engineering Division, tried to encourage them to get involved in developing turbojets. "They were very firm in their conviction that the turbine engine would never be much of a threat," he recalled. "The piston engine was going to be with us forever; it was the way to go. There might be some place for a turboprop but for a straight jet, forget it."[31]

On the military side, the Power Plant Branch at Wright Field was certainly not prepared to lead the way. First of all, in the 1920s, the NACA had very forcefully staked its claim as *the* institution responsible for fundamental aeronautical research in the United States, and it jealously guarded its monopoly throughout the 1930s. The Air Corps, by law, was to limit its activities to applied research, and throughout the 1930s, officials at Wright Field were loathe to invade the NACA's turf for fear of arousing Congress' ire. As far as Air Corps leaders were concerned, it was the NACA's job to conduct fundamental research and keep up with the latest scientific developments, and always strapped for funds throughout the 1930s, they were quite willing to defer to the NACA in this regard.[32]

The NACA had abandoned propulsion research to industry and the military, but this does not mean that anybody ever directed the Air Corps to fill the void or undertake fundamental research of any kind. The military's job was to conduct applied research, and thus, as I. B. Holley has observed, the personnel of the Power Plant Branch at Wright Field "had their goals rather clearly laid out for them: they were to strive for better engines, meaning more horsepower at less weight. They were to minimize fuel consumption, to reduce frontal area in order to reduce drag, and to achieve maximum reliability and durability."[33]

Moreover, even if given the job, there were a number of other circumstances which militated against *any* kind of serious research effort. Gen. Jimmy Doolittle once observed that research and development is like virtue; everyone believes in it but no one wants to sacrifice for it. This was certainly true for the Army Air Corps during the interwar years. Throughout the period, its entire R&D budget generally hovered between $2 million and $4 million, most often, at the lower end of the scale. More tellingly, between 1926 and 1939, R&D

expenditures as a percentage of the total Air Corps budget plummeted from 16 to just 5 percent.[34] Out of these paltry sums, no more than 30 percent was ever dedicated to propulsion systems, and virtually none was directed toward experimental research of any kind because the emphasis at Wright Field was on the procurement of systems destined for the operational inventory. The very structure of the Materiel Division mandated this kind of emphasis.

With the establishment of the Air Corps in 1926, both R&D and procurement were brought together under the new Materiel Division at Wright Field. While the merger improved coordination between the two areas, it had a number of unintended side effects. Most important, the requirements of the procurement side of the house absorbed an ever greater percentage of the available technical manpower, facilities, and other resources in support of routine specification compliance testing of aircraft and systems submitted by manufacturers. The practical consequence of this, as I. B. Holley has noted, was that experimental research fell by the wayside.[35]

Inadequate funding also translated into serious deficiencies both in the number and quality of technical personnel assigned. The Materiel Division suffered from a serious shortfall in engineering manpower throughout the 1930s. A single project officer assisted by a single civilian engineer, for example, was typically responsible for the development of all pursuit, bombardment, or trainer aircraft. Moreover, the scientific and technical competence of the staff was well below par. Lt. Gen. Laurence C. Craigie served several tours at Wright Field during the 1930s and 1940s, and he recalled that, when he arrived in late 1934, no more than a dozen individuals, out of 1,100 personnel, could be considered as "real scientists" There were fewer still who, like Kotcher, could cross disciplines. Five years later, an investigating board reported "an appalling lack of qualified personnel . . . particularly in key positions." The most serious deficiency was among the officers, only a fraction with any of the relevant scientific and technical training which had, by then, become so necessary to cope with the burgeoning complexity of aviation technology.

A handful of the most qualified officers were selected each year to attend the Air Corps Engineering School. The year-long curriculum, however, provided little more than a one- or two-week orientation into the activities of each of the labs and test organizations at Wright Field. The much larger civilian staff tended to be a cut above the officers, but low pay and limited promotion potential generally drove the best among them to higher paying jobs in industry. Thus there were, at best, never more than a few individuals at Wright Field who were sensitive to the growing interaction between fundamental and applied research and fewer, still, who were capable of crossing disciplines and perceiving the sudden convergence of thermodynamic with aerodynamic principles. The upshot of all of this was not only that the Air Corps' principal R&D organization was ill-equipped to conduct serious research but also that it put the Air Corps at a tremendous disadvantage in attempting to deal with the larger scientific and technical community from which it might have benefited.[36]

All of this made for an almost classic "who's minding the store?" scenario. Industry depended on the Air Corps for direction in terms of requirements, and the Air Corps, in turn, depended on the NACA for fundamental research. Because the piston engine appeared to be such a given, the military never called on the NACA to investigate radical new forms of propulsion and the NACA, in turn, virtually abandoned the field, leaving it up to industry and the military. However, industry did not have the incentive to take on the job and the military did not have the expertise to look in new directions or even to direct either industry or the NACA to do so.

By 1940, as noted above, Pratt & Whitney was doing some very limited, component-level work on a compound engine. The NACA was actually conducting some useful research on compressors, and one of its most brilliant aerodynamicists, Eastman Jacobs, was preparing to demonstrate the feasibility of a ducted fan concept first conceived by Italian Secondo Campini in 1930. If all had gone well, it was conceivable that this system could have been ready for inflight testing by 1943. Earlier, someone in the Engineering Section at Wright Field had produced a report in 1936 titled "The Gas Turbine as a Prime Mover for Aircraft," but like Kotcher's report three years later, it did not generate enough interest to stimulate any kind of major research program. In addition to looking at jet-assisted (really rocket) take off, the use of piston engine exhaust to provide supplementary jet thrust, and reviewing (and typically rejecting) proposals for all manner of reaction propulsion systems, the Power Plant Laboratory had launched a modest program in 1938 aimed at developing a successful compound engine by 1943. With no sense of urgency in any of the above-mentioned efforts, none of them ever evolved into successful propulsion systems.[37]

As in Europe, interestingly enough, the only projects underway which were headed in the right direction all had their genesis outside of the aeropropulsion establishment. In 1936, engineers at General Electric started publishing internal research bulletins and reports on the feasibility of employing gas turbines as a primary source of power to drive propellers, and by 1939, Dale Streid was writing optimistically about "propulsion by means of a jet reaction." These studies were ongoing right up to April 1941, when GE (Schenectady Division), Allis Chalmers, and Westinghouse were invited to join Dr. Durand's Special Committee on Jet Propulsion, and each of these turbine manufacturers ultimately began development of their own turbojet designs.[38]

Meanwhile, Jack Northrop appeared to have stolen a march on everyone. On the basis of design studies initiated in 1939, he became convinced of the superiority of a gas turbine over the conventional piston engine for driving propellers. After commencing initial development of a turboprop engine he called the Turbodyne with his own resources, he approached the Army and

Navy for support. Neither showed any interest until June 1941 when they issued a joint contract to pursue development of what was subsequently designated the XT37. Like all of the early turboprops, the project was ambitious in concept and excruciatingly slow in development. Three test engines were finally built in 1947, and though never flight tested, one of them eventually delivered an impressive 7,500 horsepower during bench tests before the project was canceled in 1949. By then, Northrop's ingenious engine had been overtaken by the turbojet.[39]

By far the most interesting development was taking place at Lockheed. Since the mid-1930s, Kelly Johnson had been well aware of the theoretical implications of compressibility phenomena, and by 1939, he and Hal Hibbard had decided to do away with the prop altogether! Unlike so many others in this country, they were capable of perceiving the sudden convergence of aero-dynamic with thermodynamic principles, and they asked Nathan Price to design a pure turbojet to power a truly radical interceptor at speeds never before envisioned in this country. Initial development of the engine, designated L–1000, got underway in 1940, and though his initial concepts were far too complex to be practicable, Price ultimately came up with a truly remarkable design — a high-compression-ratio, twin-spool, axial-flow turbojet promising a then extraordinary 5,000 pounds of thrust at takeoff. Meanwhile, Johnson led a small design team that came up with the L–133, an equally remarkable twin-engine, stainless steel airplane, featuring thin wings and canard surfaces and projected to attain a whopping 620 mph at 20,000 feet (and nearly that speed at 50,000 feet)! Much to Johnson's chagrin, officials at Wright Field considered the radical airplane to be far too risky a venture when he delivered the design and technical data in March of 1942. The engine, however, showed enough promise for Lockheed to win a contract for further development of what became known as the XJ37. The engine never got beyond the development stage, however, Kelly Johnson's knowledgeable interest in jet-propelled airplanes had made a very important impression on the Experimental Engineering Section at Wright Field.[40]

Like so many among the top Air Corps leadership, Hap Arnold had never been technically inclined, and he was probably unaware of most of these developments. But, when confronted with the palpable evidence of Whittle's achievement, he immediately grasped its implications and acted quickly to expedite America's late entry into the jet age. After promising the British he would clamp the tightest security precautions on the project, he managed to gain permission to build the Whittle engine in the United States by late summer 1941. Next, he had to decide who would produce it, but for the reasons noted above, the major engine manufactures were excluded. Brig. Gen. Oliver P. Echols, Chief of the Materiel Division of the recently redesignated Army Air Forces, and his assistant, Lt. Col. Benjamin W. Chidlaw, recommended GE because they were well aware that the company had pioneered in turbine

Lockheed chief engineer Hal Hibbard (left) and designer Nathan Price
with the experimental L–1000 (XJ37) axial-flow turbojet.

technology and, over the years since World War I, had perfected the development of turbosuperchargers which permitted piston-engined airplanes to climb to otherwise impossible altitudes. Indeed, turbosupercharging was based on many of the same principles as jet propulsion: at high altitudes, the thin air was compressed to sea level conditions by a centrifugal compressor, directed through a carburetor, where fuel was added, and through an intake valve into a piston cylinder where it was ignited. After ignition, the exhaust gases were channeled through a turbine wheel that drove the compressor. GE's extensive work with turbosuperchargers and, most important, the high-temperature alloys necessary to build them made it the logical choice to take the next step. Thus, in a meeting in Arnold's office on September 4, 1941, GE was offered a contract to reproduce the 1,650-pound thrust Whittle W.2B engine.[41]

Arnold's choice to design and build the airframe was almost as easy. His concerns about disrupting top priority existing development and production programs were a major factor in this decision. Based again on advice from Echols and Chidlaw, he selected a company which certainly was not overburdened with such work. With innovative (though not very successful) designs, such as the YFM–1 "Airacuda" and the P–39 "Airacobra," the Bell Aircraft Corporation's team of designers had at least established a reputation for inventiveness. Larry Bell's own seemingly boundless drive, Arnold and his staff believed, would guarantee that any project would be completed on time and up to expectations.[42]

Bell agreed to tackle the job on September 5, 1941. The next day, he selected a small group of six engineers and assigned them the task of creating

17

a preliminary design for the aircraft. Working with little more than a small free-hand sketch of the engine, the "Secret Six," as they were called, prepared a design proposal and a 1/20th scale model within the span of just two weeks. Arnold gave his approval, and a fixed fee contract for $1,644,431 was finalized on October 3. It stipulated that the first of three "twin-engine, single-place interceptor pursuit models," with a projected combat ceiling of 46,000 feet and a top speed of nearly 500 mph, should be delivered within just eight months. A similar $630,000 contract was negotiated with GE for fifteen engines, with the initial pair of flight-ready engines, each providing 1,650 pounds of thrust, to be available for installation on the first aircraft. Remarkably, and though Arnold doubted that it was possible, his staff was hoping that an engine-airframe combination could be designed and developed which could be rapidly transitioned into a combat-worthy production fighter. This goal was incredibly ambitious and the schedule was tight, to say the least.[43]

Chidlaw was selected by Arnold and Echols to provide overall direction for the program, and Majors Ralph Swofford, from the Experimental Aircraft Projects Section at Wright Field, and Don Keirn, from the Power Plant Lab, were assigned as airframe and engine project officers, respectively (within months Chidlaw was promoted to the rank of brigadier general and Swofford and Keirn each to the rank of full colonel). Swofford and Keirn each shouldered tremendous responsibility. In those days, a project office was responsible for all of the many functions now handled by system program offices staffed with hundreds of personnel. Due to the "Super Secret" nature of this program at its outset, no more than a dozen people at Wright Field had any knowledge of its existence. In Swofford's and Keirn's case, each was intimately involved in the design and development process on a daily basis. Each had enormous authority, and every design change required their personal approval. During the early months of the flight test program, long before official AAF flight tests got underway, each would also find himself serving as a test pilot. After every significant modification to one of the prototype airframes, for example, Swofford would always fly the airplane before approving or disapproving it for inclusion in the production design. Small wonder that after he had retired as a two-star general years later, Don Keirn recalled that he had been entrusted with far more authority as a major during the hectic early months of this program than he would ever enjoy as a general officer.[44]

In a fashion which would become a hallmark of the American aviation industry during the war years, a small design team hastily set to work at Bell with a profound sense of urgency and only a few rough drawings of the proposed engine in hand. Tasked with designing an entirely new type of airplane, they were further required to come up with a design which would also be suitable for combat service. Beyond the single stipulation to wrap an airframe around a pair of the new powerplants, they were free to improvise, but they had to work quickly and without the benefit of any outside advice or assistance. Because of security restrictions imposed by Arnold, for example, they were not permitted to make use of the NACA's full-scale wind tunnel facilities and were forced, instead, to rely on very imperfect data from the five-foot, low-speed tunnel at Wright Field. By mid-November, General Echols was already pleading with Arnold to rescind this restriction because he could already foresee boundary-layer problems with the engine inlets unless the design team could get some hard data on high-speed flow conditions. Arnold, however, was adamant and this decision, indeed, resulted in some serious miscalculations which severely limited the performance of the airplane. Nevertheless, working in haste, the design team completed its work by early January 1942, and a small select crew of Bell workers began to build it, literally by hand, on the closely guarded second floor of a Ford agency in Buffalo, New York. In the interests of secrecy, the aircraft had been given the designation XP–59A, a designation originally intended for a proposed Bell pusher-prop fighter which never got beyond the mock-up stage.[45]

Equally stringent security precautions were in force at GE's Lynn River facility in Massachusetts, where another small team headed by Donald F. "Truly" Warner labored, nonstop, on a design that for security purposes had been designated "Type I–A supercharger." With the benefit of Whittle's W.1X engine, which had been used in the taxi tests of the E.28/39 and on which they were able to run tests, and working from reportedly incomplete drawings of his W.2B design, they made some minor modifications to the diffuser, combustors, and bearings of the British design and built a prototype. On March 18, just 5-1/2 months after taking on the job, they wheeled the engine into a test cell — aptly named "Fort Knox" — for its first test run. However, the engine stalled and this attempt was unsuccessful. But, exactly one month later, on April 18, Truly Warner once again advanced the throttle, and this time, the engine successfully roared to life. With the push of a hand, he had finally lit the flame of the turbojet revolution in America.[46]

The GE Type I–A engine was a centrifugal, reverse-flow, turbojet which represented a quantum advance over the design in Frank Whittle's original 1930 patent. The GE engine featured inlets configured with guide vanes that directed air into a single-stage, double-sided impeller — a centrifugal compressor — that roughly tripled the air's pressure as it passed through the diffuser and into the ten reverse-flow combustion chambers where it was ignited. The intensely hot, expanding gases raced through the turbine , which drove the compressor, then combined to exit through a single exhaust nozzle at high speed to produce thrust.[47]

The GE team proceeded with what would become a lengthy and sometimes painful development process. The thrust performance of the test unit, for example, never came close to matching the British design predictions for the W.2B (it was not until early 1943 that they would learn that the thrust curves they were using were different than those employed by the British). When Wing

The GE I–A turbojet engine.

Commander Whittle arrived in June 1942, he found that Truly Warner and his team were struggling with excessive turbine inlet temperatures, cracked turbine blades, bearing failures, excessive carbon formation in the flame tubes due to poor combustion efficiency, and a host of other problems. Warner had found it necessary to experiment with a variety of different diffuser, combustor, and turbine bucket designs and materials. Whittle was quick to caution that, due to the decision to locate the engine nacelles alongside the airplane's fuselage (as opposed to the wing mounted pods that would be employed on the Meteor), boundary layer problems would severely reduce ram air efficiency. Despite all of these problems, Chidlaw reported to Arnold's office that "Bell and GE have both done a bang-up job in rushing this thing through" and that the XP–59A project was "well ahead" of Britain's Meteor project which had enjoyed a one-year head start. He attributed this lead principally to GE's years of experience with turbosuperchargers which put the United States well ahead in the development of high-strength, heat-resistant alloys. Nevertheless, Bell's completion of the first airframe was held up by GE's inability to deliver flight-rated engines until early August, and it was already quite apparent that the I–A powerplants would never be able to deliver more than 1,250 pounds of thrust. Indeed, Warner had already proposed major modifications to the original design which would result in an I–16 unit capable of producing the desired 1,650 pounds of thrust.[48]

20

Meanwhile, as the Bell team assembled the first airplane during the spring and summer of 1942, the construction of a small Materiel Center test site got underway a continent away on the edge of an enormous dry lake at an out-of-the-way place called Muroc in California's high desert. Six miles to the south, Muroc Army Air Field served as a training base for fighter and bomber crews preparing for overseas deployment. The site was selected by Chidlaw and Swofford in April 1942 because of its extremely remote location, the excellent year-round flying weather, and the availability of Rogers Dry Lake, with an expanse of forty-four square miles. It was obvious to them that the immense, concrete-like lakebed would provide a natural landing field from which to explore all of the unknown characteristics of the new jet. When Bell chief test pilot Bob Stanley arrived there in August, he found what could best be described as "Spartan-like" accommodations: a water tower, an unfinished portable hangar, and a wooden military barracks. These three totally unimpressive structures represented the humble beginnings of what would one day become the USAF Flight Test Center.[49]

On September 19, the engines and crated pieces of the airplane were off-loaded from box cars after a long cross-country journey on what its weary GE escorts mockingly called the "Red Ball Express." Working, quite literally, day and night, Bell and GE personnel set about to reassemble the craft. They completed the job within a week, and on September 26, the XP–59A rolled out from the hangar for the first time. In many regards, it appeared to be a fairly conventional design, but certain features caught the eye. Fully loaded, it weighed just over 10,000 pounds, and with a wing loading of 25 pounds per square foot, its immense wings (400 square feet) appeared to be optimized for high-altitude flight. The tail section swept upward very noticeably and the craft rested extremely low to the ground on tricycle landing gear. And then, of course, there was no prop, and tucked beneath the wings, along the fuselage, were a pair of nacelles housing the I–A engines.[50]

Those engines roared to life in the aircraft for the first time that day, and by September 30, just four days later, Bob Stanley and the airplane were primed for its initial taxi tests. After completing some low-speed trials, he proceeded to a series of high-speed runs to get a feel for the controls. On a couple of these runs, late in the day, the wheels of the aircraft actually lifted a couple of feet off the ground. Stanley, a brilliant engineer and a relentlessly hard-driving personality who seldom counted patience among his virtues, was all for making the first flight then and there. Larry Bell, however, overruled him; high-ranking official observers — such as Dr. Durand and Col. Laurence C. "Bill" Craigie, Chief of the Experimental Aircraft Section at Wright Field — were not scheduled to arrive for two days. On the following day, October 1, Stanley made four additional "high-speed taxis," during the first of which the aircraft lifted off and soared some twenty-five feet above the surface of the lakebed. On subsequent runs, it climbed to as high as a hundred feet. Unofficially, the

First run-up of the XP–59's engines at Muroc, California.

XP–59A had unquestionably flown, but the brass had not been there to witness the event. "Officially," it had not really happened.[51]

Finally, on October 2, the brass were on hand. At about 1 p.m., Stanley advanced the throttles, released the brakes, and the aircraft, slowly at first, moved across the hard-baked clay. After what seemed like an unusually long takeoff roll, its wheels finally left the ground and he made what he described as a "leisurely" climb to 6,000 feet. Remarkably, just one year — almost to the day — after commencing the project, the United States had finally and officially entered the jet age.[52] GE's Ted Rogers reported what he called a "strange feeling" as he watched the flight: "dead silence as it passed directly over-head, . . . then a low rumbling roar, like a blowtorch . . . and it was gone, leaving a smell of kerosene in the air."[53] After a second flight, Stanley turned to Colonel Craigie and said: "Bill, we've only got about 45 minutes left on the engines [they had to be pulled for inspection after every three hours of running time]. How'd you like to take it up?" He didn't have to ask him twice. Although he had been on hand only to serve as the AAF's official observer, he went up for a thirty-minute flight and thus, quite by happenstance, Bill Craigie became America's first military jet pilot. As he was to recall many times: "Things were a lot less *formal* in those days."[54]

Less formal, indeed! There were no safety chase airplanes that day, and the most important instrumentation — at least during the initial flights — remained the seat of the pilot's pants. It may not have been too scientific, but by latter-day standards, it was relatively inexpensive and afforded a means of real-time data acquisition which was always certain to yield immediate analyses of any

The original mission control center for the XP–59 flight test program
consisted of a two-way radio and a wire recorder. The wires
radiating out from the radio formed its antenna.

problems. There was no telemetry — indeed, the entire "mission control center"
consisted of a two-way radio and an old voice recorder set up on the lakebed
adjacent to the hangar. Although the aircraft was ultimately instrumented to
cover between 20 and 30 different parameters, the instrumentation was often
primitive, to say the least. Control stick forces, for example, were measured
with a modified fish scale, and engine thrust was originally measured by means
of an industrial spring scale attached to the landing gear and anchored to the
ground. The lack of a satisfactory means of measuring thrust on the aircraft,
especially in flight, would, in fact, hamper flight test efforts throughout the
P–59 program — making it impossible, for example, to correlate airplane drag
to net engine thrust.[55]

As the business of flight testing the airplane and engines proceeded, they
encountered more than their share of headaches. Early on, for example, they had
so much trouble starting one of the engines that they named the number one
airplane "Miss Fire." Overheated bearings, malfunctioning fuel pumps and
barometric controls, detached turbine blades, the three-hour inspection requirement and countless other problems eventually forced them to remove the
cowling panels so often that they later started calling it "Queenie," in honor of
a much-admired exotic dancer. (The designation "Airacomet" only came into
use much later as a result of a contest among Bell employees.) Indeed, persistent
engine breakdowns and lengthy delays in the delivery of replacements, spare
parts, and uprated higher thrust models of the engine caused the program to fall

way behind schedule. Program officials in the Experimental Engineering Section at Wright Field had expected to start receiving useful performance data by January 1943, but by mid-April, the airplanes had only accumulated 29 flying hours. The engine problems, plus the fact that no one really knew how to test a jet airplane, delayed the start of the AAF's unofficial performance evaluations until late September 1943, and the official tests were not completed until March 1944.[56]

Although the testing proceeded at an excruciatingly slow pace, the pilots quickly became familiar with the characteristics of the jet, gaining a lot of wisdom they would impart in the flight manual. The throttles, for example, had to be treated very carefully. Rapid acceleration caused engine surges which could burn up the combustors and turbines. The engines' extremely slow acceleration also taught them never to go low and slow on final approach. Lacking an airstart capability, the engines also had a nasty habit of flaming out, and as had been expected, they consumed enormous quantities of fuel. Experience with both of these problems bore out the wisdom of selecting the lakebed for the tests. In fact, attempting to get as much out of each mission as possible, the pilots eventually made it a common practice to fly until the tanks went dry and then glide in to dead-stick lakebed landings.[57]

Hoping to catch up in a hurry, the Army Air Forces had attempted to make the great leap from a proof-of-concept, experimental vehicle into a 500-mph combat fighter, all in one airplane. It was a bold hope, too bold. The performance of the XP–59A with the original I–A engines fell far short of expectations. In part, this was because the original thrust data provided by the British for the W.2B engine were misinterpreted, and the I–A's actual performance fell about 25 percent short of what had been very optimistic projections. Even with modified I–14 engines, each providing about 1,450 pounds of thrust, the maximum speed attained was only 424 mph at 25,000 feet. This speed was attained, moreover, only after the entire airplane's surfaces had been puttied, smoothed and sanded and its wings polished. By comparison, in its normal "dirty" configuration, the airplane's top speed was only 404 mph at the same altitude.[58]

The performance of the slightly heavier YP–59s was even more disappointing. Some of the YPs were representative of the ultimate production version of the aircraft. For example, the wingtips were clipped and squared off, reducing the span from 49 feet to 45 ½ feet and its wing area by about 15 square feet. The size of the vertical stabilizer was reduced and its tip squared off, as well. The hinge-mounted, side-opening canopy, which was flush with fuselage of the XP–models, was replaced by a new sliding canopy which protruded about two inches above the fuselage surfaces and incorporated a larger and flatter windscreen. To everyone's surprise and disappointment, the top speed achieved by the aircraft was only 409 mph at 35,000 feet, even though the YPs were configured with the uprated I–16 models of the engine (AAF designation J31)

24

rated at 1,650 pounds of static thrust, the rating for which the airframe was originally designed. This poor performance, in comparison with the XP–model, was attributed primarily to the substantial increase in drag caused by the new canopy and windscreen.[59]

The disappointing performance of the overall design, however, was blamed on a number of other factors. In September 1943, Bell engineer Randy Hall's plaintive cry to chief project engineer Ed Rhodes belabored the obvious: "We need thrust, thrust, and *more* thrust."[60] The low thrust-to-weight ratio and the oversized (scarcely laminar flow) wings were among the most obvious contributors. There were many other flaws, however, which could conceivably have been identified and remedied during the initial design process if the Bell team could have had access to reliable high-speed wind tunnel data. Their original calculations concerning boundary-layer effects and engine nacelle inlet area, for example, were way off the mark, and after the airplanes started flying, Bell was forced to experiment with various new configurations. The original inlet of 2.86 square feet was ultimately reduced to 2.08 square feet, but even then, it was scarcely optimized for peak performance. The failure to completely understand the dynamics of airflow within the nacelles led to a multitude of other problems. A lot of engineering effort was expended after the flight test program got underway, for example, attempting to reduce rear compressor inlet temperatures. The aircraft also exhibited a directional "snaking" tendency which increased in severity with speed. Repeated modifications to the vertical tail and rudder were to no avail, and the aircraft was judged "unsatisfactory" as a gunnery platform during official AAF tests.[61] The real source of the problem may actually have had little to do with the rudder, but may well have stemmed back, once again, to the failure to adequately understand nacelle inlet problems. At a symposium in late 1945, Benson Hamlin, one of Bell's key flight test engineers on the program, reported that the snaking "is believed to be due to the very large inlet scoops in which it is possible for the inlet ducts on either side to alternately stall and unstall, causing a fluctuating air flow in the scoops or nacelles producing an unstable directional stability of the airplane."[62]

Though it served as a useful testbed to explore the potential advantages — and pitfalls — of a radical new technology (and it won at least one distinction when, in February of 1944, Maj. Everrett Leach climbed to an American record of 47,700 feet), the P–59 was really, for all practical purposes, a 350-mph airplane — no faster than the prop-driven fighters of its day. And, indeed, in operational suitability tests during which it was flown in mock combat engagements against P–38s and P–47s, it was outclassed in virtually every category by the conventional fighters. Ambitious plans for a major production run were canceled. In addition to the 3 XP–59A and 13 YP–59A prototypes, only 50 production models came off of Bell's assembly line. Not suited for combat, they were used to train America's first cadre of jet pilots, a role which, indeed, made them unique among the first generation of jet aircraft.

More important, still, was the fact that America's aviation industry went to school with this aircraft, and those in it learned their lessons well.[63]

On January 8, 1944, just two days after the AAF first announced the existence of the P–59, another jet prototype was prepped for its maiden flight at Muroc. In contrast to the Airacomet, there was nothing conventional-looking about this airplane. Designed by Kelly Johnson and delivered by his fledgling "Skunk Works" in just 143 days, the sleek, single-engined XP–80 looked like it was made for jet power, and indeed, it was. It was powered by yet another British import, the British de Havilland Halford H.1B, and as he accelerated to 490 mph, Lockheed test pilot Milo Burcham put on an impressive demonstration above the lakebed that morning. Among those viewing it was Bell test pilot Tex Johnston. Immediately afterward, he fired a cable back to Bob Stanley in Buffalo: "Witnessed Lockheed XP–80 initial flight-STOP-Very impressive-STOP-Back to drawing board-STOP." Though its Halford engine was never able to deliver more than 2,460 pounds of thrust, during official AAF performance tests conducted just over a month later, the XP–80 became the first American aircraft to exceed 500 mph.[64]

The XP–80, however, was really only an aerodynamic testbed. Prior to the end of 1942, GE design engineers had already learned enough from their work with the original I–A engine that the Engineering Division at Wright Field was willing to give the go-ahead to develop an engine which would more than triple the I–A's thrust. Development of the I–40 (J33) proceeded so rapidly that, in August 1943, Johnson was asked to design a substantially larger airframe to house an engine providing 4,000 pounds of static thrust. This airplane, the XP–80A, was the prototype for America's first combat-worthy jet fighter, the P–80 Shooting Star. It first flew in June of 1944, and the first production models were accepted by the AAF in February 1945. Capable of speeds approaching 600 mph, the P–80 demonstrated how far and how fast the United States had come in just three years. The lessons learned in the P–59/I–A engine program had paid extraordinary dividends.[65]

The turbojet airplane could have been — and, but for the delusions of Adolph Hitler, might have been — a decisive weapon in World War II. But it was not, and although the United States failed to put a jet aircraft into combat, with Germany's surrender and the development of the J33-powered P–80, this country had arguably moved from the back of the pack into the forefront of the turbojet revolution within a span of just three years. How did we do it? Well, in large part, quite obviously because of tremendous advantages in terms of material, skilled manpower, and industrial know-how. But also, in part, almost ironically, because of that very same focus on applied science which, Edward Constant has argued, initially put us behind. No nation in the world was more adept at — or had more impressive facilities for — transforming the fruits of pure science into superior products. In some cases, being first is not nearly so advantageous as being a really superior second, third, or even fourth. Once

26

Lockheed XP–80 on the morning of its first flight at Muroc.

presented with a good idea, no nation was better prepared to run with it and a so-called weakness became an immediate strength.

Nevertheless, none of this would have been possible without the aid and ongoing assistance of the British, and this lesson was certainly not lost on the man most intimately involved in the process. Returning from a trip to England in August 1943, Col. Don Keirn was exasperated by the fact

> that enough emphasis has not been placed on research facilities to enable this country to keep up with developments. Our present position is largely due to the aid given us by Great Britain and our ability to sift the information and follow those lines which appear to be most immediately profitable.[66]

The implications of his report extended far beyond the turbojet, and they were not lost on any of those who had been involved in importing the new technology to the United States.

By the late summer of 1945, as the U.S. military was completing its inventory of Germany's massive R&D infrastructure, General Craigie was preparing to take over as the Chief of the Engineering Division. It would be his job to help build a new U.S. Air Force that could meet the challenges of the future. The recent war had taught him that science and warfare had become inextricably intertwined, and in the future, he was convinced, there probably would not be time to borrow, let alone to catch up. In a speech to the International Aeronautical Society, he emphasized that the United States must "tear a page from the German book of experience and use it as a warning lest we forget that research can only rarely be hurried, that it must be continuous, and that most of it must be accomplished during years of peace." This, he further emphasized, would require the creation of a massive R&D establishment "prepared to stand on its own feet" within the Air Force, and he concluded, "these feet can only be provided through adequate appropriations and the provision of adequate personnel and facilities."[67]

This was essentially the same message which Dr. Theodore von Kármán and the AAF Scientific Advisory Group were about to deliver to General Arnold. And, indeed, he would define the establishment of a comprehensive and well coordinated R&D capability which would be second to none — one which would not only encompass the NACA, industry, and the universities, but also for the first time, a major inhouse establishment, as well — as the AAF's highest postwar priority. The turbojet was the most publicized — and, therefore, embarrassing — example of the failure of the underfunded, fragmented, and uncoordinated pre-war military R&D system in this country. In that sense, it would become a useful symbol for those, like General Craigie, who were given the job of convincing an austerity-minded Congress — and, indeed, the rest of the Air Force — that being first was no longer just a matter of national pride, it was now a matter of national survival.[68]

At war's end, the turbojet revolution was still in its infancy. The AAF already had at least 19 turbojet aircraft projects underway. Most of them, however, would be relatively crude attempts to adapt existing airframe concepts to the new propulsion technology and even the most successful of them, such as the sweptwing F–86, could be considered as, at best, no more than transitional designs. G. Geoffrey Smith observed, at the time, that the turbojet revolution had precipitated a momentous turn of events:

> it is only as a result of successful development of the gas turbine and jet propulsion that engine manufacturers are able, for the first time in history, to supply more powerful units than the builders of airframes can at the moment usefully employ. The relative position [of each] has been reversed.[69]

On a very basic level, the genius of Whittle and von Ohain's vision of a high-speed airplane had been based on the perception that the engine and airframe were really two components of a single system joined together in a kind of symbiotic relationship in which the capability of each was dependent on the maximum efficiency of the other. Aerodynamicists had unwittingly brought on the demise of the reciprocating engine, and they now found themselves in the position of having to catch up with the new technology spawned by their efforts to take full advantage of its potential.

There was also, of course, a multitude of jet engine development projects underway at the time as the emphasis shifted overwhelmingly toward axial flow designs. General Electric, Westinghouse, and the erstwhile piston-engine manufacturers poured millions into a painstaking search for lighter weight, higher strength, and more heat-resistant materials as they strove to achieve higher compression and thrust-to-weight ratios and reduced fuel consumption while improving the durability and acceleration capabilities of their engines. Well before the end of the war, they had made tremendous strides in aero-thermodynamics (achieving combustion in high-speed airflow). They had also started looking into the advantages to be gained from various types of thrust augmentation, such as water injection and afterburning, and they were already well aware of the tremendous fuel economies that could be achieved with turbofan designs.[70]

The turbojet also compelled a host of developments in other fields. The tremendously high speeds and altitudes which were now within reach, for example, meant that human physiology could easily become the most critical limiting factor in the design of high-performance airplanes. Aeromedical research, a heretofore neglected field, suddenly became a top-priority endeavor, as did the development of ejection systems, g-suits, pressurized cockpits, pressure-breathing oxygen systems, and full-pressure suits. The turbojet also drove major efforts in weapon systems development. An immediate demand for dramatic improvements in lead-computing optical gunsights and bombsights gave way to a massive effort to develop radar tracking systems and, among many, to the conclusion that guns and classic dog fights had become relics of a bygone age and only guided missiles could meet the requirements of future air-to-air combat. High speeds and human limitations also compelled the development of hydraulically boosted and irreversible flight controls and stability and control augmentation systems. The development of sophisticated automated fire and flight control systems, in turn, mandated the development of compact, high-speed computers. The spin-off effects of the turbojet seemed to be endless.

Like an irresistible force, the awesome potential of the turbojet also forced designers to confront the reality of transonic flight. Aerodynamicists had long speculated on the possibility of flight beyond the speed of sound, but it was now obvious that the means were at hand to actually propel a piloted airplane into that region. Speculation and theory were one thing, but no one had any valid data on high-speed stability and control and the effects of compressibility and there was an urgent need for such information. Ezra Kotcher finally got his transonic research airplane — the Bell X–1 — and the rest, as they say, is history.[71]

The new U.S. Air Force had already made tremendous strides in all of these and many other related areas when turbojet technology finally achieved mature status with the development of the Pratt & Whitney J57. On April 15, 1952, almost exactly ten years after Hap Arnold had first witnessed the E.28/39 making short hops during its high-speed taxi tests, eight prototype J57s powered the YB–52 on its maiden flight. This engine-airframe combination was an extraordinary accomplishment. Early model B–52s could outpace an F–86E at altitude, and they demonstrated an intercontinental range capability which, only a couple of years earlier, had been thought to be impossible for jet-powered aircraft. The J57 opened the door for the development of long-range commercial airliners and supersonic fighters. Early versions of the engine provided about 12,000 pounds of dry thrust and 17,000 pounds in afterburner. In May 1953, the J57–powered YF–100, with its burner lit, became the first aircraft in history to exceed Mach 1 on its maiden flight. With the arrival of the YF–100 and the other first generation supersonic fighters, the marriage of aerodynamics to thermodynamics was, at last, successfully consummated; for they were the first airplanes to achieve the kind of symbiotic harmony which, three decades before, had inspired the visions of Frank Whittle and Hans von Ohain.[72]

Notes

1. John Golley, *Whittle: The True Story* (Washington: Smithsonian, 1987), pp 23–24; Frank Whittle, "The Birth of the Jet Engine in Britain," in *The Jet Age: Forty Years of Jet Aviation*, Walter J. Boyne and Donald S. Lopez, eds (Washington: Smithsonian, 1979), p 3; Edward J. Constant, *The Origins of the Turbojet Revolution* (Baltimore: Johns Hopkins, 1980), pp 180–82.

2. Ronald E. Miller and David Sawers, *The Technical Development of Modern Aviation* (London: Routledge & Kegan Paul, 1968), pp 47–97; Robert Schlaifer, *Development of Aircraft Engines* (Boston: Harvard Graduate School of Business Administration, 1950), pp 156–320; J. S. Butz, Jr, "General Electric: Pioneer of U.S. Jet Engines," *Flying*, vol 51 (Feb 1963), p 85; Bill Gunston, *World Encyclopaedia of Aero Engines*, 3d ed (Sparford, Nr Yeovil, Somerset, UK: Patrick Stephens, 1995), pp 122–23. For a useful brief history of the development of piston-driven aero engines, see Herschel Smith, *A History of Aircraft Piston Engines*, corrected ed (Manhattan, Kan: Sunflower University Press, 1986).

3. Richard P. Hallion, *Designers and Test Pilots* (Alexandria, Va: Time-Life Books, 1983), pp 68–78; Don Berliner, *Victory Over the Wind: A History of the Absolute World Air Speed Record* (New York: Van Norstrand Reinhold, 1983), pp 43–48, 58–65; J. R. Smith and Anthony L. Kay, *German Aircraft of the Second World War*, 7th printing (London: Putnam, 1990), pp 520–23; Gunston, *World Encyclopaedia of Aero Engines*, p 49.

4. As quoted in Smith and Kay, *German Aircraft*, p 524.

5. James J. St. Peter, ed, *The Memoirs of Ernest C. Simpson: Aeropropulsion Pioneer* (Wright-Patterson AFB, Ohio: Aeronautical Systems Div, 1987), pp 9–11; Leslie E. Neville and Nathaniel F. Silsbee, *Jet Propulsion Progress: The Development of Gas Turbines* (New York: McGraw-Hill, 1948), pp 194–95; Hallion, *Test Pilots*, p

78; Constant, *Turbojet Revolution*, pp 109, 138, 152–3, 249–50.

6. Theodore von Kármán with Lee Edson, *The Wind and Beyond: Theodore von Kármán, Pioneer in Aviation and Pathfinder in Space* (Boston: Little, Brown, 1967), pp 216–21; Richard P. Hallion, *Supersonic Flight: Breaking the Sound Barrier and Beyond* (New York: Macmillan, 1972), p 11.

7. Clarence L. "Kelly" Johnson with Maggie Smith, *Kelly: More than My Share of It All* (Washington: Smithsonian, 1985), p 95.

8. Golley, *Whittle*, pp 24, 32; Constant, *Turbojet Revolution*, pp 182–83.

9. Frank Whittle, *Jet: The Story of a Pioneer* (London: Frederick Muller, 1953), pp 24–25. See also Whittle, "Birth of the Jet Engine," p 4.

10. Golley, *Whittle*, p 248.

11. Golley, *Whittle*, pp 34–36, 51–53; Whittle, "Birth of the Jet Engine," pp 4–5; Constant, *Turbojet Revolution*, pp 138–49, 184–86; Schlaifer, *Aircraft Engines*, pp 334–35.

12. Golley, *Whittle*, pp 64–81, 248–49; Whittle, "Birth of the Jet Engine," pp 7–8; Constant, *Turbojet Revolution*," pp 186–87.

13. Golley, *Whittle*, pp 86–91, 97–98, 101–21, 250–52; Whittle, "Birth of the Jet Engine," pp 8–12; Constant, *Turbojet Revolution*, pp 191–92; Schlaifer, *Aircraft Engines*, pp 343–44, 350–51.

14. Constant, *Turbojet Revolution*, pp 194–200; Schlaifer, *Aircraft Engines*, pp 377–79.

15. Constant, *Turbojet Revolution*, pp 201–7, 210–11; Schlaifer, *Aircraft Engines*, pp 379–98, 418–28; Neville and Silsbee, *Jet Propulsion Progress*, p 8. The Junkers Jumo 004–B had an eight-stage axial compressor that produced a pressure rise of 3.1:1 at an efficiency of 78 percent. Weighing 1,590 pounds, it produced 1,980 pounds of thrust at 8,700 rpm. The engine was 12 feet 8 inches long and 31.5 inches in diameter and had a specific fuel

consumption (pounds of fuel consumed per pound of thrust produced per hour) of 1.4. Between 5,000 and 6,000 of the engines were produced before the end of the war.

16. Schlaifer, *Aircraft Engines*, pp 424–29; Constant, *Turbojet Revolution*, pp 208–13; Neville and Silsbee, *Jet Propulsion Progress*, pp 9–10, 16, 21–33; Jeffrey Ethell and Alfred Price, *World War II Fighting Jets* (Annapolis: Naval Institute, 1994), pp 42, 55. In a letter to Gen Arnold in early Sep 1944, Lt Gen Carl Spaatz conveyed his extreme alarm concerning the threat posed by the Me 262 and, in passing along his list of priorities for assistance, stated: "Most important of all is to put long-range jet fighters into the field at the earliest possible date," Spaatz to Arnold, Sep 3, 1944. His and Lt Gen Jimmy Doolittle's concerns only mounted as the Germans fielded increasing numbers of the airplane and employed them with greater tactical skill. See David R. Mets, *Master of Airpower: General Carl A. Spaatz* (Novato, Calif: Presidio, 1988), pp 246–47, 54; Richard G. Davis, *Carl A. Spaatz and the Air War in Europe* (Washington: Center for AF History, 1993), pp 512–13, 538–42, 574–75.

17. Golley, *Whittle*, pp 119–20, 122 –28, 165–71; Whittle, "Birth of the Jet Engine," pp 12–15; Neville and Silsbee, *Jet Propulsion Progress*, pp 58–63. For further details concerning the first flight and subsequent tests of the E.28/39, see Bill Gunston, "Dawn of the Jet Age," *Aeroplane Monthly*, Apr 1977, pp 184–88; John Golley, "The Whittle Revolution," *Aeroplane Monthly*, Jun 1991, pp 346–51. For details concerning jet engine development in England throughout the war and early postwar years, see Neville and Silsbee, *Jet Propulsion Progress*, pp 55–97; Schlaifer, *Aircraft Engines*, pp 332–74.

18. H. H. Arnold, *Global Mission* (New York: Harper & Row, 1949), pp 242–43; D. Roy Shoults, "Wartime Diary Tells of Exciting Days that Led to First Flight of XP–59, America's First Jet," GE Flight Propulsion Div *World*, vol I, no 26 (Oct 6, 1967), p 2; Virginia P. Dawson, *Engines and Innovation: Lewis Laboratory and American Propulsion Technology* (Washington: NASA, 1991), pp 46–49; Alex Roland, *Model Research: The National Advisory Committee for Aeronautics, 1915–1958* (Washington: NASA, 1985), 189–91; James R. Hansen, *Engineer in Charge: A History of the Langley Aeronautical Laboratory, 1917–1958* (Washington: NASA, 1987), 230–32.

19. See Schlaifer, *Aircraft Engines*, 321; Wesley Frank Craven and James Lea Cates, eds, *The Army Air Forces in World War II*, vol 6, *Men and Planes* (Chicago: University of Chicago, 1953), 246.

20. Schlaifer, *Aircraft Engines*, 489.

21. Constant, *Turbojet Revolution*, 151–77, 244, 271.

22. Capt Ezra Kotcher, "Our Jet Propelled Fighter," in "Condensed Model Specification for Twin Engine Jet Propelled Fighter Airplane," Bell Aircraft Corp Tech Data Rpt # 27–947–001–2, Apr 23, 1945, p 14. For testimony from a contemporary on Kotcher's brilliance and breadth of knowledge, see Lt Gen Ralph P. Swofford, Jr, USAF (ret) interview by Lt Col Arthur W. McCants, Jr, Apr 24–25, 1979, Oral History K239.0512–1120, AF Hist Res Agency, Maxwell AFB, Ala, pp 36–37. For comment on how widespread discussion of jet propulsion was, see Schlaifer, *Aircraft Engines*, p 486.

23. Hallion, *Supersonic Flight*, 12–13. Kotcher went on to become the primary instigator and champion of the X–1 supersonic research prog. See Hallion, *Supersonic Flight*, pp 20–26, 34–36, 40–41, 45; James O. Young, *Supersonic Symposium: The Men of Mach 1* (Edwards AFB, Calif: AFFTC History Office, 1990), pp 5–7; Hansen, *Engineer in Charge*, pp 260, 271 –73.

24. Martin P. Claussen, *Materiel Research and Development in the Army Air Arm, 1914–1915*, AAF Hist Studies 50 (Washington: AAF Historical Office, 1946), pp 98–102; Craven and Cates, *Men and Planes*, pp 228–29; Donald C. Swain, "Organization of Military Research," in Melvin Kranzberg and Carroll W. Pursell, Jr, eds, *Technology in Western Civilization:*

Technology in the Twentieth Century, vol II, (New York: Oxford, 1967), p 540.

25. Neville and Silsbee, *Jet Propulsion Progress*, pp 100–101; Constant, *Turbojet Revolution*, p 142; Hansen, *Engineer in Charge*, pp 224–26; Roland, *Model Research*, p 188.

26. *Engineer in Charge*, p 225. See also Dawson, *Engines and Innovation*, p 43; Roland, *Model Research*, p 186; Schlaifer, *Aircraft Engines*, pp 8, 33.

27. Irving Brinton Holley, Jr, *Buying Aircraft: Materiel Procurement for the Army Air Forces* (Washington: Center of Military History, 1989 ed), pp 23–25; Claussen, *Materiel Research and Development*, pp 75–80; I. B. Holley, "Jet Lag in the Army Air Corps," in Harry R. Borowski, *Military Planning in the Twentieth Century: Proceedings of the Eleventh Military History Symposium,* (Washington: Office of AF History, 1986), pp 135–36; Jacob Vander Meulen, *The Politics of Aircraft: Building and American Military Industry* (Lawrence, Kan: University Press of Kansas, 1991), pp 5–7, 44–45, 53–54, 63, 77–85. It is also significant that there were practically no research contracts issued to U.S. universities until very late in the 1930s.

28. Schlaifer, *Aircraft Engines*, pp 85, 445, 453–54; Dawson, *Engines and Innovation*, pp 1, 47. At a Jul 27, 1944, meeting of the NACA Executive Committee called to consider postwar research policy and the role industry should play in fundamental and applied research, the aeroengine industry drew the most attention. Dr. Vannevar Bush concluded, "The engine people did not do a thing on that subject [jet propulsion] or on any other unusual engine." Maj Gen Oliver P. Echols, who had served in the Materiel Div at Wright Field throughout the 1930s and was Gen Arnold's wartime chief of R&D, noted a conservatism which, in fact, was not just limited to the aeroengine manufacturers: "Industry is always looking over its shoulder at its competitors. If their research is one step ahead of their competitors they are satisfied. It has always been apparent they are not interested in the general

progress of the art." See "Notes on discussion at meeting of NACA, Jul 27, 1944," in Roland, *Model Research*, vol II, p 688.

29. Schlaifer, *Aircraft Engines*, pp 451–53; Dawson, *Engines and Innovation*, p 47; *The Pratt & Whitney Aircraft Story* (United Aircraft Corp, 1950), pp 154–55.

30. Roland, *Model Research*, p 190; Dawson, *Engines and Innovation*, pp 46–47.

31. Lt Gen Donald L. Putt, USAF (ret), interview by James C. Hasdorff, Apr 1–3, 1974, Atherton, Calif, Oral History K239.0512-724, AF Hist Res Agency, pp 31–32. Gen Putt punctuated his comments by stating that Pratt & Whitney's attempt to shift responsibility from itself was "just a bunch of malarkey."

32. Holley, "Jet Lag," pp 144–45; Claussen, *Materiel Research and Development*, pp 63, 66–67, 91–93; Roland, *Model Research*, pp 186–87; memo, Maj K. B. Wolfe, Chief Prod Eng Sect, to Col Oliver P. Echols, Asst Chief, Materiel Div, Apr 29, 1940, Records of the USAF Eng Div, 1917–1951, Box RD 3735, Washington National Records Center, Suitland, Md, (documents from this collection hereafter cited as "RD" plus digits); memo, Maj Stanley M. Umstead, Chief, Flying Branch, to Chief, Aircraft Branch, Jun 27, 1938, RD 3735; memo rpt, Maj Carl J. Crane, Chief, Flight Research Projects, to Chief, Exp Eng Sect, Mar 21, 1940, RD 3735. By the late 30s, a number of officials at Wright Field were lamenting that their dependence on NACA to conduct fundamental flight research had become so complete. In 1938, for example, the chief of the Aircraft Laboratory proposed the reestablishment of an exp flight research capability at Wright Field. In the memo cited above, Major Umstead, the chief of the Flying Branch, responded that, while his organization "recognizes the value of flight research and deplores the conditions which brought about its almost complete discontinuance as a [Materiel] Div activity," he did not believe such an action would be possible. "The first and most general reaction to such a request," he

noted, "would be a denial on the grounds of duplication of allotment of funds and delegation of work of which the NACA is now the responsible agency. Any deviation from the present policy would probably be interpreted as reflecting upon the results achieved by the NACA."

33. Holley, "Jet Lag," p 132. Though Professor Holley refers to this article as "but a modest footnote" to Edward Constant's sweeping treatise, it provides a penetrating analysis of the state of Army Air Corps R&D during the 1930s.

34. Claussen, *Materiel Research and Development*, pp 48–52; Craven and Cates, *Men and Planes*, p 177. Figures do not include funds appropriated for military pay and the construction of facilities.

35. Schlaifer, *Aircraft Engines*, p 267; Holley, "Jet Lag," pp 132–33.

36. Claussen, *Material Research and Development*, pp 42–48; Holley, *Buying Aircraft*, pp 97–98, 463–64; Benjamin S. Kelsey, *The Dragon's Teeth?: The Creation of United States Air Power for World War II* (Washington: Smithsonian, 1982), p 43; Lt Gen Laurence C. Craigie, USAF (ret), interview by Maj Paul Clark and Capt Donald Baucom, Colorado Springs, Colo, Sep 24, 1971, AF Hist Res Agency Oral History 637, pp 70–71; Lt Gen Laurence C. Craigie, USAF (ret), interview by author, Riverside, Calif, May 14, *Air Corps Engineering School* (Wright Field, Dayton, Ohio, 1935), Maj Gen Osmond J. Ritland papers, AFFTC History Office (hereafter AFFTC/HO); Holley, "Jet Lag," pp 136–37, 139, 144; Putt interview, pp 15, 23, 47–48; Swofford interview, pp 34–37, 39–40.

37. Hansen, *Engineer in Charge*, pp 222–23, 227–29, 233–44; Roland, *Model Research*, p 191; Dawson, *Engines and Innovation*, pp 48–49; *Eight Decades of Progress: A Heritage of Aircraft Turbine Technology* (Cincinnati: GE, 1990), p 28; Schlaifer, *Aircraft Engines*, pp 441–42; Neville and Silsbee, *Jet Propulsion Progress*, p 101; ltr, Robert H. Goddard to Gen George H. Brett, Chief, Materiel Div, Jul 27, 1940; memo, Maj Paul H. Kemmer, Chief, Aircraft Laboratory, to Chief, Exp Eng Sect, Aug 19, 1940; memo, Kemmer to

Chief, Exp Eng Sect, Sep 14, 1940; Maj Franklin O. Carroll, Chief, Exp Eng, to I. M. Laddon, Nov 23, 1940; memo, Carroll to Chief, Power Plant Laboratory, Oct 7, 1940; memo, Kemmer to Chief, Exp Eng Sect, Oct 29, 1940; memo, Carroll to Chief, Materiel Div, Mar 25, 1941. The Campini ducted-fan engine was employed in the Caproni-Campini C.C.2 aircraft that first flew in Aug of 1940. Attaining a top speed of only 230 mph, the aircraft—and its propulsion system — were judged failures. See G. Geoffrey Smith, *Gas Turbines and Jet Propulsion for Aircraft* (New York: Aircraft Books, Inc., 1946), pp 48–50.

38. *Eight Decades of Progress*, pp 28, 38; Schlaifer, *Aircraft Engines*, pp 88, 445; Roland, *Model Research*, pp 189–90; Dawson, *Engines and Innovation*, pp 47–48.

39. Schlaifer, *Aircraft Engines*, pp 446–48; Fred Anderson, *Northrop: An Aeronautical History* (Los Angeles: Northrop Corp, 1976), pp 39.

40. Constant, *Turbojet Revolution*, pp 223–24; Schlaifer, *Aircraft Engines*, pp 448–50; Johnson, *Kelly*, pp 72, 74–76, 79, 95; Lockheed Aircraft Corp Rpt # 2578, Manufacturer's Preliminary Brief Model Specification, Airplane, Interceptor Pursuit (Model L–133), Feb 27, 1942; Lockheed Rpt # 2581, Performance Characteristics (Model L–133), Feb 27, 1942; Lockheed Rpt # 2579, Manufacturer's Preliminary Brief Model Specification: Jet Propulsion Unit, Feb 27, 1942; Lockheed Rpt # 2571, Design Model Features of the Lockheed L–133, Feb 24, 1942. All Lockheed rpts on file AFFTC/HO. Key correspondence between Lockheed and Wright Field officials throughout the development of the L–1000 may be found in *Case History of the L–1000 (XJ37–1) Jet Propulsion Engine* on file in the Aeronautical Systems Center History Office (ASC/HO), Wright-Patterson AFB, Ohio. Bell had already been issued a contract to develop a single-engine XP–59B by that time. By late Nov of 1942, however, when it became apparent that Bell was not making satisfactory progress, Col Ralph Swofford recommended to Col Bill Craigie, chief of the Exp Eng Sect:

"Perhaps we should consider the assignment of the project to another company, which could give us expedited delivery." See memo, Col Ralph P. Swofford, Chief, Fighter Branch, Experimental Engine Section to Col L. C. Craigie, Chief, Experimental Engine Section, Nov 22, 1942, subj: XP–59B Airplanes, located in *Case History of XP–59 Airplanes*, vol I, ASC/HO.

41. Cablegram, M/A, London, England, to MILID, Jul 21, 1941 (this document, along with much of the other correspondence concerning development of the I–A series of engines cited herein, may be found in "Summary of the Whittle Jet Propulsion Engine," ASC/HO); unsigned memo to Gen [Oliver P.] Echols, Chief, Materiel Div, subj: Summary of Sep 4, 1941 Meeting with GE, Sep 5, 1941; cablegram, Arnold to [Lt Col J. T. C.] Moore-Brabizon, Ministry of Aircraft Prod, Sep 4, 1941; notebook, Maj B. W. Chidlaw, "Notes on Conferences," Sep 9, 1941; Gen Benjamin W. Chidlaw, USAF (ret), interview by Murray Green, Colorado Springs, Colo, Dec 12, 1969, Murray Green Collection, Gimble Library, USAF Academy (hereafter Green Collection); Gen Benjamin W. Chidlaw, USAF (ret), to Ronald Neal, Aug 19, 1965, AFFTC/HO; Maj Gen Franklin O. Carroll, USAF (ret), interview by Murray Green, Boulder, Colo, Sep 1, 1972, Green Collection. For more on superchargers, see Sanford A. Moss, *Superchargers for Aviation* (New York: National Aeronautics Council, Inc., 1944 ed) and Dorothy L. Miller, "Case History of (selected) Turbosuperchargers," Air Materiel Command Historical Office, Wright-Patterson AFB, Ohio, Mar 1951, ASC/HO. The latter is a compendium of key documents — or summaries of same — covering the period 1918 through 1950.

42. Chidlaw to Neal; Chidlaw interview; "Notes on Conferences."

43. Memo, E. P. Rhodes, Chief of Eng, Bell Aircraft Corp, to M. M. Henchan, PA Office, Bell, Jan 7, 1944; Capt D. T. Tuttle, "Final Rpt of Development of XP–59A and YP–59A Model Airplanes," Air Corps Tech

Rpt # 5234, Jun 28, 1945, p 3; Chidlaw to Neal; Maj Gen H. H. Arnold to Col Alfred J. Lyon, Office of Special Army Observer, London, UK, Oct 2, 1941; David M. Carpenter, *Flame Powered: The Bell XP–59A Airacomet and the General Electric I–A Engine* (Boston: Jet Pioneers of America, 1992), pp 13–14; Ronald D. Neal, "The Bell XP–59A Airacomet: The United States' First Jet Aircraft," *Journal of the American Aviation Historical Society*, vol 11, no 3 (Fall 1966), pp 156–57; Maj R. N. Wheat, Air Corps Cost-Plus-Fixed-Fee Contract, Contract number W 535 ac-21931 (5748) with Bell Aircraft Corp for Three Airplanes, Air Corps Model XP–59A, Sep 30, 1941 (this document along with much of the other key documentation concerning development of the XP–59A may be found in "Case History of the XP–59A, YP–59A...[etc.]," ASC/HO); Maj D. J. Keirn, War Dept Contract W 535 ac-22885 (6111) with GE for 15 Air Corps Type I Superchargers & Data, Sep 10, 1941. For XP–59A specifications, see memo, Maj Gen O. P. Echols, Materiel Div, to Lt Gen H. H. Arnold, Jun 5, 1942.

44. Craigie interview by author; memo, Col R. P. Swofford to CG, Materiel Command, Mar 23, 1943; Maj Gen Donald J. Keirn, USAF (ret), interview by Murray Green, Delaplane, VA, Sep 25, 1970, Green Collection; D. J. Keirn, "Pilot-Engineer-Scientist," *Daedalus Flyer*, Mar 1981, p 19; Grover Heiman, *Jet Pioneers* (New York: Duel, Sloan and Pearce, 1963), pp 52–64.

45. Carpenter, *Flame Powered*, pp 19–21; Neal, "Bell XP–59A," 160; Bell Aircraft Corp, "America's First Jet-Propelled Fighter," *ca* 1947, on file at National Air & Space Museum (hereafter NASM); memo, Brig Gen Oliver P. Echols, Materiel Div, to Gen Arnold, Nov 13, 1941; Bell Aircraft Corp Tech Data Rpt # 27–943–011: Test of 1/16-Scale Bell XP–59A Interceptor Pursuit, Feb 13, 1942, RD 813. For details concerning the design of the XP–59A, see Benson Hamlin, "Twin Aircraft Gas-Turbine Jet-Propelled Fighter," *Aircraft Gas Turbine Engineering Conference* (West Lynn, Mass: GE, 1945), pp 165–68. The papers in this volume were

presented at the Swampscott Conference hosted by GE, May 31–Jun 2, 1945.

46. Carpenter, *Flame Powered*, pp 15–17; *Eight Decades of Progress*, p 49; HQ AAF Routing and Record Sheet, Echols rpt to Arnold, Dec 13, 1941; J. C. Miller, Aviation Div, GE, to Lt Col D. J. Keirn, Apr 16, 1942, RD 3021.

47. Neal, "Bell XP–59A," p 159.

48. Memo, Col Ralph P. Swofford to Col B. W. Chidlaw, Jan 22, 1943; Whittle's complete rpts may be found attached to a series of letters from Maj J. N. D. Heenan, British Air Commission, to Col D. J. Keirn, Jun 26 and Jul 8–9, 1942, RD 3777; memo, Col B. W. Chidlaw, Chief, Exp Eng Sect, Materiel Div, to Col Beebe, Jun 5, 1942; memo, D. F. Warner, GE, to Col Keirn, *et al*, Jun 22, 1942; memo, Col F. O. Carroll, Chief, Eng Sect, to Materiel Command Asst Chief of Staff, Jul 11, 1942; memo, Col D. J. Keirn to CG, Materiel Command, Aug 13, 1942; Maj Rudolph C. Shulte, "Design Analysis of GE Type I–16 Jet Engine," *Aviation*, Jan 1946, pp 44.

49. Swofford interview, pp 44–45; memo, Col A. E. Jones, Chief, Contract Sect, Materiel Center, to Bell Aircraft Corp, May 12, memo, R. P. Swofford for Col F. O. Carroll, Chief, Exp Eng Sect, to Administrative Executive, Wright Field, Jun 27, 1942, RD 2784; Capt M. J. Dodd, "Origin of Desert Testing Station: Muroc Flight Test Base," Jul 27, 1945, on file in the USAF Flight Test Center History Office, Edwards AFB, Calif (hereafter AFFTC/HO); Robert M. Stanley, "My Thirty Years in Aviation" (unpublished manuscript), pp 26–27; Donald J. Norton, *Larry: A Biography of Lawrence D. Bell* (Chicago: Nelson-Hall, 1981), p 122; memo, Col F. O. Carroll to Asst Chief of Staff, Materiel Command, Jul 11, 1942; memo Carroll to Asst Chief of Staff, Materiel Command, Aug 12, 1942; Craigie interview by author.

50. Carpenter, *Flame Powered*, pp 28–30; Neal, "Bell XP–59A," p 162; Ted Rogers and Frank Burnham, "We Piloted the First Jet Airplane Coast-to-Coast," reprint of 1952 interview published by GE, *ca* 1967; Jack Russell interview by author,

Feb 14, 1992 (Mr Russell was one of the Bell employees involved in the project from the outset).

51. R. M. Stanley, Pilot's Rpt: XP–59A, Oct 1, 1942; Edgar P. Rhodes, "America's First Jet," *Rendezvous*, Sep–Oct 1962, p 12; Neal, "Bell XP–59A," p 161; Carpenter, *Flame Powered*, pp 30–31; Norton, *Lawrence D. Bell*, pp 122–23; Bell Corp, "America's First Jet-Propelled Fighter," p 5; Russell interview.

52. R. M. Stanley, Pilot's Rpt: XP–59A, Oct 2, 1942; E. P. Rhodes, Materiel Center Flight Test Base, to R. F. Hall, Buffalo, New York, Oct 4, 1942.

53. T. J. Rogers, "The First American Jet," *Flying*, Mar 1947, p 74.

54. Stanley, Pilot's Rpt, Oct 2, 1942; L. C. Craigie, Pilot's Rpt: XP–59A, Oct 2, 1942; Craigie interview by author. See also Lt Gen Laurence C. Craigie, USAF (ret), "P–59," *Proceedings of the XXIII Annual Symposium of the Society of Experimental Test Pilots*, Sep 26–29, 1979, pp 156–57.

55. Shulte, "Design Analysis of Type I–16 Jet Engine," pp 47–48; Craigie interview by author; 1st Lt G. J. McCaul, Materiel Center Test Base, to Col D. J. Keirn, Power Plant Lab, Aug 1, 1941, RD 2784; Donald Thomson interview by author, Edwards AFB, Calif, Feb 12, 1992 (Mr Thomson worked as a Bell instrumentation technician on the XP–59A); Harry H. Clayton telecon interview by author, Sep 14, 1992 (Mr Clayton was the AAF's lead propulsion engineer throughout the XP–59A test prog).

56. Rogers, "The First American Jet," p 68; memo, Col F. O. Carroll, Chief, Eng Div, to Asst Chief of Staff, AAF Materiel Command, Jul 11, 1942; memo, Carroll to Brig Gen B. W. Chidlaw, Chief, Materiel Div, AC/AS MM&D, Apr 24, 1943; Capt Wallace A. Lein, Materiel Center Flight Test Base, to Chief, Flight Test Branch, Sep 29, 1943, RD 1061; Capt Nathan R. Rosengarten, Materiel Center Flight Test Base, to Lt Col Osmond J. Ritland, Flight Test Branch, Wright Field, undated [ca Oct 1943], RD 1061. For details concerning all of the problems encountered with the airplanes, especially the engines, see AAF

Materiel Center [and Command] Flight Test Base Power Plant Research Group Daily Log: XP–59A [and YP–59A], Sep 26 1942–Jan 13, 1944, AFFTC/HO; daily correspondence between Randolph P. Hall and Edgar P. Rhodes, Mar 1–Oct 29, 1943 (Hall and Rhodes took turns supervising Bell's test prog at Muroc throughout this period), "Muroc Letters: Trip 1," NASM; Bell Aircraft Corp Special Projects Eng Dept Weekly Rpts, Jan 1–Oct 20 1944, "Muroc Letters," NASM. Although much of the delay in testing was caused by development problems encountered with the engines, Bell's conduct of the tests was severely criticized by AAF flight test personnel. In the letter to Lt Col Ozzie Ritland cited above, for example, Rosengarten complained: "The uncorrected data [from the AAF's initial "unofficial" performance tests] was turned over to Bell and, of course, when they worked up the data, they cried about us being about 25–35 miles lower in speed than what they expected. This, of course, is an old story with them and I didn't let it bother me because I knew they never ran tests on this particular airplane [I–16 powered YP–59A] and, like all their test work, it is strictly theoretical calculations." He was subsequently directed to prepare an evaluation of every contractor flight test organization. Lockheed was rated highest and Bell at the bottom of the scale. See memo, Rosengarten to Col Ernest K. Warburton, Chief, Flight Sect, Mar 27, 1944. In a rpt to the Dir of the Air Tech Service Command in Oct 1944, Col Mark Bradley, Chief of the Flying Sect at Wright Field, complained that, over a nearly two-year period, Bell had yet to complete a full performance evaluation of the airplane and that the sparse results that were obtained were "erroneous and misleading." See memo, Bradley to Dir, ATSC, Oct 20, 1944.

57. "Pilot's Flight Operating Instructions for Army Models P–59A–1 and P–59B–1," USAAF AN 01–110FF–1: May 5, 1945; Capt Nathan R. Rosengarten, "Complete Performance Tests on YF–59A Airplane," Eng Div Memo Rpt #

ENG–47–1739–A, AAF # 42–108777, Apr 15, 1944; Russell interview.

58. Memo, R. F. Hall, Bell Flight Test Unit, to E. P. Rhodes, Eng, Bell Aircraft Corp, Mar 29, 1943, NASM; Carroll to Chidlaw, Apr 24, 1943; memo, Brig Gen Chidlaw, Chief, Materiel Div, AC/AS MM&D, to Gen Arnold, Apr 29, 1943; B. Hamlin, "Performance Summary for X– and YP–59 Airplanes," Bell Aircraft Corp Rpt # 27–923–013, Nov 30, 1944, RD 813.

59. AAF Materiel Command Memo Rpt on YP–59A #5 Airplane, AAF # 42–108772, Dec 14, 1943, NASM; F. Crosby, "Preliminary Speed Tests of the YP–59 Airplane with I–A Units," Bell Aircraft Corp Tech Data Rpt # 27–923–008, Dec 12, 1943, RD 813; Capt Wallace A. Lein, "Pilot's Comments," AAF Memo Rpt on YP–59A Airplane # 42–108722, Jan 7, 1944, RD 813; Rosengarten, Complete Performance Tests on YP–59A Airplane.

60. Hall to Rhodes, Sep 4, 1943, "Muroc Letters: Trips," NASM.

61. Memo, Col H. Z. Bogert, Tech Staff, Eng Div, to Chief, Aircraft Laboratory, Jul 10, 1943; memo, Melvin Shorr to Col R. P. Swofford, Aug 28, 1943, RD 813; R. H. Wheelock, "XP–59A Nacelle Entrance and Sealing Performance Investigation," Bell Aircraft Corp Tech Data Rpt # 27–923–006, Aug 28, 1943, RD 813; memo, Melvin Shorr to Aircraft Laboratory, Jan 21, 1944, RD 813; P. Largustrom, Investigation of Various Nacelle Modifications on the YP–59A Airplane, Bell Aircraft Corp Tech Data Rpt # 27–923–018, Dec 9, 1944, RD 813; Lt Col O. G. Celline, Pilot's Rpt, YP–59A Gun Firing Tests, Apr 16, 1944; memo, Lt Col Oliver G. Celline to CG, III Fighter Command, Apr 25, 1944.

62. "Twin Aircraft Gas-Turbine Jet-Propelled Fighter," p 174. For difficulties encountered in the testing of first-generation American turbojet engines, see W. J. King, "Flight Testing of Aircraft Gas Turbines," *Aircraft Gas Turbine Engineering Conference*, pp 241–54.

63. Rosengarten, Complete Performance Tests on YP–59A Airplane; Gen H.

Technology and the Air Force

H. Arnold to Maj Gen Howard C. Davidson, Mar 20, 1944; AAF Proving Ground Command, Final Rpt on the Test of the Operational Suitability of the P–59A–1 Airplane, Apr 17, 1944; Brig Gen E. L. Eubank, "Tactical Suitability of the P–59A Airplane," AAF Board Project # (M–1) 46A, May 3, 1944; memo, AC/AS OC&R to C/AS OC&R and AC/A R&S, Oct 5, 1944; transcript of telecon, Ray Whiteman, Bell Aircraft Corp, to Brig Gen Orval R. Cook, Oct 11, 1944.

64. 1st Lt Bastian Hello, Final Rpt of Development, Procurement, Performance, and Acceptance [of] XP–80 Airplane, Air Corps Tech Rpt # 5235, Jun 28, 1945, pp 3–8, 13; C. L. Johnson, *et al*, XP–80 Log Book, Jun 18 1943–Jan 28, 1944, AFFTC/HO; Lockheed Aircraft Corp Rpt # 4592: Preliminary Flight Tests on Lockheed XP–80 Airplane, Feb 1, 1944, AFFTC/HO; Lockheed Aircraft Corp Rpt # 4732: Performance Flight Tests of the Model XP–80 — Tests 6 through 32, May 5, 1944, AFFTC/HO; A.M. "Tex" Johnston with Charles Barton, *Tex Johnston: Jet-Age Test Pilot* (Washington: Smithsonian, 1991), p 69.

65. Memo, Col H. Z. Bogert, Chief, Tech Staff, Eng Div, to Brig Gen F. O. Carroll, Chief, Eng Div, Dec 21, 1942; Col D. J. Keirn to R. C. Muir, GE, Mar 13, 1943; memo, Carroll to CG, AAF, Jul 26, 1943; draft memo rpt, Col D. J. Keirn, subj: Jet Propulsion Aircraft Engines, Aug 13, 1943; Dale D. Streid, "Design Analysis of GE Type I–40 Jet Engine," reprint of article published in *Aviation*, Jan 1946; Clarence L. Johnson, "Development of the Lockheed P–80A Jet Fighter Airplane," *ca* late 1945 (unpublished manuscript); Jay Miller, *Lockheed's Skunk Works: The First Fifty Years* (Arlington, Tex: Aerofax, Inc., 1993), pp 22–27; E. T. Woolridge, *The P–80 Shooting Star: Evolution of a Jet Fighter* (Washington: Smithsonian, 1979), pp 26, 36–42, 45; Swofford interview, pp 43–44.

66. Draft memo rpt: Jet Propulsion Aircraft Engines, p 12.

67. "Research and the AAF," reprinted in *Aeronautical Engineering Review* vol 4 (Oct 1945), pp 1, 3. For an eyewitness account of the excellence of the German R&D facilities, especially in comparison with those at Wright Field, see Col "Toby" Tobiason, Instrument Lab, to Col Theodore B. Holliday, Chief, Instrument Lab, Jun 14, 1945, attached to memo, Holliday to Brig Gen Laurence C. Craigie, Dep Chief, Eng Div, Jul 20, 1945, Lt Gen Laurence C. Craigie Papers, AFFTC/HO (hereafter Craigie Papers). Tobiason commented: "Everything is done on a big scale, nothing cheap. In other words, they have or had the best of facilities while we are still in the planning stage to a certain degree." He was also surprised by "the number of highly technical scientists employed by the G.A.F. [German Air Force]" in comparison with the AAF.

68. Theodore von Kármán, *Where We Stand* (USAAF, 1946) in Michael H. Gorn, ed, *Prophecy Fulfilled: "Toward New Horizons" and Its Legacy* (Washington: AF History and Museums Program, 1994); Theodore von Kármán, "Science, the Key to Air Supremacy," in *Toward new Horizons* (USAAF: 1946); Robert Frank Futrell, *Ideas, Concepts, Doctrine: Basic Thinking in the United States Air Force, 1907–1960* (Maxwell AFB, Ala: Air University, 1971), pp 275–76; Craigie interview by author; Swain, "Military Research," pp 543–48; A. Hunter Dupree, *Science and the Federal Government: A History of Policies and Activities to 1940* (New York: Harper & Row, 1964), pp 373–74. Craigie spoke and testified frequently on the urgent need for funding, adequately trained manpower, and state of the art laboratory facilities comparable to those developed by the Germans. In particular, he pushed hard for funding ($500 million) for the development of the complex of wind tunnels and lab facilities that would be developed at Tullahoma, Tenn (Arnold Eng Dev Center). For an example of Gen Craigie's dealings with Congress on these issues, see his testimony before the U.S. Senate Subcommittee of the Committee on Interstate Commerce, "To Establish a National Air Power Board, Jun 6, 1946, pp 5, 7–27, Craigie Papers. For an

example of a speech addressing the same issues, see "Future of Flight," delivered to the Engineer's Club, St. Louis, Mo, May 8, 1947, Craigie Papers. Old habits of mind persisted and the battle for funding and influence within the new USAF would be long and hard fought. In 1949, for example, Maj Gen Donald L. Putt, who was soon to become commander of the new Air Research and Development Command, complained: "There are those in high positions in the AF today who hold that research and development must be kept under rigid control by 'requirements' and 'military characteristics' promulgated by operational personnel who can only look into the past and ask for bigger and better weapons of World War II vintage. . . . They have not yet established that partnership between the strategist and the scientist which is mandatory to insure that superior strategy and technology which is essential to future success against our potential enemies." See Futrell, *Ideas, Concepts, Doctrine*, pp 275–76.

69. Smith, *Gas Turbines*, p 33. See also Neville and Silsbee, *Jet Propulsion Progress*, p xi.

70. For the status of turbojet R&D and projections for the future, as of 1945, see Lt Col R. B. Clevering, Chief, Aero Equipment Sect, Procurement Div, to Col [D. J.] Keirn, Eng Div, Sep 25, 1945, ASC/HO; von Kármán, *Where We Stand*, pp 38–50; von Kármán, "Science, the Key to Air Supremacy," pp 21–26; Frank L. Wattendorf, "Gas Turbine Propulsion," in *Toward New Horizons*, vol IV; Maj Langdon F. Ayers, "Present and Future Trends in the Power Plant Laboratory," (unpublished manuscript) Jan 1, 1950, ASC/HO.

71. See, for example, memo, Brig Gen F. O. Carroll, Chief, Eng Div, to CG, AAF, Mar 29, 1944; memo, Carroll to CG, AAF, Jun 26, 1944 (this memo was drafted by Kotcher).

72. Gunston, *World Encyclopaedia of Aero Engines*, pp 124–25.

Michael H. Gorn is an aerospace historian and the author of *The Universal Man: Theodore von Kármán's Life in Aeronautics,* published by Smithsonian Institution Press. He has also written *Harnessing the Genie: Science and Technology Forecasting for the Air Force, 1944-1986* and is completing a biography of Dr. Hugh L. Dryden of NASA, eminent scientist and government administrator. Dr. Gorn received a doctorate in history from the University of Southern California. In 1994, he was the Alfred V. Verville Fellow in Aerospace History at the National Air and Space Museum.

Technological Forecasting and the Air Force

Michael H. Gorn

Harnessing the Genie described the five major scientific studies undertaken by the U.S. Army Air Forces and the U.S. Air Force since the end of World War II. These included Toward New Horizons in 1945, the Woods Hole Summer Studies in 1957-1958, Project Forecast in 1964, New Horizons II in 1975, and Project Forecast II in 1986.

Since that time, new forces have gathered to produce a sixth report entitled *New World Vistas*. I first thought these studies spread out over fifty years, and with no apparent connection to one another, represented nothing more that casual attempts by the Air Force to predict the technological future. But shortly after initiating research on the subject, it became clear that several things linked the five reports and, indeed, also applied to the forecast being prepared at present. Rather than a collection of unrelated analysis, common threads ran through them. This pattern surprised me. At first appearance, the studies seemed to be entirely random, without connection to one another. They occurred without prior plan; no one organization produced them; their participants varied greatly; their methodologies were not at all uniform; their conclusions varied significantly; and, in fact, they did not even share common purposes.

Gen. Hap Arnold initiated Toward New Horizons to survey the most advanced air power technologies of World War II and project them into the future. The Woods Hole Summer Studies organized hundreds of university scientists to predict the short- and long-term military applications of space. Project Forecast attempted to revitalize Air Force thinking by linking national policy issues to advanced scientific concepts and weapons systems. New Horizons II endeavored to point the way toward incremental technological improvements in a period of expected scarcity, that is, the period following the Vietnam war. Finally, Project Forecast II sought to jolt the Air Force laboratories out of suspected complacency. Thus, for a variety of internal and external reasons, at roughly ten-year intervals since the Second World War, the Air Force has launched major science and technology forecasts.

Despite their diverse aims, the five studies did have several factors in common. First, they reflected an increasing reliance on in-house science and a steady decline in the role of independent scientists for long-range forecasts.

Technology and the Air Force

Moreover, as the importance of outside scientists slowly diminished, the Air Force's Scientific Advisory Board lost its influence over the process of predicting the future of technology. Parallelling and hastening this trend, military scientists and engineers trained in R&D came gradually to dominate science forecasting. Finally, severed from the scientific advisory board in the 1950s, the practice of periodically reporting the future of science and technology found itself an institutional orphan, unattached to any particular Air Force organization and redefined according to the imperatives of each new study director. These events developed almost absent-mindedly, with so little notice that neither military nor civilian scientists and engineers fully appreciated their occurrence or understood their significance.

The four studies that followed Toward New Horizons increasingly diverged from the pattern established by its director, Dr. Theodore Von Kármán. This eminent Hungarian physicist and mathematician did not consciously intend to present the USAF with a model for scientific forecasting. He only sought to draft a comprehensive yet practical analysis of the breakthroughs resulting from World War II aeronautics. But, in large part because of von Kármán's reputation and Arnold's patronage, it won converts at headquarters USAF and Wright Field, contributed greatly to the service's image of itself as a technically-oriented force, and established the practice of long-range planning.

Toward New Horizons operated on four principles: to endure fresh disinterested views, advice should be given by people outside the Air Force; senior university scientists, especially those equipped by temperament and experience to be generalists, should populate the panels; the report should be comprehensive, the product of sufficient time to allow serious reflection; and the findings should place scientific and technological possibilities in the context of usefulness to national defense, air power requirements, and technical practicality.

As the most influential aeronautical scientist of the century, von Kármán selected the participants of Toward New Horizons — thirty-three academicians chosen mainly from the California Institute of Technology and the Massachusetts Institute of Technology. This project originated with a request to von Kármán from Arnold to search the world for the most advanced aeronautical ideas generated by wartime research and project them far into the future.

After a year of wide-ranging study in America, Europe, and Asia, the von Kármán team, known then as the Scientific Advisory Group, issued a fourteen-volume summary of the scientific lessons of World War II and the technical implications likely to result from these breakthroughs. The product principally of physicists and mathematicians, it related advanced theoretical concepts to practical military objectives. Von Kármán delivered his study with two chief recommendations: first, scientific inquiry must be pursued, as Jim Young pointed out, constantly and applied quickly to support air power; and second, a single, distinct Army Air Force's organization should be developed and devoted exclusively to aeronautical research and development.

42

Dr. Theodore von Kármán

Ultimately, von Kármán and his report proved persuasive. The Air Force established a permanent scientific advisory board in 1947 and the Air Research and Development Command three years later. But the need remained for comprehensive long-term scientific advice for the Air Force. In 1957 the ARDC commander, Gen. Thomas S. Power, initiated a sequel to Toward New Horizons, held in Woods Hole, Massachusetts during the summers of 1957 and 1958. Von Kármán, with great reluctance, again chaired the meetings. But this time the National Academy of Sciences acted as host and, under its auspices, attracted the nation's most able scientific talent from think tanks, academia, industry, and government.

The Woods Hole Summer Studies shared a basic kinship with Toward New Horizons. University scientists dominated the proceedings, led the panels, and decided for themselves the subjects for discussion. But in their mechanics, the two differed sharply. An army of participants almost ten times the size of Toward New Horizons descended on Cape Cod during the warm months of 1957 and again the next year. Over 300 people — 198 participants and 105 consultants — appeared at the conference site. The contributors to the Woods Hole Summer Studies, too many for long-term residence or coherent group discussion, remained at the Massachusetts location for only a few days at a time, leaving the Hungarian's personal assistants to weave the committee findings into thirteen summary volumes.

Unlike Toward New Horizons, which was organized along the lines of the applied sciences, the Woods Hole leaders chose to group the recommendations into weapons system families. As a result, von Kármán's second attempt at air power forecasting yielded a broad but cautious report. Air Force brass responded to it with little enthusiasm. They did not find its conclusions untrue or invalid, merely irrelevant. It failed to answer a question of profound national importance: how to meet the defense crisis implicit in the October 1957 launch and orbit of the Soviet satellite, Sputnik? Consequently, the influence of Woods Hole proved to be nil.

At an hour when Air Force officials desperately sought measures to overcome the apparent Soviet lead, this omission in the Woods Hole Summer Studies, based on von Kármán's unyielding belief that long-range reports must provide balanced coverage of new technologies, had serious ramifications for the forecast which followed. The experience persuaded USAF authorities that civilian scientists should be subject to greater military oversight in future technology analysis. As a consequence, starting in the 1960s, the scientific advisory board found itself much less autonomous and independent than in the founding years under its famous leader.

No one did more to harness science to air power objectives than Gen. Bernard A. Schriever. As Commander of Air Research and Development Command and its successor, Air Force Systems Command, Schriever had demonstrated great capacity during the 1950s in bringing the American ICBM force to fruition. Then, directed in March 1963 by Secretary of the Air Force Eugene M. Zuckert, he undertook a major review of technologies applicable to USAF needs through the mid-1970s. Called Project Forecast, it enlisted almost 500 participants, balancing blue-suiters who understood the requirements of war with some of the most eminent civilian scientists and engineers from the universities, manufacturers, institutes, and government. In fact, Schriever drew his team from an unprecedented variety of sources — from the USAF and sixty-three other federal agencies, from twenty-six institutions of higher learning, from seventy corporations, and from ten nonprofit organizations. The selection of Schriever and his project manager, Maj. Gen. Charles Terhune, in itself suggests a maturing of the forecasting process. Both men not only understood the scientific world, but represented a growing number of engineers in uniform able to grasp the technical and military aspects of weapons development. As a result, Schriever and Terhune structured Project Forecast so that all ideas produced by the technical panels were assessed in relation to factors of cost and military requirements. In addition, evaluations of the predominant threats to American security and broad foreign policy objectives further narrowed the field of candidate technologies.

Finally, the capability panels translated the concepts which survived this screening process into actual weapons systems. Far more structured than Toward New Horizons, Project Forecast, nonetheless, incorporated truly independent scientific advice and invited the widest possible participation. Also, like Toward New Horizons, it strove for comprehensiveness, producing twenty-five volumes which related new air power technologies to the world in which the Air Force found itself. Project Forecast enjoyed widespread influence throughout the USAF and many of its recommendations, such as huge intercontinental transports and lightweight composites for aircraft and engine design, were fulfilled.

Could General Schriever's success be duplicated in the next long-range forecast? The answer would wait a decade. Almost ten years after his milestone work, the Air Force undertook yet another long-range study. Known ambitiously as New Horizons II it began in August 1974, at the direction of Air Force Chief of Staff, Gen. David C. Jones. Its Executive Director, Maj. Gen. Foster Lee Smith, the Headquarter's USAF Deputy Chief of Staff for Plans and Operations, led a steering group of Air Staff major generals. Unlike Toward New Horizons, Woods Hole and Project Forecast, civilian science advice had little weight in the deliberations. Indeed, independent scientists and members of the Scientific Advisory Board functioned only as expert consultants — not as recognized participants in the study process. As an in-house survey, all of the forty-nine study members of New Horizons II but one, the Chief Scientist of the Air Force, wore the service uniform, and almost half worked in the Offices of the Deputy Chief of Staff for Plans and Operations.

In its functioning, New Horizons II lacked both the global view and the technical scope of Toward New Horizons and Project Forecast. Its five technology panels oriented themselves toward mission rather than scientific objectives, and the study process lacked a crucial feature of Project Forecast: that of filtering candidate technologies through cost, capability, and threat assessments. In its final report to Gen. Jones, the New Horizons staff recommended a number of initiatives: advanced data processing for command and control survivable military satellites, laser weapons in the atmosphere and in space, and aircraft upgrades for night and all-weather flying. Despite the constriction of defense spending after the U.S. withdrawal from Vietnam, the report also suggested a heavy lift global-range transport of even greater capability than the C–5 aircraft. But lacking Gen. Schriever's prestige and contacts in the scientific, industrial and political worlds, New Horizons II exercised only limited influence over the course of Air Force technology. Yet it did foster, perhaps unintentionally, the idea of limiting independent civilian participation in long-term science studies.

During the mid-1980s, once more a decade after the last one, another long-term forecast occurred. Begun by AFSC commander Gen. Lawrence A. Skantze, it continued, and in some ways contributed, to the tradition of blue suit leadership in USAF forecasting. Project Forecast II, initiated in August 1985, modeled itself on Project Forecast in attempting a systematic and comprehensive survey. It utilized a similar filtering or matrix process which accounted for thread and cost factors in its analysis panels — scientific possibilities had ten technology panels and military requirements had five mission panels. Some 200 people contributed as panelists or consultants to Project Forecast II, but there the similarities ended.

Distinctly different from the varied institutional affiliations of the initial Forecast participants, all 200 Forecast II participants were Air Force employees. Although a majority of the 107 panelists were civilians, most worked in the

Technology and the Air Force

AFSC laboratory structure. Indeed, Gen. Skantze undertook Forecast II partly as a means of infusing the systems command laboratories with new ideas. While independent civilian advice may have been solicited by the Forecast II staff, little found its way into the final report. Even the Air Force Scientific Advisory Board, the mother institution of USAF forecasting, had almost no impact on its ultimate contents. Altogether, about 2,000 technical ideas flowed from the Forecast II process. Nine hundred originated in the Forecast II offices and 1,100 came from outside sources, including universities, industries, and think tanks. While all of the 900 received full consideration in the project screening process, 90 percent of the 1,100, that is, the outside suggestions, were rejected without any formal review. Thus in-house science and engineering reached its zenith in Project Forecast II.

The principal feature of the von Kármán model of applying independent civilian talent to long-range advising all but disappeared. Gone, too, was the Toward New Horizons practice of relating the technological future to the institutional life of the entire Air Force and the nation's defense needs as a whole.

Eventually seventy candidate systems and technologies emerged from the rigorous Forecast review system. Unlike its namesake, Forecast II did not relate them to national security policy or overall military objectives, but simply presented them as the technological champions of the future. They included such highly advanced concepts as knowledge-based computer systems, ultrastructured materials, antiproton technology, the transatmospheric vehicle, widely distributed phased radar in space, and the so-called super cockpit.

Attempts at implementing the massive 1,700-page final report began almost immediately with significant AFSC laboratory funding devoted specifically to further exploration of these selected technologies. Nonetheless, by 1985, responsibility for Air Force technology forecasting had devolved on a single major command, leaving the process a corporate orphan in the USAF as a whole. After yet another ten years, the Air Force discovered the need for yet another technology forecast. But this time the old imperatives no longer implied. The demise of the USSR transformed the nation's international objectives in ways not yet fully understood.

The entire U.S. defense posture underwent changes of heroic size, as did the required technical capability of the armed forces. In the wake of falling budgets and overturned priorities, the USAF has reassessed its future. No less than its operational basis, the Air Force's scientific and technical substructure also underwent intense scrutiny. The renewal of Air Force science began in 1994 at the behest of the Chief of Staff, Gen. Merrill A. McPeak. He told the Scientific Advisory Board to look toward the year 2020 and "stretch beyond the evolutionary and make sure we don't miss the leap frog technologies, the breakthroughs that are our best guarantee that the Air Force will remain the world's dominant air and space power."

At the same time, the Secretary of the Air Force, Dr. Sheila Widnall, seized the occasion of the board's fiftieth anniversary year to reinvigorate it. Her initiatives began with a symposium late in 1994 commemorating the board's half century of service to the Air Force by looking back at its major achievements. On November 29, 1994, Dr. Widnall continued her offensive on behalf of the board by directing its chairman to focus the energies of the SAB over the following year on a long-range forecast she called New World Vistas. The Secretary wanted New World Vistas first, to predict those scientific fields likely to be at the forefront of technological change; second, to predict their impact on affordability of systems and operations; third, to review dual use and commercial opportunities; fourth, to identify technology issues of special interest to the military; fifth, to advise on science and technology infrastructure; and finally, to consider joint service applications. Not accidentally, New World Vistas appears to be a hybrid of Toward New Horizons and Project Forecast I. Von Kármán's original report has been studied closely in recent years and its influence is clear both in the structure and objectives of New World Vistas.

As in 1945, the last time a new world system began to materialize, the Scientific Advisory Board is in a pivotal position in the process. The board will consider structural changes in the Air Force laboratories and the study is guided, ultimately, not by a search for novel technical solutions, but by von Kármán's old objective of identifying broad offensive and defensive capabilities for future aeronautics. New World Vistas borrows equally from Project Forecast, the other great Air Force study of the past, in its broad inclusion of industrial and university scientists and engineers and in its clear organizational distinction between technical ideas on the one hand and practical Air Force applications on the other. But, unlike either of the earlier reports, New World Vistas is conceived as part of a broad servicewide planning process. Indeed, the Scientific Advisory Board canvassed missionary and planning staffs at the major commands, as well as the so-called revolutionary planning team in the Office of the Deputy Chief of Staff for Plans for insights about which new technologies best suited long-range Air Force plans.

Once the report is completed in December 1995, fifty years to the month since *Toward New Horizons* appeared, the senior Air Force leadership will review its recommendations and integrate the most desirable into the services corporate strategic plan. New World Vistas' stated objective to foster "a more capable, flexible, and less expensive Air Force" both reflects the upheavals the world has undergone since 1989 and suggests how different the technological choices will be in this latest instance of Air Force forecasting.

47

Richard P. Hallion is the Air Force Historian. He earned his Ph.D. in the history of aerospace technology from the University of Maryland in 1975. He has held numerous positions in government and academe, including the Harold Keith Johnson Visiting Professorship at the U.S. Army War College and the Charles A. Lindbergh Visiting Professorship at the Smithsonian Institution's National Air and Space Museum. Dr. Hallion has written fourteen books on military history and technology. His latest work is *Storm Over Iraq: Air Power and the Gulf War.*

The Air Force and the Supersonic Breakthrough

Richard P. Hallion

The turbojet revolution, one of the two great revolutions affecting aeronautics at midcentury, promised flight at speeds faster than 500 mph. But it was a second and essentially contemporaneous revolution that enabled the fulfillment of this promise. This latter revolution was the achievement of practical transonic and supersonic flight: flight around or in excess of the speed of sound.[1]

The accomplishment of supersonic flight involved a multiagency and industrial partnership, development of creative ground and inflight research methods and tools (including specialized research airplanes), the design of experimental prototypes conceived to meet perceived operational requirements, and, eventually, the production and deployment of operational military systems. Some of the latter were great successes, while others were far less so. Indeed, the story of the Air Force and supersonic flight is one that has strong elements of both great success and nagging disappointment, at once both an encouraging and cautionary tale.

Background to a Breakthrough

The birth of supersonic flight in the United States stemmed from a generalized "compressibility crisis" encountered by high-performance fighters, propeller-driven, jet-propelled, and rocket-propelled, in the late 1930s and on through the Second World War. Essentially, as an airplane dove at speeds exceeding Mach 0.7, the accelerated airflow around the aircraft (which could, in places, exceed the speed of sound) generated a series of disturbing performance anomalies including shockwave formation, abrupt drag rise, a marked reduction in lift, an abrupt reduction (for propeller-driven aircraft) in propeller efficiency, pronounced airframe buffet, occasional flutter of flight control surfaces, and unsettling control characteristics. Depending on the nuances of a particular aircraft, the latter might consist of undamped or poorly damped longitudinal (pitching); lateral (rolling); and directional (yawing) oscillations; "wing drop" as the lifting characteristics of a wing abruptly changed; abrupt pitch-up or pitch-down tendencies; or dramatic changes in control feel and function, ranging from extreme oversensitivity, to a sense that the control stick was fixed immovably in concrete or even, in some cases, to total control reversal.

Many aircraft were lost as a result of structural weaknesses, inadequate understanding of transonic phenomena and their influence on aircraft controllability, adverse handling qualities, and poor pilot training. Virtually all high-performance fighters of World War II were susceptible to compressibility effects, and some , such as the American Lockheed P–38 Lightning, the British Hawker Typhoon, and the German Messerschmitt Me 163 Komet, were notoriously so. Thus one powerful impetus for developing an understanding of high-speed aerodynamic phenomena was flight safety — to enable the design of safer high-performance fighter aircraft.

A second impetus driving the supersonic breakthrough was to improve predictive methodologies; in the era before slotted-throat wind tunnels, no truly adequate method existed to furnish reliable transonic design information. Further, the information that did exist was, in most cases, suspect.[2] Based on tunnel tests in closed-throat tunnels, designers as early as 1920 knew that drag and lift experienced a marked inverse relationship as velocities approached that of sound, with drag rising alarmingly.[3] In 1935, British aerodynamicist William F. Hilton stated that this drag rise phenomenon "shoots up like a barrier against higher speed as we approach the speed of sound." The resulting news accounts shortened Hilton's cautious statement to the far more lurid and popular "sound barrier."[4]

But the exact magnitude of these phenomena was not precisely understood, for shockwave reflection across the test section of closed throat wind tunnels limited their usefulness precisely where researchers were most interested in understanding what was happening: Mach 0.75 to Mach 1.25. Below Mach 0.75, shock waves did not form. Beyond Mach 1.25, shock waves assumed such a sharply raked cone form that they generally did not interfere with reliable tunnel measurements. Clearly the intervening transonic region was one that required close study, if for no other reason than drag rise plots which hinted at an asymptotic curve as aircraft velocity approached the magic Mach 1 mark.

The third impetus promoting the supersonic breakthrough was military necessity and, related to it, the impact of postwar technical sifting through the ashes of Nazi Germany's aeronautical research establishment. During the war, the dive limitations of conventional fighter aircraft seriously impacted combat effectiveness of all the combatant nations. The United States was fortunate that its major air superiority fighters, the Republic P–47 Thunderbolt and North American P–51 Mustang, had generally acceptable high-speed dive characteristics. The Lockheed P–38 Lightning, however, was quite a different matter, and required extensive study and the addition of dive recovery flaps before it became a fully satisfactory airplane.[5] During World War II, the United States, Great Britain, and Nazi Germany all ran major investigation programs on high-speed fighter performance and handling qualities in a quest to improve the

The Messerschmitt Me 262.

performance of existing aircraft such as the P–47, Spitfire, and Bf 109, as well as to develop a generalized knowledge base applicable to the design of future high-speed jet airplanes.

Though it did not serve as either the catalyst or the primary reason for postwar American supersonic research and development efforts, the opportunity to examine Nazi Germany's research establishment at the end of World War II nevertheless constituted a major force behind such work, particularly given the unexpectedly strong postwar threat posed by a powerful and technologically advancing Soviet Union. In truth, Nazi Germany badly managed its scientific and technological establishment, to the great relief of the Allied war effort. But, if unfocused, nevertheless there had been a surprisingly fecund quality to German research, spawning a dizzying series of projects and ideas, including operational jet fighters and bombers (e.g., the Messerschmitt Me 262 *Schwalb* and Arado Ar 234 *Blitz*), pilotless weapons (e.g., the *Fritz-X* glide bomb, the V–1 cruise missile, and the V–2 ballistic missile), rocket fighters (e.g., the Messerschmitt Me 163 *Komet* and Bachem Ba 349 *Natter*), surface-to-air missiles (e.g., *Wasserfall, Rheintochter, Feuerlilie,* and *Enzian*), fanciful future projects (e.g. the Lippisch P–13 ramjet fighter and the Messerschmitt P.1101 sweptwing testbed), an air-launched rocket-propelled sweptwing supersonic research airplane (the DFS 346), a long-range Mach 4+ winged missile based on the earlier V–2 (the sweptwing A–4b, the first winged vehicle to ever exceed Mach 1 — in February 1945 — though it subsequently broke up during atmospheric entry), and even a proposed rocket-propelled hypersonic orbital skip-bomber (the sled-launched Sänger-Bredt antipodal aircraft). Much of this German work would serve as inspiration and/or confirmation for ideas

Technology and the Air Force

expressed in two of the most influential postwar Army Air Forces reports, Theodore von Kármán's *Where We Stand* (August 1945) and *Toward New Horizons* (December 1945), to be discussed subsequently. By that time, however, the first steps in the Air Force's steps to the supersonic breakthrough were already well in hand.[6]

The Kotcher Initiatives

Much of the success that accompanied the early American efforts at supersonic flight are directly attributable to a remarkably prescient and dedicated Army Air Forces engineering officer, Maj. Ezra Kotcher.[7] Kotcher, a prewar civilian engineering instructor at the Air Corps Engineering School at Wright Field, became interested in supersonic flight while listening to a lecture in the mid-1930s by Lt. Col. Heinz Zornig, the chief of ballistics research for the Army's Aberdeen Proving Ground. Artillery shells obviously moved faster than sound, and Zornig's lecture showed that they experienced the same abrupt transonic drag rise characteristic of wings. But their transonic drag rise did not exceed beyond a factor or three or four the subsonic drag value. There were daunting differences; if nothing else, artillery shells were symmetrical bodies, not constrained by wings and other awkward protuberances. Nevertheless, as he listened, Kotcher developed an intuitive feeling that the "sound barrier" was "not necessarily a permanent flight barrier, but rather a wind tunnel technique barrier or a psychological barrier."[8]

In August 1939, Kotcher submitted a report on future aeronautical research to the Kilner-Lindbergh board established the previous May by Army Air Corps Chief Maj. Gen. Henry H. "Hap" Arnold. In his report, Kotcher called for "comprehensive flight research programs" to achieve supersonic flight and advocated development of gas turbine and rocket propulsion systems to over-come the obvious disadvantages of propeller-driven designs. Kotcher's boldness was remarkable; in 1939, the leading American fighter, the Curtiss P–36, could barely fly half as fast as the speed of sound at altitude, while the previous year, a U.S. Navy engineering board had concluded that gas turbines were utterly unsuited as a means of propulsion for aircraft.[9]

Kotcher's calls went unanswered by the AAC, though another federal agency, the prestigious National Advisory Committee for Aeronautics, had a number of leading aerodynamicists — such as Eastman Jacobs and, foremost, John Stack — who were pursuing studies aimed at extending the flight frontier beyond Mach 1. But the NACA had its own blind-side: the potential of the gas turbine airplane. In April 1941, after seeing the rapid progress of British gas turbine research, Hap Arnold returned to the United States determined never to let the AAC develop a dependency on such outside organizations as the NACA for its future capabilities. The Arnold trip to England marked the birth of his interest in creating within the service its own scientific forecasting and research

52

capabilities. This led, in time, to his strong reliance on Theodore von Kármán, the émigré Hungarian scientist who directed the Guggenheim Aeronautical Laboratory at the California Institute of Technology, eventually triggering the creation of the postwar Air Force Scientific Advisory Board.[10]

After Pearl Harbor, Kotcher shed his civilian suit for a uniform, and by 1944, he was deep in the midst of project management for various AAF jet and rocket fighters, his interest in transonic flight unabated and unrequited. By late 1943, a burgeoning interest in supersonic flight had emerged within the Federal scientific establishment. Thus, in mid-January 1944, the Materiel Division at AAF Headquarters issued a confidential technical instruction authorizing the initiation of a study for a transonic research airplane to explore flight conditions from 600 to 650 mph.[11] The chief of the Engineering Division at Wright Field, Brig. Gen. Franklin O. Carroll, asked Arnold's (as yet unofficial) scientific advisor von Kármán if it were possible to develop an airplane to fly at Mach 1.5. The distinguished scientist assembled a small study team and generated a quick report over a weekend favorably endorsing the feasibility of a ramjet-powered airplane with a gross weight of 10,000 pounds and a small wing of 125 square feet that would be capable of reaching Mach 1.5 at 40,000 feet and flying at that speed for five minutes. With ramjet technology in its infancy (even more so than gas turbines and liquid-fuel rocket technology), such a design had little chance of winning development approval. Nevertheless, it was an important psychological encouragement, as was a virtually equivalent Mach 1.6 proposed ramjet-powered research airplane study by two NACA engineers, Macon Ellis and Clinton E. Brown, in midsummer 1945, by which time, of course, the XS–1 and the D–558–1 were well underway.[12]

In mid-March 1944, a series of meetings at the NACA's Langley Memorial Aeronautical Laboratory drew together transonic research aircraft partisans within the NACA, the Navy, and the AAF and resulted in the NACA assigning personnel to coordinate with the two services for the possible development of such a craft. Having been slow to awaken to the turbojet revolution, the NACA now conservatively favored the lower risk turbojet over the more exotic liquid-fuel rocket. However, in April 1944, on his own initiative, Kotcher had directed a comparative study that clearly demonstrated the superiority of the rocket. Kotcher had the Design Branch of the Aircraft Laboratory at Wright Field evaluate two similar configurations for a "Mach 0.999" design — a whimsical reference to the supposed "impenetrable sonic barrier" — one powered by a TG–180 (J35) turbojet and the other by a proposed 6,000 lb. thrust Aerojet rocket engine. The rocket powered variant — which bore a generalized similarity to the subsequent X–1, with the exception of its horizontal tail location — clearly offered superior performance. Kotcher showed the design to von Kármán, who concurred in its potential, and then Kotcher took it to General Carroll who endorsed Kotcher's design approach, effectively putting the service on record behind the Kotcher position.[13]

Technology and the Air Force

From this point onwards, the AAF envisioned a rocket-powered research airplane as the best means to substitute for the lack of suitable ground-based research methods, while the NACA persuaded the Navy to support the less challenging turbojet alternative; in July 1944, in fact, the AAF rejected outright a NACA proposal for a jet-propelled research airplane as too conservative. By mid-1944, two development paths had already diverged from common Air Force-NACA-Navy-industry desires to build a Mach 1 testbed, spawning by early 1945 the clearly supersonic Air Force-sponsored rocket-propelled Bell XS–1 along with the just as clearly transonic Navy-sponsored turbojet-powered Douglas D–558–1 Skystreak.[14]

In the summer of 1944, as V–1s rained down on London and other English cities, Kotcher was sidetracked from transonic research by a crash program to produce an American copy (the Republic JB–2) of this Nazi buzz bomb, but with more precise guidance. Kotcher did not return to transonic planning until fall. Then, on November 30, 1944, came the catalytic event that triggered the birth of the Bell XS–1, the first supersonic airplane: a casual conversation between Kotcher and the chief engineer of the Bell Aircraft Corporation, Robert J. Woods. That day, as Woods visited Wright Field on other matters, the two men discussed the problem of transonic flight. Kotcher broached the idea of a transonic rocket-propelled research airplane; Woods was interested. Kotcher went further, asking if Bell might be interested in building such a craft, capable of attaining 800 mph at 35,000 feet for at least two minutes. Woods, on his own, committed Bell on the spot to developing the plane. The following month, in a joint conference held at Langley on December 11–12, 1944, the AAF, Bell, and NACA collaborated on developing final development specifications for the aircraft, stipulating that it carry a research instrumentation payload of 500 lbs. Thus was born the XS–1, first of the famed postwar X-series aircraft. Its subsequent development consumed the next year, with the AAF issuing a formal contract to Bell on March 16, 1945, for a rocket-powered straight-wing research airplane. Though the first of two XS–1s began gliding trials in January 1946, it was not ready for its assault on the "sound barrier" until the fall of 1947.[15]

The Foreign Dimension

As soon as Nazi Germany collapsed in May 1945, Allied technical intelligence teams began poring over captured German research and development facilities. The intensive effort that Germany had put into studying high-speed flight, particular the development of high-speed wind tunnel complexes and the widespread use of sweptwing planforms for high-speed aircraft and missile designs, surprised technical intelligence assessors and triggered immediate efforts to turn the fruits of this work to postwar advantage. Virtually all the major combatant nations exposed to Nazi technical work incorporated portions of it in their postwar aeronautical development schemes.[16]

54

Immediately, three shortfalls in previous Allied work appeared: the failure to emphasize the development of the turbojet, the failure to capitalize on the postulation of transonic sweptwing theory, and the failure to begin, at an earlier date, a serious investigation of the problems of supersonic flight.

Of all of these, critics leveled their greatest criticism at the failure to appreciate the sweptwing, first postulated as a means of overcoming the problems of transonic flight by Adolf Busemann at the seminal 1935 Volta Conference on High Speeds in Aviation held at Campidoglio, Italy.[17] Immediately after the war, two aircraft projects then under development — the North American XP–86 jet fighter and the Boeing XB–47 jet bomber — were radically restructured to make use of sweptwing planforms. In the almost hysterical climate surrounding exploitation of German wartime research, the NACA came under sharp criticism from the Air Force's Production Division for not emphasizing a sweptwing planform for the XS–1 then well under development. In fact, primary responsibility for the decision to build the XS–1 with a straight wing rested with the AAF's own Air Technical Service Command, which at the time of the March 1945 contract award, was well aware of the indigenous American sweptwing research of NACA aerodynamicist Robert T. Jones. The ATSC rejected a swept XS–1 at the time because the sweptwing was still an unproved concept, and the NACA strongly concurred; had the contract been awarded six or eight months later, it might have been a very different story.[18]

In any case, the discovery of German sweptwing research data had two important results in addition to the impact it had on the F–86 and B–47 efforts. First, it was largely responsible for accelerating much of the subsequent "Round One" research aircraft program, including the Bell X–2, the Northrop X–4 semitailless research airplane (inspired in part by the Me 163), and the variable sweep Bell X–5 (based outright on the Messerschmitt P.1101 project).[19] Secondly, and beyond this, the German revelation created its own myth — namely, that America only discovered the sweptwing when it sifted the ashes of Nazi Germany, a myth that further weakened the already battered image of the NACA. In fact, such was not the case, for, independently of German work, NACA's Jones had postulated, tested, and reported on the potentialities of both delta and swept wings for transonic and supersonic flight well in advance of any confirmatory data coming from the rubble of the Reich. Indeed, on the basis of reports of Jones' work, von Kármán had arranged for comparative tests in April 1945 (before he went overseas) of an experimental sweptwing and a conventional straight wing in a supersonic tunnel at the Army's Aberdeen laboratory, at speeds to Mach 1.72; the sweptwing proved superior.[20]

By the late summer of 1945, an international race to be the first to exceed the speed of sound had essentially already begun; the players were the United States, with the AAF's rocket-propelled X–1; Great Britain with the jet-powered

The Bell XS–1.

Miles M.52 and the de Havilland D.H. 108 Swallow; and the Soviet Union, with the Mikoyan I–270 rocket-powered experimental interceptor and its own version of the Nazi DFS 346 rocketplane. Much further behind was France, pursuing development of an air-launched ramjet designed by René Leduc. On October 14, 1947, with the first supersonic flight of the Bell XS–1, the United States emerged victorious.

But it had been, as Wellington said of Waterloo, "a close-run thing." In mid-1946, citing financial and possible safety reasons, the British government foolishly cancelled the M.52 program, the first of a long series of questionable postwar development decisions affecting British aviation. Then, in September of that year, Geoffrey de Havilland, the son of Great Britain's oldest and most distinguished aircraft firm, was killed in the crash, due to loss of control near the speed of sound, of a semitailless D.H. 108 Swallow, a possible X–1 rival inspired in part by the Me 163. It would be two years before a Swallow would again approach the speed of sound, this time successfully.[21]

The steps the Soviet Union took to exceed the speed of sound are still shrouded in mystery. In 1947, the Soviets flew two Mikoyan I–270 rocket-propelled interceptors patterned on the captured Junkers Ju 248, which was an outgrowth of the Me 163. Though potentially capable of exceeding the speed of sound, both I–270s were lost in accidents and thus to history as well. The Soviet Union made some flight attempts in 1947 with the DFS 346 they had captured in eastern Germany. In XS–1 fashion, they air-launched it with a German test pilot using one of three B–29 bombers acquired in 1945 when AAF crews made emergency landings following bombing raids on Japan. Again, no claim of supersonic flight resulted, and the fate of the airplane is unknown. In the late 1940s, the Soviets test-flew the Bisnovat B–5, an X–1 lookalike with a modestly swept-back wing, but without any apparent supersonic success. Then, in September 1949, they succeeded in broaching Mach 1 for the first time, reportedly with a developmental aircraft for the MiG–17.[22] By that time, the "sound barrier" was a thing of the past, having been broken by the XS–1 (October 1947), the XP–86 (April 1948), the D.H. 108 (September 1948), the straightwing D–558–1 (September 1948), and the sweptwing D–558–2 (February 1949).

The von Kármán Reports

In November 1944, Hap Arnold had instructed von Kármán to prepare a detailed report on the state and future of aviation to be used as a basis for long-range Air Force planning, research, development, and acquisition. In his mandate to von Kármán, Arnold stated, among other assumptions, that, in the postwar world, "supersonic speed" was a "requirement."[23] In Europe to assess Nazi technical developments immediately after the war, von Kármán wasted little time in preparing two seminal reports on aeronautical development. The first of these was *Where We Stand*, a assessment of the current state of aeronautical development. The second was the multipart *Toward New Horizons*, which forecast the future of aviation and suggested bold courses of action. Both appeared in 1945, the former as Japan was atom-bombed, and the latter in December, as Bell was rolling the first XS–1 out of its Buffalo, New York, plant.

Where We Stand addressed supersonic flight, noting that it had appeared as "a remote possibility" before 1940, but that as the result of "bolder and more accurate thinking . . . this stone wall . . . will disappear in actual practice if efforts are continued."[24] Von Kármán went on to note that "we were slow in recognizing the necessity of supersonic wind-tunnel research;" by 1945, few American supersonic test tunnels existed, while Germany had no less than eight in service in four research complexes, six of which could exceed Mach 3, and one of which could exceed Mach 4. He concluded,

> It seems to me that the Air Forces have to recognize the fact that the science of supersonic aerodynamics is no longer a part of exterior ballistics but represents the basic knowledge necessary for design of manned and unmanned supersonic aircraft. The Air Forces have to provide facilities and include this field in their research, development, and training programs.[25]

He also noted the potential value of the sweptwing as a supersonic aircraft planform, and went on to make four recommendations involving development of large supersonic wind tunnels, transonic and supersonic research airplanes to substitute for the lack of tunnel measurement capabilities in the transonic region, vertical take-off rocket-boosted fighters, and use of forward-firing rocket thrusters for deceleration of high-performance aircraft prior to landing. He concluded by emphatically stating that "We cannot hope to secure air superiority in any future conflict without entering the supersonic speed range."[26]

It was his *Toward New Horizons*, however, that drew the greatest attention, particularly the opening essay "Science, the Key to Air Supremacy." This document confidently predicted a future for the Air Force built around supersonic manned and pilotless aircraft and missiles, atomic weapons and atomic energy, operations over global ranges, the ability to fly and navigate with greater precision and safety, the ability to attack in all-weather conditions with devastating force and greater accuracy, and the ability to defeat and counter enemy efforts to defend against aerial attack. Overall, the report reflected the climate of the age after Hiroshima and Nagasaki, that in an era of atomic warfare (and this was written four years prior to the Soviets acquiring an atomic bomb),

> All we can hope is that absolute air superiority, combined with highly developed and specialized warning and homing devices, will help us erect an impregnable aeroelectronic wall, which will reduce to a minimum the possibility of any enemy device [i.e., atomic weapon] slipping through undetected and undestroyed.

Von Kármán and his research team accepted as a given:

> the necessity for a powerful air force, which is capable of:
> a. Reaching remote targets swiftly and hitting them with great destructive power.
> b. Securing air superiority over any region of the globe.
> c. Landing, in a short time, powerful forces, men and firepower, at any point on the globe.
> d. Defending our own territory and bases in the most efficient way.[27]

Transonic and Supersonic Flight: The Early Years

By the end of 1945, then, the United States Army Air Forces, together with other Federal organizations and private industry, had set for itself the task of exploiting the transonic and supersonic speed regimes. The AAF had a planning document prepared by some of the most outstanding American scientists and engineers that clearly forecast a supersonic future, and it also had funded development of a specialized rocket-propelled research airplane on the verge of its first flight. Two other supersonic research aircraft were also under consideration: the proposed rocket-propelled air-launched sweptwing Bell XS–2, and the jet-propelled Douglas XS–3. A further two transonic configuration testbeds (the semitailless Northrop XS–4 and a proposed Bell variable wing-sweep research airplane that would emerge as the X–5) were edging towards development. Two advanced sweptwing aircraft were undergoing design development, the North American XP–86 and the Boeing XB–47. A third program, for an ambitious rocket-boosted ramjet-powered delta interceptor, the XP–92, was gestating at Convair. Over the next five years, all these aircraft would fly and, with others, would shape the future of American aviation.

A key question involved forging the best possible flight-testing partnership between the AAF and the NACA, at the time the nation's premier aeronautical research organization, for as will be seen, tensions between the service and the agency had grown in the wartime years and immediately afterwards. The NACA had a long heritage of creative flight and ground-test research work, and maintained laboratory complexes at Langley, Ames, and Lewis that were uniquely valuable to the supersonic and turbojet breakthroughs. By the spring of 1947, with two XS–1s in flight test, the time had obviously come to strike an agreement between the soon-to-be independent Air Force and the NACA on how to best run the research aircraft programs. The choice of test site was not a question: Muroc Dry Lake (now Edwards AFB), California. On June 30, 1947, representatives of the Army Air Forces' Air Materiel Command and the NACA met at Wright Field to discuss the future research administration of the XS–1 program; they agreed that the AAF would undertake accelerated flight testing of the XS–1 to take it through the speed of sound as quickly as possible, while the NACA would retain the second aircraft for a more detailed exploration of the sonic regime.[28]

This became the pattern followed on other research aircraft as well, up to the time of the X–15: rapid exploration of the airplane over its performance envelope by the Air Force, and then a more detailed and systematic study by the NACA. (Beyond this, it inaugurated the traditional "MOU" arrangements that have characterized the partnership of the Air Force Flight Test Center and the NASA Dryden Flight Research Center at Edwards on programs such as the XB–70A, YF–12A, lifting bodies, AFTI F–16, the X–29, and, more recently, the X–31). Within the Air Force, program management oversight for the early X-series aircraft resided within the Air Materiel Command, and, subsequently, the ARDC; within the NACA, it resided within the agency's Research Airplane Projects Panel. Subsequent research airplane coordination between the service and the NACA was good to excellent at all levels, particularly at Edwards itself, thanks in part to a strong bond of mutual respect and support between the onsite Air Force installation commander, Brig. Gen. Albert Boyd, and his NACA counterpart, Walter C. Williams.[29]

On October 14, 1947, piloted by Air Force Capt. Charles E. "Chuck" Yeager, the Bell XS–1 #1 achieved Mach 1.06 (approximately 700 mph) at

The Northrop X–4.

43,000 feet over the Mojave desert, the first manned supersonic flight in aviation history, dramatically fulfilling the expectations of research aircraft partisans. By the next spring, it had exceeded Mach 1.4 (960 mph). But such was the pace of aircraft development that, even before the XS–1 made its historic flight, the prototype F–86 Sabre had arrived at Muroc for testing. It was another aircraft capable of exceeding Mach 1, albeit in a dive, and did so in April 1948, thanks to its sweptwing. While the XP–86 was no substitute for the heavily instrumented XS–1, it is ironic that the lengthy contractor test period on the Bell rocket plane — from January 1946 through the spring of 1947 — nearly robbed the first of the X-series of its opportunity to make its mark in aviation history.

Thereafter, supersonic flight at Muroc was virtually commonplace, and soon, after the introduction of the F–86A into squadron service in 1949, so were sonic bangs from supersonic dives by enthusiastic fighter pilots around the country. Even at this early point North American had plans underway for an advanced sweptwing fighter called the "Sabre 45" which would emerge in due course as the YF–100. It reflected lessons already learned from the XS–1 by the late summer of 1948: relocating the horizontal tail from midfin to low on the aft fuselage, and changing the adjustable horizontal stabilizer and elevator combination (which the F–86 shared with the XS–1, and which worked much better than a fixed horizontal stabilizer and elevator, as MiG pilots would discover to their sorrow in Korea) to a genuine slab all-moving tail.

But in many ways, what was happening on the ground was of far more importance to the future of supersonic flight than what was taking place in the air. One of the most interesting (of many) aspects within the final von Kármán report was its matter-of-fact acceptance that supersonic flight was a practical

The North American XP–86 during its flight tests in October 1947.

probability, even though the XS–1 had not yet flown and test Chuck Yeager's first supersonic sojourn was still nearly two years in the future. Among other recommendations, the report suggested creation of a supersonic and pilotless aircraft research center, and a specialized research and development command.[30] These two were, in fact, acted on, with the establishment of the Arnold Engineering Development Center at Tullahoma, Tennessee, and (following further recommendations by an advisory group chaired by Dr. Louis N. Ridenour) the eventual creation of Air Research and Development Command.[31]

The AEDC story is an often neglected component of the Air Force's contribution to the national supersonic breakthrough. As early as June 1945, AAF members of the Allied technical intelligence teams roaming Germany had suggested to higher headquarters that German developments warranted creation of new research laboratory facilities. By December, with the issuance of *Toward New Horizons*, the work of Wright Field partisans and von Kármán had crystallized. He envisioned an ambitious "Center for Supersonic and Pilotless Aircraft Development," equipped with Mach 3 tunnels, a hypersonic tunnel, combustion research facilities, and laboratories for studying flight control systems, medical aspects of high-speed/high altitude flight, as well as an actual flight test facility, on the model of such great German research institutions as Braunschweig and Peenemünde.[32]

Such plans threatened the position of the NACA as the nation's premier aeronautical research authority, but the NACA had, unfortunately, lost much support within both the military and industrial community. In a bid to retain primacy, the NACA's leadership first called in the fall of 1945 for a "unitary" facilities plan meeting the needs of the services, the industry, and the NACA. When this failed to trigger any enthusiasm, they next announced, in March 1946, their own intention to develop a "National Supersonic Research Center." The two plans came into sharp conflict with the Congress and the aircraft industry for funding and support. An industry-led review panel attempting to produce a unitary plan satisfactory to all parties failed to "deconflict" the two proposals, despite three years of study; at one point, the proposed plan envisioned multiple centers with no less than thirty-three transonic, supersonic, and hypersonic tunnels, and a total cost of almost $3 billion — inconceivable in the post-World War II fiscal environment.[33]

Exasperated, Congress finally stepped in and structured its own plan, costing approximately $250 million. Though postwar budgetary considerations nearly caused cancellation of the plan, the strong personal intervention of the first Secretary of the Air Force, W. Stuart Symington, and, ironically, the availability of captured German equipment, test facilities, and plans of proposed high-speed test installations, helped preserve much of von Kármán's vision, and the core of the Air Force's proposal. In 1949, Congress passed the National Unitary Wind Tunnel Plan Act. The NACA received $136 million for three supersonic wind tunnels, and an additional $10 million for tunnel construction at various educational institutions but — significantly — received no support for a new national supersonic center. The Air Force, on the other hand, received $100 million (as but a first step) for construction of a new center — which became Arnold — though nowhere near as expansive as von Kármán had envisioned. Construction began on the Arnold center in 1950, and it began its first research operations two years later. Arnold has subsequently played a major role in both the history of supersonic flight, and the history of spaceflight as well, working productively with industry and other Federal agencies such as NASA, as well as for the Air Force.[34]

Too Much of a Good Thing?

The acceptance of supersonic flight as a practicality, which was accompanied by a remarkable casualness of attitude, did create its own set of problems. Among these was a tendency by many technologists to view supersonic flight as merely an extension of subsonic flight with annoying "transonic tangles and traps" (as one engineer termed it) in between. Put another way, many perceived the transition from subsonic to supersonic flight as the great challenge of supersonic flight when, in fact, "breaking" the "sound barrier" was but a means of entering a whole new aerodynamic world. It was here that the more deliberative approach to testing and development characteristic of the NACA (after the 1958 Space Act, NASA) paid off.[35] Once beyond Mach 1, an

62

entire subset of related problems demanded resolution, including greatly reduced directional stability (and consequent development of stability augmentation systems), aircrew escape requirements, external stores separation, engine-airframe matching, and, beyond Mach 2, aerodynamic heating. No better examples of the results which such overconfidence could produce exist than the North American F–100A Super Sabre, America's first supersonic "on the level" jet fighter, and the Bell X–2 rocket research airplane.

The preproduction YF–100 began flight testing in 1953 and, with its impressive Mach 1+ performance, quickly won enthusiastic support from pilots of the Air Force's Tactical Air Command assigned to evaluate it. TAC's pilots minimized cautions from the project test pilot (himself a distinguished fighter pilot) that the plane required careful study, and, at TAC behest, HQ USAF ordered the F–100 into full production too soon. Shortly after entering service, a disastrous series of accidents resulted in the loss of six aircraft and their pilots. The F–100, with its long fuselage and relatively short wing, had fallen victim to the phenomenon of inertial coupling, essentially diverging and tumbling out of control during supersonic maneuvering. It required extensive redesign to be made into a safe and satisfactory fighter, with changes in fin and wing area, and the addition of yaw and pitch dampers.[36] Tragically, lack of a full appreciation of how lateral-directional stability tendencies would continue to deteriorate as Mach number increased, together with questionable flight test management and mission planning, led to the loss of the Bell X–2 #1 research airplane at Mach 3.196 on September 27, 1956, even though its pilot (who died in the accident) had extensive experience on the F–100 inertial coupling program.[37]

A second problem was a tendency to favor the radical at the expense of the practical; this resulted in "Buck Rogers"-type proposals (such as von Kármán's own suggestion of vertically launched fighters and rocket deceleration systems). Common sense usually prevailed, but in any case, time, effort, and often great sums of money were lost in the process. Several notable programs typified this will for the fanciful, such as the dolly-launched rocket-boosted ramjet-powered XP–92 which, in addition to its other peculiarities, also would have had its pilot sitting within the inlet duct of the ramjet! The program spawned the small XF–92A research testbed; by the time it flew, cooler heads had cancelled the radical ducted rocket-ramjet XP–92 outright. Another was the Republic XF–103 turboramjet-powered Mach 3 interceptor, cancelled in 1957. (Other services were far from immune to such thinking as well, with such projects as the Navy's attempts to develop turboprop-powered tail-sitting vertical-takeoff-and-landing fighters — the Lockheed XFV–1 and the Convair XFY–1 — for shipboard use, and the Army's proposed "flying infantry" platform, the Davy Crockett battlefield nuclear weapon, tanks with detachable flying gun turrets, and, wildest of all, proposed use of Redstone missiles to deliver squads of troops into battle.)[38]

A third was a tendency to minimize developmental problems by an overreliance on extrapolative approaches. The so-called "1954 Ultimate Inter-

The North American YF–100 over the North Base at
Edwards AFB during its flight tests in May 1953.

ceptor," the Convair YF–102A Delta Dagger, typified this problem. Developers overestimated the value of the XF–92A testbed experience, thinking that the YF–102 could, in effect, represent a simple "scaling up" of the smaller plane. Accordingly, the Air Force made major production commitments to this aircraft under the aegis of the Cook-Craigie concurrency development approach even as significant unknowns about the magnitude of its transonic drag rise characteristics existed. Following its initial flight testing, the YF–102 had to be extensively redesigned to incorporate area ruling, with other changes to leading edge configuration, fin size, and the aft fuselage, resulting in virtually none of the early production aircraft having any commonality with post-change aircraft. Worse, two-thirds of the production tooling purchased for the aircraft had to be scrapped and replaced by new sets. In fact, the problem continued beyond the F–102; Convair and the Air Force both underestimated the challenges and difficulties of going from the F–102 to the Mach 2 F–106 (which began as the so-called "F–102B"), causing delay, complications, and rapid cost escalation on that program as well.[39]

A fourth — and most serious of all — was an unfortunate tendency to see the future almost exclusively in terms of supersonics and atomic weaponry. After the Korean War, Air Force research and development stressed developing supersonic jet fighters and bombers, supersonic cruise missiles, supersonic air defense interceptors — some even possibly atomic-powered.[40] The service made an extensive effort to investigate hypersonic orbital boost gliders as well. While some of these projects represented reasonable approaches to the defense challenges facing the nation, others did not, and, along the way, lessons from earlier conflicts ranging from World War II through the Korean and Indochinese experiences were not incorporated in the acquisition of future Air Force aircraft. This was dramatically highlighted in the Vietnam war when the Air Force, to its discomfort, had to acquire no less than three Navy-developed aircraft to meet its wartime needs: the Douglas A–1 Skyraider attack aircraft, the Ling-Temco-Vought A–7 Corsair II strike aircraft, and — most notably — the McDonnell F–4 Phantom II jet fighter. (In part, this was because after the Korean War — in which the Air Force and its aircraft had performed well — the service turned away from conventional war and focused virtually exclusively on a nuclear war future. The Navy, in contrast, had serious challenges meeting the requirements of the Korean conflict and afterwards generally used the lessons learned to good advantage when it made its post-Korean acquisition plans. It dropped dubious ideas — such as the tail-sitting VTOL fighters — and concentrated on producing practical supersonic and transonic aircraft such as the Vought F8U–1 Crusader, the McDonnell F4H–1 Phantom II, and the Douglas A4D–1 Skyhawk).

In particular, the Century Series fighters offer an instructive lesson in how the emphasis on speed — not merely low supersonic speed, but near or above Mach 2 as well — and atomic warfare predominated. By 1960, the Air Force's conception of the fighter's role had evolved from air superiority (typified by the F–86 in Korea) to either an interceptor of enemy atomic bombers (F–101B, F–102A, F–104A, F–106A) or a deliverer of nuclear weapons (F–101C, F–104C, F–105D). Indeed, of the 5,525 Century Series "fighters" (the F–100, F–101, F–102, F–104, F–105, and F–106), only 1,274 (the F–100D family of the mid-1950s) were truly multirole "classic" fighter-bombers capable of undertaking multiple mission taskings. These 1,274 represented only 23 percent of the fighters procured by the Air Force from 1952 through 1964. Some, while newer, overemphasized speed at the expense of doing much else, as perhaps was best exemplified by the Lockheed F–104 Starfighter (dubbed "The Missile with a Man in It"), essentially a more powerful straight-forward adaptation of the aerodynamic design approach exemplified by the Douglas X–3 research airplane.[41] Poor design features limited the maneuverability of some aircraft, such as the F–101 and particularly the F–104, which were seriously constrained by the pitch-up vulnerabilities induced by their T-tail configurations (with which, astonishingly, they were designed, even though, at the time of their design, the weaknesses of such a configuration were already well-appreciated). Most required long and extensive development and upgrade programs to be truly useful, while one — the F–105, designed as a nuclear strike fighter — was called on to act as a conventional iron-bomb dropper in the Vietnam War.

The same tendencies that influenced Century Series fighter development affected bomber development as well; proposed bombers in the 1950s increasingly reflected unrealistic expectations of sustained Mach 2.5+ or even Mach

Five of six "Century Series" fighters. Clockwise from bottom: Lockheed
F–104 Starfighter, North American F–100 Super Sabre, Convair F–102
Delta Dagger, McDonnell F–101 Voodoo, and Republic F–105
Thunderchief. The aircraft not shown is the Convair F–106.

3+ flight, with complex "chemical" or nuclear propulsion. One only slightly less
radical aircraft, the Mach 2 Convair B–58A Hustler, did enter service, but
proved difficult to maintain, accident-prone, and of only marginal strategic
value.[42] The drive for even more exotic supersonic fighters and bombers
climaxed with the proposed (but never built) North American F–108 Rapier
interceptor and the gargantuan North American XB–70A Valkyrie bomber.
While technically possible, they were increasingly divergent from the real
military needs of the United States in the late 1950s; at the time of their
cancellation, they had gobbled nearly $1.7 billion in taxpayer funding.[43]

In Retrospect

Historians always have the luxury of 20/20 hindsight, and one must sympathize at the least with the challenges and decisions that the Air Force leadership of the 1940s and 1950s had to make. They had to steer a course between the too conservative and the too fanciful, and by and large, they did so quite well, certainly at first. Aircraft such as the F–86, B–47, F–100, and B–52, and obviously the X-series themselves, are testimonials to the basic wisdom of acquisition decision making during the early years of transonic and supersonic aircraft development. Certainly the Air Force deserves great credit for fostering a supportive climate for supersonic research and for undertaking development of the X-series aircraft, which as airborne research tools, enabled aeronautical science to move forward without having to wait for ground research methodologies to catch up. Without Air Force money, even if military prototypes had rapidly exceeded Mach 1 (as was the case with the XP–86), the comprehensive body of knowledge generated by the supersonic X-series and related programs would have been missed, making far more difficult the task of industry in the 1950s as it tried to come to grips with the challenge of designing transonic and supersonic aircraft. Indeed, so great was the expansion of military (and industrial, thanks to military interest) knowledge within supersonic and related fields of aircraft design that, by the 1970s, Hap Arnold's vision of a service largely independently pursuing its technological future without a need to rely (as opposed to consult) on outside organizations was generally fulfilled.[44] The success of contemporary systems, such as the F–15, F–16, and F–117, all attest to the wisdom of his intentions.

But that success came only after the spoiled fruits of poor choices had already been sampled. Within a decade of the first supersonic flight, the same strengths that gave forth comprehensive knowledge of supersonic flight were giving individuals and major commands entranced more with technological opportunity than with military necessity the chance to pursue acquisition choices sadly distant from what the nation and the service really needed at the time. The impact of those choices would be felt all too soon as war broke out in Southeast Asia.

The climate that produced this situation was not inherently a bad one; there was a healthy vibrancy to aeronautics at midcentury that, today, seems sorely lacking. But that same optimistic and unquestioning climate nurtured most of the problems discussed previously. Those problems also illustrate more serious difficulties which were not per se connected to the supersonic breakthrough but which have, at other times, also afflicted both military and civilian acquisition.[45] One was the overreliance on contractor-based ideas for new aircraft development programs. Many of the least satisfactory aircraft undertaken by the Air Force in this time frame began as projects initiated within industry and then, as "paper airplanes," were offered up to the service. Industry is undoubtedly a source of fruitful ideas, and a service needs to be aware of what industry has to offer, but in this case, the price was very high indeed.[46] Related to this was the lack of a strong doctrinal underpinning to post-World War II weapon system development so that technological capability more than requirements necessity became the deciding factor.[47] The weakness of not matching technological developments with appropriate doctrinal shifts is a long standing one; as one of the most distinguished students of air doctrine, Dr. I. B. Holley, has noted, "New weapons when not accompanied by correspondingly new adjustments in doctrine are just so many accretions on the body of an army."[48] If nothing else, the story of Air Force fighter and bomber acquisition in the 1950s and 1960s illustrates the importance of relating acquisition to clearly defined military doctrine, national needs, and appropriate technology. The history of systems acquisition by the Air Force since Vietnam, for all the challenges and difficulties it has experienced, offers an equally important lesson of what can be accomplished when military doctrine, national needs, and appropriate technological choices, as opposed to mere technological opportunities, are placed first and foremost before service decision-makers. The two stories are the twin sides of the supersonic coin forged by Ezra Kotcher, Theodore von Kármán, and all the others who envisioned flight beyond Mach 1, and then worked to make it a reality.

Notes

1. The speed of sound — approximately 760 mph at sea level, but which varies with altitude, being approximately 660 mph at 40,000 feet — is commonly referred to as Mach 1, after Ernst Mach, a nineteenth-century Austrian physicist and philosopher. Mach number, as commonly applied to aircraft velocity, is the speed of the aircraft divided by the speed of sound at the altitude the aircraft is flying. As originally described in the technical literature of the 1940's, transonic flight was generally considered to be flight at speeds from Mach 0.75 to 1.25; supersonic flight was flight from Mach 1.25 to Mach 5. (Mach 5 to Mach 25 is considered hypersonic flight). In broad usage today, supersonic flight is now defined as flight faster than Mach 1, with transonic generally reserved for vehicles moving above Mach 0.75, but less than Mach 1.

2. For an excellent account of transonic tunnel development, see John V. Becker's *The High-Speed Frontier: Case Studies of four NACA Programs, 1920–1950*, NASA SP–445 (Washington: NASA, 1980), pp 61–118. Becker was one of NACA/NASA's most distinguished aeronautical engineers, and the driving force behind the agency's X–15 concept studies as well as other hypersonic work.

3. For example, see Frank W. Caldwell and Elisha N. Fales, *Wind Tunnel Studies in Aerodynamic Phenomena at High Speeds*, NACA Tech Rpt No. 83, 1920.

4. James R. Hansen, *Engineer in Charge: A History of the Langley Aeronautical Laboratory, 1917–1958*, NASA SP–4305 (Washington: NASA, 1987), has a good account of this on p 253.

5. Edwin P. Hartman, *Adventures in Research: A History of Ames Research Center, 1940–1965*, NASA SP–4302 (Washington: NASA, 1970), pp 97–99.

6. For a discussion of the von Kármán effort, see Michael H. Gorn, ed, *Prophecy Fulfilled: "Toward New Horizons" and Its Legacy* (Washington: AF History and Museums Program, 1994).

7. Much of this discussion is based on extensive correspondence with the late Ezra Kotcher in 1971–72 and on documentation that he sent to me. I wish to acknowledge his assistance with the greatest respect and affection.

8. Kotcher ltr, Jan 23, 1972.

9. Theodore von Kármán with Lee Edson, *The Wind and Beyond: Theodore von Kármán, Pioneer in Aviation and Pathfinder in Space* (Boston: Little, Brown, 1967), p 225.

10. The tortuous story of American development of the jet engine is well summarized in Robert Schlaifer and S. D. Heron's seminal *Development of Aircraft Engines and Fuels* (Boston: Harvard University, 1950), particularly Chapter 17, "Why Was the United States Behind in Turbojet Development?", pp 480–508. The impetus for creation of the SAB is discussed in Michael H. Gorn, *Harnessing the Genie: Science and Technology Forecasting for the Air Force, 1944–1986* (Washington: Office of AF History, 1988) and Thomas Sturm, *The USAF Scientific Advisory Board: Its First Twenty Years, 1944–1964* (Washington: USAF SAB, 1967).

11. Much of the subsequent discussion is drawn from my book *Supersonic Flight: Breaking the Sound Barrier and Beyond — the Story of the Bell X–1 and Douglas D–558* (New York: Macmillan in association with the Smithsonian Institution, 1972). It is possible the CTI issued by HQ AAF (CTI 1568) resulted from a meeting held on Dec 18, 1943 at NACA HQ in response to a lecture by W. S. Farren, director of the Royal Aeronautical Establishment on the future of aeronautics. At that meeting, a young Bell Aircraft Corp engineer, Robert Wolf, advocated building a jet-propelled transonic research airplane; but, in any case, by this time the AAF was well aware of German interest in rocketry, jet propulsion, and high-speed aircraft

69

design, and this was generating its own pressures.

12. Details of this study project are drawn from von Kármán's statement in his *Where We Stand*, reprinted in Gorn, *Prophecy Fulfilled*, p 19. There is some confusion in von Kármán's mind about when this request occurred. In *Where We Stand* (written in 1945), he states it was early in 1944. In his later *Wind and Beyond* (1967, pp 233–34) he states 1943. I have accepted the 1944 figure because it was recollected closer to the event and also because it ties more closely and logically with other events and activities surrounding AAF interest in supersonic flight than an earlier 1943 date. Information on the Ellis-Brown study can be found in Becker, *High-Speed Frontier*, pp 93–95, 98. I have also benefited from discussions with the late John Stack, Macon Ellis, and Clinton Brown on this study effort. As an aside, ramjets always offered far more promise than they actually delivered, and it was not until the late 1950's that the practical winged ramjet-powered aircraft became a reality, as exemplified by the Lockheed X–7 pilotless testbed, the Boeing IM–99 Bomarc SAM, and the French Nord Griffon technology demonstrator.

13. Ltrs, Kotcher to Hallion, Sep 10, 1971, Jan 23, 1972, and Feb 22, 1972; ltr, Kotcher to William Lundgren, 4 Nov 1953; Kotcher statement at a historical session of the American Institute of Aeronautics and Astronautics, San Francisco, Calif, Jul 28, 1965, copy in the NASA Historical Archives, NASA HQ, Washington, DC; Air Materiel Command Correspondence Summary of Project MX–653 History, Jan 14, 1947, and "Beginnings of the X–1," nd, p 3, copy in NASA Historical Archives.

14. *Ibid*

15. The best summary account from the manufacturer's standpoint of the XS–1's subsequent development is in R. M. Stanley and R. J. Sandstrom's "Development of the XS–1 Airplane," in *Air Force Supersonic Research Airplane XS-1 Report No. 1* (HQ USAF, Jan 9, 1948), a copy of which is in the NASA Historical Archives. A shorter and less extensive version of this appeared in Stanley and Sandstrom's "Development of the XS–1 Supersonic Research Airplane," *Aeronautical Engineering Review*, vol 6, no 8 (Aug 1947), pp 22–26, 72. Though Kotcher retained an interest in the subsequent program, his role essentially ended at this point. His work had been profoundly influential and, given his rank at the time, bold in the extreme. After the war he remained at Wright Field, introducing generations of students to the nuances of aeronautical engineering, before retiring in the 1960's. Though justly revered for his accomplishments, it is nevertheless unfortunate that he did not share in the subsequent 1948 Robert J. Collier Trophy, for no one deserved it more, particularly since his own role in formulating the XS–1 was clearly greater than two of the three joint recipients, Larry Bell and John Stack. (The third was test pilot Chuck Yeager). To his own great credit, the late John Stack was always unstinting in his praise of Kotcher, and, for his own part, the ever-modest Kotcher never once grumbled at what is arguably the greatest oversight in the history of Collier trophy presentations.

16. Two such summary reports that offer good views of Nazi aeronautical research are Ronald Smelt, "Notes on Discussion with the Staff of Focke-Wulf Ltd.," (Farnborough, Eng: Royal Aircraft Establishment, May 1945), indexed as "E/R.A.E./Tech Note Aero 1644, R3477 F758," (Wright Field T–2 HQ AMC, USAAF), copy in the microfilm files of the National Air and Space Museum archives, Washington, DC; and Fritz Zwicky, "Report on Certain Phases of War Research in Germany," Summary Rpt F–SU–3–RE (Dayton, Ohio: HQ AMC, Jan 1947), copy in the archives of the Aeronautical Systems Center History Office, Wright-Patterson AFB, Ohio.

17. Adolf Busemann, "Aerodynamische Auftreib bei Überschallgeschwindigkeit," *Luftfahrtforschung*, Oct 3, 1935, pp 210–20.

18. Kotcher ltr., Nov 7, 1971; Kotcher AIAA stmt; see also Joseph A. Shortal, *History of Wallops Station: Origins and*

Activities Through 1949 (NASA, comment edition, nd), p 74A, copy in the NASA Historical Archives.

19. By the late 1950's, researchers distinguished three "rounds" to postwar research aircraft programs. In "Round One" were the early transonic and supersonic research aircraft, including the X–1 family (XS–1, X–1–3, X–1A, X–1B, X–1D, and X–1E), the X–2, X–3, X–4, X–5, the D–558 family (D–558–1 Skystreak, D–558–2 Skyrocket), and the XF–92A; "Round Two" was the hypersonic X–15, and "Round Three" was a proposed orbital boost-glider that became the X–20 Dyna-Soar, cancelled in 1963. See James A. Martin, "The Record-Setting Research Airplanes," *Aerospace Engineering*, vol 21, no 12 (Dec 1962), pp 49–54.

20. Inspired by the earlier work of emigre Russian engineer Michael Gluhareff transmitted to the NACA via glide bomb contractor Robert Griswold, Jones conceptualized the low aspect ratio, low thickness/chord ratio delta (triangular) wing as a potential solution to the problems of supersonic and transonic flight. In Jan 1945, he discussed his work with Ezra Kotcher, who brought it to the attention of von Kármán and one of von Kármán's chief assts, Hsue-shen Tsien. By May, Jones had flight tested a model mounted on the wing of a diving P–51 Mustang. He next explored the sweptwing configuration, with a test model of it undergoing study late in May 1945, still well in advance of German sweptwing data reaching the United States. The myth of German omnipotence in swept and delta-wing research, unfortunately, continues to thrive. For the record, Jones personally took the position that the XS–1 should be a sweptwing aircraft, not a straight-wing design. For a survey of this story, see Richard P. Hallion, "Lippisch, Gluhareff, and Jones: The Emergence of the Delta Planform and the Origins of the Sweptwing in the United States," *Aerospace Historian*, vol 26, no 1 (Mar 1979), pp 1–10. For Jones' reports, see his "Properties of Low-Aspect Ratio Pointed Wings at Speeds Below and Above the Speed of Sound," NACA Rpt No. 835, May

11, 1945; and his "Wing Planforms for High-Speed Flight," NACA Tech Note No. 1033, published in Mar 1946, but issued at Langley laboratory on Jun 23, 1945. See also R. T. Jones, "Recollections from an Earlier Period in American Aeronautics," *American Review of Fluid Mechanics*, 1977, pp 1–11; and Theodore von Kármán, *Aerodynamics: Selected Topics in Light of Their Historical Development* (New York: McGraw-Hill, 1963), pp 50–57, and 132–34. The von Kármán Aberdeen story is from his *Where We Stand*, reprinted in Gorn, *Prophecy Fulfilled*, pp 23, 25.

21. Charles Burnet, *Three Centuries to Concorde* (London: Mechanical Engineering Publications, 1979) is an excellent survey of British supersonic flight research.

22. Yefim Gordon with Bill Sweetman, *Soviet X–Planes: Experimental and Prototype Aircraft, 1931 to 1939* (Osceola, Wisc: Motorbooks, 1992) is a useful guide to Soviet work on these projects. The title is somewhat misleading, however, for, unlike the U.S., the Soviets never had a comprehensive and coordinated "X–series" program as did the United States. I have also benefited from a discussion in 1979 on this subject with the late Carl Duckett, the former director of science and technology for the CIA.

23. Ltr, Arnold to TvK [Theodore von Kármán], Nov 7, 1944, copy in the files of the AF Scientific Advisory Board, HQ USAF, Washington, DC.

24. This and subsequent quotes from *Where We Stand* are drawn from the reprinted version found in Gorn, *Prophecy Fulfilled*, pp 19–20, 23, and 26, as this version is more readily accessible to readers than the original archival copies.

25. *Ibid*, p 20.

26. *Ibid*, p 26.

27. *Ibid*, p 98.

28. Hartley A. Soulé, Memo for Chief of Research (NACA), Jul 21, 1947, in the historical files of the NASA Langley Research Center, Hampton, Va.

29. I have benefited from discussions with the late Dr. Williams and other early X–series program management officials on

this point.

30. *Ibid*, pp 176–78.

31. See "Weapons," in *Air Force Magazine*, vol 40, no 8 (Aug 1957), p 343.

32. Von Kármán, *Science*, in Gorn, *Prophecy Fulfilled*, p 177. It can hardly be presumed to be coincidental that the proposed center acronym was SPAD, which was, of course, the acronym of a famous French aircraft company at the time of World War I.

33. There is an excellent brief discussion of this in Alex Roland, *Model Research: The National Advisory Committee for Aeronautics, 1915–1958*, NASA SP–4103 (Washington: NASA, 1985), pp 213–21. See also Donald D. Baals and William R. Corliss, *Wind Tunnels of NASA*, NASA SP–440 (Washington: NASA, 1981), p 65.

34. Symington's role is discussed in George M. Watson, Jr., *The Office of the Secretary of the Air Force, 1947–1965* (Washington: AF History and Museums Program, 1993), pp 66–67; see also Roland, pp 218–19; Baals and Corliss, pp 65–66; there is a useful (if brief) summary of AEDC history in Steve Greenhut's "40 Years of Aerospace Ground Test Leadership," *High Mach* (a monthly AEDC publication), vol 38, no 5 (Jun 1991), pp 6–10.

35. For examples of this, see Hansen, *Engineer in Charge*, and Richard P. Hallion, *On the Frontier: Flight Research at Dryden, 1946–1981*, NASA SP–4303 (Washington: NASA, 1984).

36. Marcelle Size Knaack, *Post-World War II Fighters, 1945–1973*, (Washington: Office of AF History, 1986 ed) pp 115–16, 119. The best technical summary of the F–100's problems is found in NACA Research Memo RM–H55A13, "Flight Experience with Two High-Speed Airplanes Having Violent Lateral-Longitudional Coupling in Aileron Rolls," (Edwards AFB, Calif: NACA High Speed Flight Station, Feb 1955). (The second aircraft this report refers to is the X–3). The ignored AF test pilot was Frank K. Everest; see his autobiography (as told to John Guenther), *The Fastest Man Alive* (New York: Pyramid Books, 1959 ed), pp 12–13, 19–20.

37. The pilot was, unfortunately, making his first powered flight in the X–2, and was thus unfamiliar from personal experience with its performance quirks. See Ronald Stiffler, *The Bell X–2 Rocket Research Aircraft: The Flight Test Program* (Edwards AFB, Calif: AF Flight Test Center, Aug 12, 1957).

38. There are archival materials on the Army programs in the library holdings of the U.S. Army Military History Institute, Carlisle Barracks, Pennsylvania. I wish to acknowledge the assistance of Linda Brenneman and Judith Meck to my research in these materials.

39. "Scaling up" only rarely works as planned, and has often been accompanied by real disaster (viz the Langley Aerodrome of 1903). Even "straightforward" attempts in recent times (such as the F–16 from the YF–16, or the F–18 from the YF–17) have had more than their share of problems. For the F–102, see Knaack, *Fighters*, p 164, and pp 207–21 for the F–106. This story is also well told in Thomas A. Marschak, *The Role of Project Histories in the Study of R & D* (Santa Monica, Calif: Rand Corp, Jan 1964).

40. For example, the atomic-powered airplane program, which lasted into the early 1960's, and the PLUTO program for a supersonic atomic-powered ramjet cruise missile.

41. Ironically, the F–104 had begun as a lightweight "hot rod" approach based on the recommendations of fighter pilots back from Korea who had fought the MiG–15. The connection between the F–104 and the X–3 is explored in Marschak, *Project Histories*, pp 85–86, 90; I also benefited from discussions with the late Melvin B. Zisfein. The overemphasis on brute speed (and little else) in fighter development is explored in greater depth in Richard P. Hallion, "A Troubling Past: Air Force Fighter Acquisition Since 1945," *Airpower Journal*, vol 4, no 4 (Winter 1990), pp 4–23.

42. Marcelle Size Knaack, *Post-World War II Bombers, 1945–1973* (Washington:

Office of AF History, 1988), pp 351–98. Two excellent essays on B–58 development are R. Cargill Hall's "To Acquire Strategic Bombers: The Case of the B–58 Hustler," *Air University Review*, vol 31, no 6 (Sep–Oct 1980); and "The B–58 Bomber: Requiem for a Welterweight," *Air University Review*, vol 32, no 1 (Nov–Dec 1981).

43. Knaack, *Fighters*, pp 330–31; Knaack, *Bombers*, pp 559–73. The F–108 was cancelled in 1959, and the XB–70A in 1962. There are good technical discussions of these two projects and the supersonic cruise issue in T. R. Parsons, "B–70 and F–108 Perspectives on Supersonic Cruise," in John Chuprun, Jr, and Wayne M. O'Connor, eds, *Proceedings of the Conference on the Operational Utility of Supersonic Cruise* (Wright-Patterson AFB, Ohio: ASD/XR, May 1977), pp 3–40; and in F. Edward McLean, *Supersonic Cruise Technology*, SP–472 (Washington: NASA, 1985), pp 23–77.

44. For example, in 1976, just over three decades after von Kármán's *Toward New Horizons* report, Milton O. Thompson, a senior NASA engineering administrator exposed to the AF/industry-developed General Dynamics YF–16 lightweight fighter reflected that the days where designers had to rely on the agency for ideas had clearly come to an end. "NASA," he concluded, "no longer enjoys [an] esteemed position in the aeronautics world, largely due to default," ltr, Thompson to David Scott, Dir, NASA Dryden Flight Research Center, Jan 2, 1976, in the DFRC archives.

45. In the civilian case, consider nuclear-powered merchant ships, nuclear power stations, gas turbine-powered automobiles and trains, and SSTs.

46. To be fair, one must also recognize those aircraft primarily privately initiated that were great successes: notably the outright-private KC–135, T–38, and F–5; the near-private F–86, U–2, SR–71, and C–130; and the akin to private F–4 and B–52. For more of this discussion, see Richard P. Hallion, "Girding for War: Perspectives on Research, Development, Acquisition, and the Decision-Making Environment of the 1980's," *Air University Review*, vol 37, no 6 (Sep–Oct 1986), pp 46–62.

47. For an elaboration of this, see Richard P. Hallion, "Doctrine, Technology, and Air Warfare: A Late Twentieth Century Perspective," *Airpower Journal*, vol 1, no 2 (Fall 1987), pp 16–27.

48. I. B. Holley, Jr, *Ideas and Weapons: Exploitation of the Aerial Weapon by the United States During World War I: A Study in the Relationship of Technological Advance, Military Doctrine, and the Development of Weapons* (Washington: Office of AF History, 1983 ed), p 14. See also Holley's "Of Saber Charges, Escort Fighters, and Spacecraft: The Search for Doctrine," *Air University Review*, vol 34, no 6 (Sep–Oct 1983), pp 2–11.

Thomas A. Julian was graduated from the United States Naval Academy, but elected to be commissioned in the Air Force. After earning his Ph.D. from Syracuse University, he served for seven years on the history department faculty at the USAF Academy, rising to Associate Professor and Deputy Head of the Department. A command pilot, his service included assignments as an air commando in Vietnam, Deputy Commander for Operations and Vice Commander of a C–141/C–5 Wing, Chief of the Nuclear Policy Section at SHAPE, and Chief of the NATO Initiatives Division in the Plans Directorate, Headquarters USAF. Since retiring, he has consulted on defense issues and recently completed four years with the National Defense University Staff. Currently, he is revising his dissertation and pursuing research in Soviet-American military relations.

The Origins of Air Refueling in the United States Air Force

Thomas A. Julian

J. F. C. Fuller, the respected British theorist of mechanized warfare and military historian, noted in 1945 that range throughout military history had been "the characteristic that dominated the fight," and with the experiences of the Second World War fresh in his mind, he voiced his belief that "the fulcrum of combined tactics" in the new airpower era had to be the airplane.[1] However, even with the vast advances in aircraft design and propulsion systems since the Wright Brothers first flew some 43 years before, in 1946, the issue of range for both bombers and fighters had not yet been solved in ways that met the national security requirements of the United States as they were then perceived.

The ultimate solution, air refueling,* was to involve what might well be called reverse technology transfer: the techniques and basic equipment were pioneered by the United States and then adopted, further developed and applied by the British; the British-developed air refueling system was then adopted by the United States Air Force, and in turn, modified and applied by the USAF as an interim system to meet its postwar requirements while it developed a new American system. The latter would incorporate elements of the improved British system and drew upon the collective experience with both systems.

The earliest technology was by today's standards relatively crude. Initially what was involved was merely the translation of standard ground refueling equipment, i.e., refueling hoses, storage tanks, fuel tanks on the aircraft, and procedures, into a vertical dimension. However, it was a translation which called for a considerable amount of ingenuity and pilot skill and courage, particularly if one considers that there was the ever-present possibility that a hose might foul the receiver aircraft's propeller or become wedged into one of the control surfaces.

The evolution and development of air refueling as a system was driven at first by the activities of fliers and aeronautical engineers intent on pushing the limits of this new field, later by the attempts to apply air refueling to commercial activities. Air refueling's adoption as a standard procedure by the

---

*Air refueling will be used throughout this paper, although the terms flight refueling, in-flight refueling, and aerial refueling are used interchangeably in the literature of refueling.

new United States Air Force in 1948, however, represents a rather clear case study of how operational requirements drove the development of new technology rather than vice versa.[2]

The earliest American military experiments with air refueling were in 1923 at Rockwell Field, San Diego, then commanded by the future Commanding General of the Army Air Forces, Henry H. "Hap" Arnold. In a series of flights beginning in April, Lieutenants John Richter and Lowell Smith demonstrated the ability to transfer fuel between aircraft by manually grasping a hose hanging down from the aircraft serving as tanker, connecting it to a fuel tank aboard their aircraft, and letting gravity flow occur. The origin of the tests is not clear

In the 1923 refueling experiments at Rockwell Field, a de Havilland DH-4 dropped a hose from the rear cockpit (above) to another DH-4 (left).

but Richter was quoted at the time as having wished he had had a refueling source during his participation in the World War I St. Mihiel Offensive. He had flown nine sorties but had to return to his home base to refuel after each sortie because his Spad could only stay aloft 20 to 40 minutes in combat. Richter and Smith's efforts culminated in November in a nonstop flight of a little over 12 hours from the Canadian border to Tijuana, Mexico, a distance of some 1,280 miles, during which their de Havilland DH–4B was refueled twice.[3]

A tragic crash the next month caused by the hose becoming entangled in the wings of the participating aircraft brought further experiments to a halt until the famous flight in 1929 of *The Question Mark*. This aircraft, piloted by Carl Spatz (as he then spelled his name) and Ira Eaker, who were to be Arnold's closest collaborators in the later creation of the independent Air Force, and Lieutenants Harry Halverson and Pete Quesada, stayed aloft over Los Angeles, California, for 150 hours and 40 minutes — over six days, in the course of which 5,660 gallons of gasoline and 245 gallons of oil as well as meals, water, and other supplies were transferred, during more than 50 air refuelings.[4]

More significant than the undoubted publicity which the airmen's feat generated was the technical lesson that it demonstrated, namely, that air refueling allowed an aircraft with sufficient structural strength, after becoming airborne, to be overloaded with fuel to a gross weight at which it could not have lifted off the ground because the wing would not generate sufficient lift at the low speeds associated with take off. In more technical terms, as the aircraft increased its airspeed once aloft thereby generating increased lift from the wing, it could fly with wing loadings (expressed in pounds per square foot of wing lifting surface) which were considerably higher than the maximum wing loading at which a loaded aircraft could takeoff from the ground. This extra fuel could mean greatly increased range with a heavier payload.[5]

The then Major Spatz was so impressed with the implications of the *Question Mark*'s extended flight that he recommended that projects be set up to apply air refueling to bombardment, pursuit, and observation aircraft. He also recommended that the Air Corps Engineering Division study whether provision for refueling could not be included in all aircraft during manufacture.[6] However, the War Department did not choose to act on these recommendations, and while a series of commercial fliers successively bettered *The Question Mark*'s record the very next year and there was experimentation in other countries including Germany, Russia, and Japan, serious consideration of air refueling shifted to England.[7]

There were two parallel British efforts. The first, by the Royal Aircraft Establishment, was one in which Royal Air Force Squadron Leader Richard Atcherley figured prominently. Atcherley had participated in the American National Air Races in 1930 and observed a number of the endurance flights of that year. The other, with more of a commercial focus was directed by Sir Alan Cobham, who in 1934 incorporated Flight Refuelling, Limited, for experimental

In 1929, *The Question Mark*, a Fokker C–2, received fuel from a Douglas C–1 (above). The hose was lowered from the C–1 and caught by a crewmember in *The Question Mark* (left), usually Spaatz.

and development work in the field of air refueling. Cobham secured the sponsorship of British Imperial Airways during his early years, and later, that of the Air Ministry.

At the RAE, which in 1931 demonstrated an air refueling system similar to that employed by Hines and Richter in 1923, Atcherley developed a safer and simpler method of contact between tanker and receiver aircraft, consisting of weighted cables with grappling hooks trailed by each aircraft which maneuvered to have the cables cross and the grapples lock the cables together.

The refueling hose attached to the tanker cable was then pulled in for attachment to a refueling receptacle. Simple in concept, but difficult to execute, this method was still being used in the 1950s by B–29 aircraft. Atcherley also developed refinements to this hose method, including a powered reel which permitted the hose to be hauled in more rapidly, an automatic coupling which opened and closed the fuel valve on the receiver as the hose nozzle entered, and as a safety measure, a guillotine which was activated by an explosive charge to sever the refueling hose in an emergency.[8]

Cobham, meanwhile, was successful in demonstrating the potential of air refueling to make air service between England and the United States commercially viable; and in 1938, he received a contract to refuel an Atlantic mail service which would operate specially modified flying boats. The key to commercial viability was reducing the aircraft fuel load at takeoff in favor of mail but overloading the aircraft once in the air with sufficient fuel to enable it to fly against the prevailing westerly winds, from Shannon, Ireland, to Botwood, Newfoundland, the terminal points of the transatlantic crossing. The two specially built flying boats used, the Caribou and the Cabot, were strengthened to carry an overload in flight and had air refueling equipment installed. They were given Certificates of Airworthiness for a maximum take-off weight of 46,000 pounds with a maximum flying weight after refueling of 53,000 pounds. A fuel-dumping system was developed concurrently to deal with the possibility of an emergency at gross weights above the allowable landing gross weight.[9]

The Air Ministry was now also interested — Cobham's Flight Refuelling company took over the development work which Atcherley had previously done — and provided four modified Handley Page Harrow bombers to serve as tankers, two of which were shipped to Newfoundland. Sixteen such flights were made with such success that Imperial Airways planned to expand the service in 1939, but the outbreak of World War II caused cancellation of these plans. As with most flights conducted in the 1930s, Cobham experiments were conducted at relatively low altitudes, none higher than 3,000 feet and most at only 1,000 feet.[10]

Cobham had introduced the so-called Ejector Method of contact between lines from tanker and receiver aircraft in which, using a line throwing gun like those used for life saving at sea, a line from the tanker was fired across a weighted line suspended from the aircraft to be refueled. The two lines again secured to one another by grapples, and the now established procedures beginning with hauling a hose attached to one of the lines to a fill point on the receiver aircraft were followed. Cobham's procedure added purging the hose lines with nitrogen to eliminate the danger of explosive vapors and other safety features.[11]

With war clouds hovering over Europe and concern over Japanese expansion growing in the United States, extending the range of bombardment

A B–50 receives fuel from a hose trailed from a KB–29.

aircraft became a serious issue for Air Corps leaders. In August 1939, Hap Arnold, now Chief of the Air Corps, requested and received a description of Cobham's system from Jimmy Doolittle, who had left the Air Corps and was then with the Shell Oil Company. Doolittle also cited several articles and studies which would explain Cobham's system and the aerodynamic principles involved.[12]

Doolittle's letter generated action on the Air Corps Staff including a short study of air refueling by the Materiel Division and a request for a more comprehensive comparative study of various methods of increasing the range of bombers to provide a basis for a Division policy on how to best achieve the result. The methods to be compared included catapulting, air refueling, and the construction of large airports with longer and possibly even sloping runways to increase the speed at which takeoff could be made with commensurately greater wing loadings.[13]

When the United States entered the Second World War, deficiencies in the combat radii of action of existing bombers and fighters posed major problems for the U.S. desire to strike directly at the heartlands of its German and Japanese enemies. Even before the United States became a belligerent, fears that Hitler might conquer Britain leaving no base area from which Germany might be attacked, led the Army Air Corps in April 1941 to open a design competition for a truly intercontinental bomber. The initial design specifications for what was to become the B–36 called for the bomber to have a 12,000-mile range.[14] After the Japanese attack on Pearl Harbor, the War Department seriously considered air refueling as a means to extend the range of American heavy bombers so that they could bomb Japan from Pacific island bases.

Various schemes were considered. One, discussed at Wright Field in early January 1942, proposed the use of B–24s based in Hawaii to bomb Tokyo, refueling en route from Navy PBY flying boats configured as aerial tankers. The PBYs, in turn, would be refueled at sea from a mobile base consisting of surface tankers, escorts, and security forces provided by the Navy. Conferees at this meeting included Brig. Gen. George C. Kenney, Assistant Chief of the Materiel Division, who would later command Fifth Air Force for General MacArthur in the Southwest Pacific and be the first commander of the Strategic Air Command, and Lt. Col. G. F. Shulgen, representing G–3 of the War Department General Staff.[15] Greater favor was given to the idea of adapting the British air refueling system to extend the range of a B–17E using a B–24 as a tanker and Midway Island as a base from which to bomb Tokyo. While the Doolittle Raid in April 1942, the course of the war, and the ongoing development of the much longer range B–29 and B–32 aircraft took attention away from the Midway plan, experimentation with the B–17E and B–24 tanker continued.

Contrary to the assertion of one writer that wartime tests of air refueling "had been less than satisfactory," the Air Materiel Center at Wright Field conducted a series of successful tests in the spring of 1943. In mid-1942, the Center obtained a virtually complete set of refueling equipment, fabrication and installation drawings, and instructions from England; and the hose system was installed on a B–24D which served as a tanker for a series of tests in which a B–17E was refueled in flight. As a consultant, the Center used a Royal Canadian Air Force squadron leader who had served as Flight Refuelling's chief test pilot.[16]

Some modifications to the British equipment were necessary occasioned by the higher airspeeds required by the American aircraft with their different wing loadings. However, based on some seven test flights, the Center reported that the refueling equipment as installed on the two aircraft was practical. Still employing gravity feed, but with carbon dioxide purging systems, an electrically driven reel to pull the bomber's cable to the tanker, and a hydraulically operated reel to pull the refueling hose back to the bomber, the system could refuel the B–17 with 1,500 gallons of gasoline in about 18 minutes while in flight at an indicated air speed of approximately 150 mph. Other modifications of the British system necessitated by the higher air loadings on hoses and cables from the higher air speeds included substitution of a 2,000-pound-test cable for the 1,300-pound-test cable used to pull the nozzle into the refueling coupling on the bomber, changing the angle at which the harpoon gun fired the tanker-linked cable to engage the receiver's trailing cable, extension of the hose length to 235 feet, and substitution of a stronger fitting on the end of the hose nozzle to ensure positive contact during fuel flow. In its report of June 30, 1943, the Center recommended that careful consideration be given to using this method of refueling in flight to extend the present range of B–17s.[17]

Demonstration flights of air refueling by the two bombers were provided to the Army Air Forces Board in January 1944, but its members indicated verbally to the Materiel Command representatives that there was no tactical requirement for the system. However, they indicated that the refueling equipment should be sent to Fifth Air Force, that is, to General Kenney in the Pacific, for further tests in a combat theater. Although Kenney, in fact, requested five sets of refueling equipment for his Fifth Air Force the following month, none were apparently sent.[18] Underlying the AAF Board's view was almost certainly the fact that the system was obviously not applicable to the massed bomber formations employed by the AAF in the European war.

By early 1944, various other schemes to extend aircraft range for both fighters and heavy bombers that had been actively pursued during the previous years of the war had also assumed a much lower priority. These included air refueling P–38 fighter aircraft from bombers and B–17s towing fuel-filled gliders. The successful wartime solutions to the problem of inadequate range — a combination of acquiring bases closer to enemy homelands and deploying longer range aircraft such as the B–29 and P–51B — were about to be realized.[19] Developmental efforts to extend the range of fighter aircraft continued into 1945, however, at the specific direction of Lt. Gen. Howard Craig, AC/AS, Operations, Commitments, and Requirements. In the summer of 1944, the concept of extending the range of B–29s using B–24s as tankers was also studied briefly, but the idea was dropped because of the marginal extension to be obtained and the length of time needed to modify the B–29.[20]

With regard to bases, the United Kingdom had served as an "unsinkable" aircraft carrier for heavy bombers to attack Axis targets throughout the war, but the cost of other bases, particularly in the Pacific War, had sometimes been very high in American and Allied casualties. The bloody seizure of the Marianas for use as B–29 bases and the equally costly invasion of Iwo Jima, halfway from Guam to Japan, in large part to provide a base for fighter escorts for the B–29 bomber stream on its way to the Japanese Home Islands, were cases in point.[21]

Ironically, it was negative experiences during the briefly successful use of Soviet airbases in 1944 to increase coverage of Axis targets by General Spaatz's strategic air forces in Europe that helped shape the emerging requirements for the Air Force. Coupled with Soviet refusal to provide such bases in the Soviet Far East for attacks on Japan, apparent Soviet attitudes tended to condition the Army's airmen to view their wartime ally warily.

One of the airmen who had been involved was Maj. Gen. Lauris Norstad, the Assistant Chief of the Air Staff for Operations. In 1944, Norstad had been Director of Operations for the Mediterranean Allied Air Forces, whose commander, Maj. Gen. Ira Eaker, was heavily involved personally in the shuttle-bombing project. As early as October 29, 1946, Norstad presented a Top Secret briefing to President Truman, in which he named the Soviet Union as the potential enemy against which the United States must prepare. The general identified the United States and Soviet Union as the two major military powers that had emerged from World War II and stated that Soviet actions around the world showed that "a fundamental conflict of purpose" existed between them.

He concluded that the possibility of war with the USSR was "the *only* probable source of trouble in the foreseeable future" and that this eventuality should be the basis of U.S. planning.[22]

Several recent writers have too easily ascribed cynical motives to the AAF leaders' identification of the Soviet Union as a potential enemy. The earliest chronicler of the AAF's planning for the postwar world opined that

> The problem of postwar enemies [to justify a large Air Force] was solved by identifying Russia as the one long-range postwar threat, based on its assumed air power capability. Apparently, enemies were to be identified on the basis of what states had large air forces or might be expected to develop them.[23]

One might well assert by way of an answer that, following a global war in which the ability of airpower to transcend traditional barriers of space and time had been amply demonstrated, such nations were in the best position to physically threaten the United States and deserved to be viewed warily, a perspective reinforced by Soviet actions in Iran and those parts of Central and Eastern Europe occupied by the Red Army. The top U.S. airmen were also well aware that the Soviet leaders had consistently but unsuccessfully sought to obtain American heavy bombers during the Lend-Lease period and that a number of American bombers, including several B–29s, had not been returned following emergency landings on Soviet territory.[24] Those closest to the project were also aware that a substantial number of Soviet personnel had been trained to maintain American aircraft — something that the Soviet commandant of the joint base complex was to brag to his superiors in Moscow in an after-action report.[25] Ironically, in 1945, the Joint Intelligence Staff estimated that it would take at least five years for the USSR to develop a sophisticated long-range bomber like the B–29, when, in fact, Tu–4s, almost exact copies of the B–29, were coming off five Soviet assembly lines by early 1946.[26]

General Eaker's former aide and later Mediterranean Allied Air Forces Historian, who was a close associate of the AAF officers directly involved with the Frantic project and, as the historian, also had access to all the relevant documents, wrote well after the fact that the negative attitudes toward the Soviet Union of those airmen who were in a position to observe the project stayed with them the rest of their careers. Many became three- and four-star generals in the postwar era.[27] Notably, according to his biographer, the second Chief of Staff of the newly independent Air Force, Gen. Hoyt S. Vandenberg, also took away an abiding distrust of the Soviet Union from his experiences in Moscow as the Air Officer on General Deane's Military Mission during the initial negotiations for the Frantic bases.[28]

Soviet conduct during the Warsaw Uprising of 1944 appears to have had a particularly sobering impact on the AAF leadership. Even Arnold, who had written admiringly about Stalin in his diary after first seeing him at Teheran in late 1943, registered a far different feeling toward him less than a year later when he reviewed the communications about the uprising at the Quebec Conference in September 1944. He was obviously startled by the ruthless cynicism revealed by Stalin's words and his refusal to allow the British and Americans to assist General Bor-Komorowski's forces in the city.[29]

The JCS files for 1945 and 1946 also reveal a growing concern about Soviet conduct, particularly in Eastern Europe. This was shown most openly by the reaction to the memorandum from Maj. Gen. John Deane, Chief of the U.S. Military Mission in Moscow, in early April 1945, to the Joint Chiefs. In it, General Deane, with Ambassador Harriman's concurrence, recommended revising policy with relation to the Soviet Union. Deane proposed a policy of greater firmness in securing quid pro quos where Western interests were involved, while striving to maintain a relationship that preserved wartime cooperation. The Joint Strategic Survey Committee cautiously agreed, but coupled their agreement with a warning not to jeopardize wartime cooperation. However, by October 1945, on its own initiative and against the background of the American rush to demobilize and Soviet actions, the committee expressed serious concern with the current and prospective military position of the United States because of the "recent aggressive and uncompromising attitude of the Soviet Union." The committee asked for an evaluation of present and prospective United States military capabilities and a determination of those areas of the world where such capabilities "would suffice to resist successfully an attempted Russian aggression."[30]

In short, there was ample reason for Norstad's statement to the President that had nothing to do with an institutional or ideological bias, and the Soviet Union as a potential enemy posed a very serious operational and planning problem.

As the Army's airmen considered what characteristics the bomber force of the soon-to-be independent U.S. Air Force should have, the key requirement was obviously great range. However, the heavy losses suffered by American heavy bombers on unescorted deep penetrations of German airspace in late 1943 were burned deeply into their collective memories. The Air Force leaders entered the postwar era initially believing bomber missions would need to be escorted by long-range fighters as in the European air war, presenting an even greater technical challenge than extending bomber range.

As late as August 1945, Lt. Gen. Hoyt Vandenberg, Assistant Chief of the Air Staff for Operations, urged his Air Staff counterpart responsible for materiel to have the matter of increasing fighter range without sacrificing speed or other necessary characteristics investigated thoroughly. As the reason for his request, he pointed out that the war had demonstrated the "unquestioned necessity" of a large fighter escort for bombardment operations against an enemy over whom the AAF did not enjoy complete air supremacy, and he asked that two specific projects be initiated as part of the investigation. The first was a study of what performance capabilities would have to be sacrificed in order to realize an unaided 2,500-mile combat radius for a fighter aircraft; the second, another study of refueling fighter aircraft in the air. Vandenberg acknowledged that the latter idea had been previously tested and abandoned as impractical, but indicated his belief that new techniques and methods might yet make it possible.[31] However, the ultimate answer was to be a bomber program that eliminated the requirement for fighter escort by focusing on high speed as a way to reduce bomber vulnerability and air refueling to provide the needed range capability.

The seminal event in the history of aerial refueling — and, in many ways the postwar development of the independent Air Force itself — was the report of the ad hoc Heavy Bombardment Committee that General Spaatz created in early September 1947, which consisted of representatives from Headquarters, Army Air Forces, SAC, Air Materiel Command, and the Air University. Spaatz, who soon became the first Air Force Chief of Staff, directed the HBC to investigate methods of delivering an atomic attack on an enemy 4,350 nautical miles from continental U.S. bases and to "recommend actions in order of priority" for consideration by the USAF Aircraft and Weapons Board on key issues, including aircraft characteristics, tactics and techniques, the research and development program, an interim solution to the problem, and the continuation or alteration of the existing B–52 program which was developing a very heavy aircraft powered by turbine-driven propellers.[32]

The Aircraft and Weapons Board, initiated by Maj. Gen. Curtis E. LeMay, Deputy Chief of the Air Staff for Research and Development, and chaired by General Vandenberg, soon to be Spaatz's Vice Chief of Staff, was concerned with the research and development program begun during the war for heavy and medium bombers. Its deliberations had to take into account the prospect of sharply reduced defense budgets as well as the technical challenges posed by requirements to conduct atomic warfare against a distant enemy. The Air Force leadership soon also had to take account of interservice rivalry with the U.S. Navy and attacks on the Air Force bomber program.

Sixteen people attended the first meeting of the board (August 19–22, 1947), including the Commanders of the Tactical Air Command, Air Materiel Command, Air Defense Command, Air University, and the Vice Commander of the Strategic Air Command. Perhaps the most interesting thing about the verbatim transcript of this meeting is that, while there was concern expressed about the deficiencies in range of the candidate bombers (even the B–36 was not seen as part of a solution, but because of development problems, as part of the problem), there was a very strong, perhaps even greater current of concern about the ability of any unescorted bomber to penetrate hostile airspace in the face of hostile fighters. In terms of budget realities, the AAF leadership worried about the cost of large aircraft which the military characteristics seemed to indicate were necessary to attain long range.[33] Given the choice of lowering the weight and hence, the cost of the B–47, the new "workhorse" medium bomber, by reducing either its range capability or its speed, the Board opted for speed. Studies showed that high speed limited the possibility of attacks to a 55-degree arc of the bomber's tail cone by a forward firing interceptor.[34]

On this basis, even Curtis LeMay, who at the meeting had initially seemed willing to sacrifice speed for additional protective armament, ultimately agreed that higher speed was more desirable. In contrast to his hesitations at the meeting, just a month earlier he had requested that the RAND Corporation be given access to Restricted Data by the Atomic Energy Commission in connection with RAND's study of nuclear-propelled aircraft. His reasons were that decisions on the future bomber program had to be made by October 1, 1947, and he believed it "debatable" whether chemically fueled bombers could achieve sufficient speed over the long ranges the AAF had to be prepared to fly in the event of war "to provide the requisite immunity from anticipated enemy countermeasures."[35]

The new heavy bomber, the B–52, exemplified the weight/cost problem, since to meet its range, speed, and load-carrying specifications, it was estimated that it would weigh around 400,000 pounds, making it bigger and more costly than the B–36. Interestingly, there was absolutely no reference during the board's discussions to air refueling as a possible solution to the problem of range — or aircraft weight. The interim solution, pending the development of better engines with lower specific fuel consumption, was to lay on one-way missions for atomic delivery. In a discussion which would not have warmed the hearts of any aircrewman, the Operations Chief, Maj. Gen. Earl Partridge, expressed the thought that such a mission with an atomic bomb was acceptable, and that given the small number of atomic bombs that would be available (a number he could not discuss in that forum), a very large (and expensive) nuclear striking force was not necessary. In his view, the country could afford to build eight bombers for every bomb there was, and while it might sound cold-blooded, the economically best thing for the country was to "expend the crew, expend the bomb, expend the airplane all at once. Kiss them goodbye and let them go."

Brig. Gen. Thomas Power, who would be LeMay's successor as SAC Commander, disagreed, pointing out that crew reliability might decline when it became obvious crews were to be dispatched on one-way missions, and the subject was left for the moment with the Air University representative's comment that there were large remote areas of the target nation where crew pickup might be effected.[36] For whatever reason, Gen. George Kenney, the SAC Commander, did not attend this first meeting.

The HBC's Report of November 7, 1947, recommended modifications in phases to the fleet of existing bombers and those in development to meet the range requirements for striking targets in the Soviet Union and to realize the board's decision to emphasize high speed as a way of penetrating Soviet air defenses successfully. Air refueling was the key. For tactics, the committee gave top priority to refueling conventional bombers (B–29s, B–50s, and B–36s) prior to or after departing enemy territory during 4,350-mile-radius missions. For research and development, they gave first priority to AMC developing air-to-air, high-capacity, single-point refueling systems and a method of satis-factory rendezvous and refueling under all-weather conditions. They recom-mended discontinuing development of the turbo-propeller powered B–52 and identified four actions to be taken by the command as an interim solution. These were to modify expeditiously significant numbers of B–29s and B–36s as tankers for refueling bombers; to procure enough refueling equipment, range-extending bomb bay tanks, and releasable wing tanks, where applicable, to enable B–50 and B–36 bombers to reach their target areas; to develop the B–36 program to the maximum performance obtainable by the use of new engines and equipment as they came available; and to develop methods and determine the feasibility of towing airplanes under all-weather conditions and for long ranges and also "consider other methods of range extension." Contrary to the board's apparent belief that aircrews on a one-way mission to the Soviet Union could be recovered from remote areas of the country, the committee deemed such escape and evasion "improbable."[37] The board accepted the committee's recom-mendations in January 1948, and General Spaatz formally approved them on March 3, 1948.

The problem of providing fighter escort to conventional bombers still existed, however, and would continue to do so at least until the first sweptwing jet powered medium bomber, the B–47, entered the Air Force inventory in operational numbers. The Committee projected that this would not occur until January 1951. Consequently, a search for ways to extend fighter range continued to be made by AMC, including the development of so-called parasite fighters, like the P–85, which was designed to be carried by the B–36 internally and provide fighter protection at the maximum combat radius of the bombers.[38]

Even before the HBC Report, SAC had apparently been considering refueling as a means to meet its mission requirements, although, curiously, General Kenney's representative at the August Aircraft and Weapons Board meeting, Maj. Gen. Clements McMullen, had not raised the topic.[39] At SAC's request, AMC had initiated a program of experimentation in late 1947 using two B–29s assigned to the Command for the purpose and another on loan from SAC which was to be configured as a receiver aircraft. The program was to evaluate the problems of refueling in the shortest possible time and develop optimum means for in-air contact, the mechanics of refueling, and optimum equipment including single point refueling systems.[40]

It is evident from the documents of this time that developing an air refueling capability became imbued with a sense of increasing urgency as a result of the series of contemporary events that marked the formal beginning of

A B–36 carrying an F–84.

the Cold War. The Moscow Foreign Ministers' Conference of late 1947 had once again failed to establish a unified approach to administering a defeated Germany; the Czech Communists' coup of February 1948, and Eduard Benes' "suicide" put Czechoslovakia behind the "Iron Curtain," in Churchill's phrase; and in April 1948, Soviet forces in Germany took the first steps to block Western access to Berlin and, by June 1948, had instituted a full-scale blockade of the city.

To achieve the HBC's interim solution for creating a refueling capability in the new bomber force, the relatively modest AMC project had its priority upgraded to 1A in late February 1948, and in late March, it became the GEM Program. Purchase Requests for procuring modifications or equipment for GEM were to be given first priority both in contracting and in the scheduling of the contracted for deliveries, and cost-plus-fixed fee contracts were authorized whenever an Air Force procurement official deemed such contracts necessary to expedite the Program. A further measure of the importance accorded the GEM Program was that contractors were authorized unlimited overtime and were to be furnished Government facilities whenever such action was necessary to meet the required delivery schedules.[41]

Boeing was designated the supplier of all air refueling equipment and given a threefold task. First, it was to fabricate and install forty sets of equipment on forty tankers and forty receivers which had been modified to carry atomic weapons, subordinating in the process refinements such as high rate of flow, low temperature operation, and high speed at which refueling could be accomplished in favor of the earliest possible completion of the first installations. Second, Boeing was to develop and install on later B–50 airplanes an improved system incorporating a single-point refueling direct to tanks, which would permit higher rates of flow (300 to 400 gallons per minute) at higher airspeeds than possible with the gravity flow system. Finally, Boeing was to study methods of air refueling other than hose-type systems, which were inadequate for high-speed aircraft such as the projected B–52.[42]

Completion date for the GEM Program was set as December 15, 1948, but in March the development of the required equipment was still lagging. To expedite the program, it was decided to purchase a set of the refueling equipment which the RAF had developed for refueling its Lancaster Heavy Bombers and obtain technical assistance from the British in adapting it for American B–29s and B–50s. Almost immediately, the purchase requirement was expanded to include up to 40 sets.[43]

The Buy America Act was duly waived, and Lt. Col. H. E. Warden of AMC's Engineering Division was sent to England where, in late March, he contracted for the purchase of thirty-four sets of British hose-type refueling equipment together with technical support from Flight Refuelling, Limited. The latter was to include the services of technical representatives in the United States; sets of specialized tools, test equipment, test reports, fittings, a harpoon gun, etc.; and handbooks of operations and maintenance instructions for the equipment. The delivery schedule Warden negotiated called was to be stretched out over nine months with the last eighteen sets to be delivered six at a time in late October, November, and December, respectively.[44]

This delivery schedule was clearly unacceptable in view of the urgency with which the program was viewed. Consequently, AMC decided that immediate production of refueling equipment in the United States was necessary and gave Boeing a contract to procure from American sources fifty sets of tanker equipment and one hundred sets of receiver equipment built to the British design.[45] Noteworthy of the concern felt in the Pentagon about the possibility of war with the Soviet Union was the interest shown in the refueling tests by James Forrestal, the Secretary of Defense, who wanted to know where and when these tests were going to be held so that he could have a representative present.[46]

In May 1948, the initial pair of B–29 tankers and receivers was ready for testing. A series of tests during the next several months, including a demonstration for Charles Lindbergh, led to dropping the use of the harpoon gun in establishing contact between tanker and receiver aircraft in favor of a modified flying cross-over method as well as the identification of other needed improvements.[47] Also in May, development of the American System, a flying boom proposed by Boeing, was well underway, involving the efforts of AMC, Boeing, and the National Advisory Committee on Aeronautics. The system employed a rigid telescoping tube fastened to a multidirectional swivel on the tanker fuselage bottom, through which a fuel hose with a standard quick coupling on the other end could be extended and dispense fuel under pressure. The boom was equipped with control surfaces which allowed the boom operator to fly it into contact with a fixed probe on the nose or top of the receiver aircraft's fuselage, automatically couple the fuel hose inside the boom with a

A KB–29 testing the British probe and drogue refueling system.

single point refueling fitting on the probe, and transfer fuel. This new method of air refueling was viewed enthusiastically by the Air Force because its simplicity would allow installation in nearly any type of aircraft and promised to be much more operationally suitable than the hose system, particularly for jet aircraft.[48]

AMC sought to impress Boeing with the importance the Air Force attached to early development of the flying boom system, urging that every effort be expended toward "early completion of tests of the mechanical boom, since satisfactory operation there will most likely be sufficient to release the units to production."[49] While the Air Force briefed the Joint Chiefs of Staff in August 1948 that it expected the flying boom to be operational by the summer of 1949, it was not until September 1, 1950, that the first boom-equipped B–29 tanker, a KB–29P, was received by the 97th Air Refueling Squadron at Biggs AFB, Texas.[50] Even then, tests begun that same month at the Air Proving Ground concluded that the system was not satisfactory for operations at temperatures of +20 °F and below.[51] It would be the American version of the British hose system with some refinements that would provide the core of SAC's air re-fueling capability in the form of KB–29Ms during the crucial early years of the Soviet-American confrontation. The refinements, which the Air Force originally gave a lower priority than achieving the most rapid deployment of a refueling capability, included such improvements as higher rates of transfer (300 to 500 gallons per minute), transfer hoses that remained operative to -65 °F, and a bet-ter method of engaging the tanker and receiver lines.

The original program had called for the installation of 40 sets of refueling equipment on B–29s. On April 13, 1948, an additional 36 B–29s and 36 B–50s were designated to be modified as receivers (now code named Ruralists), and on April 19 USAF Headquarters directed the Commanding General AMC to

A KB–29 testing the flying boom refueling system.

add 80 more B–29s to the number of bombers to be modified as tankers (now code named Supermen) and 208 B–50s to the list of bombers to be used as receivers in air-to-air refueling. By June 26, a contract change directed Boeing, which was conducting the aircraft modifications, to deliver 76 modified B–29s by October 27, 1948, and 72 modified B–50s on or before December 15, 1948. The October target was essentially met, modifications on 75 B–29s as either tankers or receivers either having been completed or in progress at the Boeing plant in Wichita. By the autumn of 1949, training missions by B–50s refueled by KB–29 tankers were being flown at distances over 5,000 nautical miles using bases other than launch bases for recovery.[52] The HBC had projected that only after July 1950 would improved B–50 and B–36 bombers using refueling be able to make round-trip rather than one-way missions to Soviet targets, but Project GEM's success helped give the Air Force such a capability months earlier at a time when months seemed critical.[53]

The decision to develop and employ air refueling routinely for bomber missions was a true watershed. From being an expedient to overcome limitations in the heavy bomber force with which the United States emerged from World War II, air refueling's role evolved to that it played in Desert Storm when it was the limiting factor for air operations and was the key Coalition air capability, without which the Gulf War Airpower Survey concluded that the air campaign could not have been conducted successfully.[54] That evolution was marked by significant milestones, including the gradual phase-out of the converted heavy bomber KB–29 and KB–50 tankers in favor of the KC–97, a convertible tanker-transport, and after 1957, the introduction of first jet tanker, the KC–135, into the SAC inventory. It was also marked by the gradual introduction of air refueling capability into fighter aircraft, first, in 1951, a wing receptacle for use in the F–84G with the Boeing-developed and SAC-endorsed flying boom system, then, in 1955, the British-developed probe-and-drogue system for use in the F–100C. This capability permitted more rapid responses to overseas crises in support of U.S. foreign policy objectives through direct deployments from continental U.S. bases to areas of tension. SAC KC–97s tankers supported the deployment of 58 F–84Gs from Turner AFB, Georgia, to Misawa, Japan, in mid-1952, and the longest nonstop deployment in history to that time of 17 F–84Gs in August 1953, from Turner to Lakenheath, in the U.K., a distance of 4,485 miles.

From a force dedicated to the support of SAC's war plans, the SAC tanker force now had developed broader responsibilities, and in 1961, the Air Force designated SAC as the single manager for its own and TAC forces, a position which it also fulfilled during the Gulf War with regard to AWACS and JSTARS aircraft and transport aircraft from the Military Airlift Command. Since October 1993, as a result of Air Force reorganization and the elimination of SAC, all KC–10 and KC–135 tankers based in the continental United States except for KC–135s assigned to the Air Combat Command's 366th Wing at Mountain Home, Idaho, have been assigned to MAC's successor, the Air Mobility Command.[55]

The SAC tanker force's capabilities were a product of effective technical cooperation among SAC, AMC (and its successor organizations Air Force Logistics Command and Air Force Systems Command), and Boeing, along with the development of organizational structures and procedures, coupled with intensive and effective training that allowed it to fulfil many roles. When Soviet intercontinental attack capability became credible in the 1950s, KC–97s stood ground alert with the relatively short-ranged B–47s on continental SAC bases and overseas reflex bases on Guam, the North African littoral, and in Canada and Greenland. SAC tankers supported SAC bombers armed with nuclear weapons on airborne alert, and provided refueling for B–52 Arc Light missions from Guam during the Vietnam War.

Air refueling also played a new and essential role in operations by tactical aircraft against North Vietnam. Until 1964, jet tankers had been used to refuel tactical aircraft only during deployments. However, during the war, pre- and post-strike refueling of fighters engaged in combat operations by SAC dedicated tankers in Young Tiger operations were routine, and in fact, were central to the air operations as they were conducted in Southeast Asia. For example, Thailand-based tankers between August 1966 and March 1969, accomplished more than 208,500 refuelings and off-loaded 1,510,900 pounds of fuel. SAC tankers also provided emergency fueling capabilities to many U.S. aircraft leaving their targets that otherwise would have been lost.[56]

The value of air refueling capability to the achievement of U.S. military and foreign policy objectives has been demonstrated repeatedly since. Air refueling was the key to rapid reinforcement planning to support the North Atlantic Treaty Organization and Crested Cap, Creek Bee, and other exercises in which

A KC–135 refuels an F–4 over Southeast Asia as four other fighters wait.

USAF aircraft deployed to European bases and helped reassure the U.S. Alliance partners. The United States even provided C–135s and refueling technology to the French to provide an inflight capability for its Force de Frappe as a means of augmenting NATO nuclear deterrence.

The salience of an air refueling capability was in some ways most vividly demonstrated during the 1973 war between Israel and the Arab states when there were difficulties in securing landing rights for MAC airlifters carrying supplies to Israel from countries whose support the United States had previously received. This led to an expansion of refueling capability in the MAC fleet, the MAC workhorse, the C–141A, being modified to increase its carrying capacity and receive an air refueling capability, and the C–5s hitherto unused air refueling capability becoming routinely exercised along with that of the new C–141B in MAC training and airlift operations. During the 1980s, replacement of a substantial number of C–135s' engines with GE/SNECMA CFM 56 turbofan engines that produced almost twice the thrust of the original C–135 J57 engines with much reduced fuel consumption, together with the acquisition of KC–10 cargo/tanker aircraft, helped provide the tanker capability to support this expanding lien on SAC's air refueling capability.

Shortly before the Gulf War, the potential value of these improvements was demonstrated vividly and the actual value of the basic capability was once again reaffirmed in a particularly significant way. In 1986, F–111s based in the U.K. traversed with multiple air refueling a flight path over international waters to carry out the U.S. punitive attack on Libya rather than flying the much shorter direct route across France, which refused to give overflight rights.

The history of air refueling is rich in the mixture of imaginative concepts coupled with pragmatism, the application of new technologies and unusual engineering and production skills, and organizational flexibility and profes-

A KC–97 refuels a B–47.

sionalism in planning, training, and operations that have generally characterized the postwar U.S. Air Force. As the historian of the Air Force's postwar bomber force remarked, air refueling from KC–97 tankers "transformed the B–47 into an intercontinental bomber"; more important, refueling from a succession of tankers has been the means whereby the U.S. Air Force has been able to give substance to its slogan, Global Reach, Global Power.[57]

Notes

1. J. F. C. Fuller, *Armament and History: A Study of the Influence of Armament on History: From the Dawn of Classical Warfare to the Second World War* (New York: Scribner's, 1945), p 7, quoted in unpublished lecture, "The Future of Air-power," Richard P. Hallion [1994].

2. The AF's Forecast Projects and the US Army's High Technology Test Bed (HTTB) Program established at Ft Lewis, Wash, with the 9th Infantry Div, exemplify the recent tendency to identify specific technologies for possible military applications. For the Forecast Projects, see Michael H. Gorn, *Harnessing the Genie. Science and Technology Forecasting for the Air Force: 1955–1986* (Washington: Office of AF History, 1988). For a brief description of the HTTB, see "Tomorrow's infantry: More lethal and much swifter," *Business Week*, Oct 18, 1982, p 189. Admittedly, the issue whether a particular military capability is the result of "requirements pull" or "technology push" is not always clear cut.

3. Henry H. Arnold, "The History of Rockwell Field, 1923," 168.65041, White Papers, USAF Historical Research Agency, Maxwell AFB, Ala. Arnold, who observed part of the flight from another aircraft, notes that the actual record flight was delayed by a stop in San Francisco to demonstrate the new technique to Gen "Billy" Mitchell, Asst Chief of the Air Service, and delegates to the American Legion Convention then being held there. Richter is quoted in Dennis Casey and Bud Baker, *Fuel Aloft: A Brief History of Aerial Refueling* (Copy in AF Hist Support Office Library; no publisher, place, or date of publication), p 2.

4. Carl Spatz, Final Rpt, "The Flight of the Question Mark, Jan 1–7, 1929," nd, Carl A. Spaatz Papers, Library of Congress, Manuscript Div (LC/MD), Box 110; Casey and Baker, *Fuel Aloft*, p 3.

5. Within the limits established by power of the aircraft engine (or engines) to move the aircraft through the air and the resistance or "drag" induced by the passage of the wing through the air, an increase in airspeed generates an increase in the lift produced by the wing. The increase in a wing's lift is proportional to the density of the air and the square of the air speed. To maintain level flight, the lift produced by the wing must equal the gravity force of the aircraft's weight. The important corollary, with important implications for minimum airspeeds at which air refueling must be conducted, is that at a higher wing loading, the wing will stop flying ("stall") at a higher airspeed than at a lower wing loading.

6. Spatz, "Flight of the Question Mark." For a summary analysis of the flight and specific problems on which the project participants made recommendations, see Materiel Div Memo Rpt on Refueling of Airplanes in Flight, Feb 23, 1942, in Historical Office, *Case History of Air-to-Air Refueling*, vol II, *Supporting Documents* (Dayton, Ohio: Air Materiel Command, 1949), 202.2–59, Historical Research Agency, Maxwell AFB, Ala. (Vol I of the Case History is a brief narrative based upon analysis of these documents and hereafter will be cited as *Case History*; the supporting documents will be cited as *Case History Supporting Documents*.)

7. Vernon B. Byrd, *Passing Gas: The History of Inflight Refueling* (Chico, Calif: Byrd, 1994), p 47. The team of Dale Jackson and Forest O'Brine set a record over St Louis of 647 1/2 hours in Jul 1930. C. H. Latimer-Needham, *Refuelling in Flight* (London: Pitman, 1950), p 3.

8. Latimer-Needham, *Refuelling in Flight*, p 3; Byrd, *Passing Gas*, p 48.

9. Latimer-Needham, *Refuelling in Flight*, p 9.

10. Byrd, *Passing Gas*, pp 53, 59. Most of the air refueling flights in the 1930's were conducted at or lower than 5,000 feet.

11. *Ibid*, p 55.

12. Ltr, J. H. Doolittle to Maj Gen H. H.

Arnold, Aug 23, 1939, *Case History Supporting Documents.*

13. Memo, Maj H. Z. Bogert, Chief, Exp Eng Sect, AMC, to Chief, Aircraft Lab, AMC, Nov 6, 1939, *Case History Supporting Documents.*

14. Marcell Size Knaack, *Post-World War II Bombers, 1945–1973* (Washington: Office of AF History, 1988), p 5. The initial set of specifications also called for a top speed of 450 mph at 25,000 feet, a 275 mph cruising speed, and a service ceiling of 45,000 feet. These were reduced in Aug 1941 to the more realistic but still demanding specifications of a minimum overall range of 10,000 miles and an effective combat radius of 4,000 miles with a 10,000 bomb load. The required cruising speed was also reduced to between 240 and 300 mph and the service ceiling was reduced to 40,000 feet.

15. Memo for the Record: Rpt of Refueling Conf, Jan 5, 1942, *Case History Supporting Documents.*

16. Quotation from Perry McCoy Smith, *The Air Force Plans for Peace, 1941–1945* (Baltimore, Md: Johns Hopkins, 1970), p 83. See Capt D. L. Yeager, AAF Eng Div, Memo Rpt, subj: Refueling in Flight, Jun 25, 1943, *Case History Supporting Documents.*

17. Yeager rpt.

18. Ltr, Chief, Eng Div, to CG, AAF, Jan 28, 44, subj: Proj MX 204, Refueling in Flight, *Case Study Supporting Documents.* Only the test set was available to meet Kenney's request, and the Air Staff agreed to have the equipment in the test aircraft removed and installed in two B–24Js for dispatch to Fifth AF. However, this does not seem to have occurred. Ltr, Chief, Eng Div to Asst Chief of the Air Staff (AC/AS), Material, Maintenance, and Distribution, Mar 2, 1944, with 1st Endorsement, Mar 31, 1944, in *Case History Supporting Documents.* As described above, Kenny had participated in the original discussions in Jan 1942, regarding the use of air refueling to bomb Tokyo.

19. Matterhorn, the B–29 bombing offensive against Japan from Chinese bases approved at the Cairo Conference in late 1943, began with an attack on the Yawata steel works on Jun 15. Wesley Frank Craven and James Lea Cate, *The Army Air Forces in World War II,* vol V, *The Pacific: Matterhorn to Nagasaki, June 1944 to August 1945* (Chicago: University of Chicago, 1953), pp 13–27. By Jun 1944, the P–51B was ranging deep into Germany including round-trip escort missions as far as Berlin. The P–51 became a true long-range fighter with the "marriage" of the basic North American airframe with the British Merlin 61 engine and the application of droppable centerline fuel tanks. Craven and Cate, vol VI, *Men and Planes* (Chicago: University of Chicago, 1955), pp 218–20.

20. See memos, Brig Gen B. W. Childlaw for Maj Gen O. P. Echols, subj: Flight Refueling Proj, Apr 10, 1944, and Col H. Z. Bogert for AC/AS, M & S, subj: Status Rpt, Proj MX–522, Mar 2, 1945, RG 341, 171, Box 39, NARA; *Case History,* pp 8–9.

21. While the P–51's range capability had improved by mid-1944 to the point that it could escort Spaatz's bombers to Berlin and back, the Marianas were too far away from Japan for American fighters flying from Marianas bases to escort the B–29s to their Japanese targets and return. Additionally, Japanese fighters based on Iwo Jima posed a menace to the B–29s necessitating a dogleg course from Saipan to Japan which reduced bomb load and complicated navigation. Craven and Cate, *Matterhorn to Nagasaki,* pp 586–87.

22. "Presentation Given to President by Maj Gen Lauris Norstad on Oct 29, 1946, 'Postwar Military Establishment,'" Soviet Union folder, Box 63, Hoyt S. Vandenberg Papers, LC/MD. Norstad's successor at MAAF was to be Charles P. Cabell, later Dep Dir of the CIA. For a description of the shuttle-bombing project, codenamed Operation Frantic, see Thomas A. Julian, "Operations at the Margin," *The Journal of Military History* Oct 1993, pp 627–52. Norstad's presentation is discussed in Harry R. Borowski, *A Hollow Threat: Strategic Air Power and Containment before Korea* (Westport, Conn: Greenwood, 1982), pp

94–97.

23. Smith, *The Air Force Plans for Peace*, p 69. Also see James S. Sherry, *Preparing for the Next War: American Plans for Postwar Defense, 1941–45* (New Haven, Conn: Yale University Press, 1977), p 168. Part of the problem is that Smith, a political scientist, offers this comment with regard to the Postwar Planning Div of the Air Staff, but leaves the implication that this was the considered view of the AAF's top leadership when, in fact, the latter tended to share the view Norstad expressed to the President.

24. Otis Hays, Jr, provides an account of these aircraft and the vicissitudes suffered by their crews during internment until their prearranged "escape" from the Soviet Union in *Home from Siberia: The Secret Odysseys of Interned American Airmen in World War II* (College Station, Tex: Texas A&M University Press, 1990).

25. Julian, "Operations at the Margin," p 636.

26. "Air Capabilities of the USSR Generally and by Areas," Joint Intelligence Staff (JIS) 80/11, Oct 31, 1945, RG 218, CCS 092 USSR, 3–27–45, Box 208, NARA; Air Chief Marshal Sir Philip Joubert, "Long Range Air Attack," in Asher Lee, ed, *The Soviet Air and Rocket Forces* (New York: Praeger, 1959), p 107.

27. James Parton, *"Air Force Spoken Here": General Ira Eaker and the Command of the Air* (Bethesda, Md: Adler, 1986), p 407. The list of those senior AAF officers in addition to Vandenberg (future Chief of Staff) influenced negatively by their association with Soviet officialdom during the war is a lengthy one. It includes Eaker, postwar Dep CG of the AAF; Norstad, later NATO Supreme Allied Commander, Europe; Charles P. Cabell, Norstad's successor as MAAF's Ops Dep; Laurence Kuter, the wartime AC/AS for Plans, participant in the Yalta Conference in a final attempt to secure Soviet cooperation for American bomber bases in Siberia, and later North American Air Defense Commander; Carl A. Spaatz, commander of the personnel who both manned and operated from the Frantic bases, and first AF Chief of Staff; as well as a number of others who rose to senior leadership positions in the AF.

28. I am indebted to Col Phillip Meilinger, USAF, for this information. Telephone conversation, Sep 29, 1995.

29. At Teheran, Arnold admiringly called him truly "a man of steel"; at the second Quebec Conference, which Stalin did not attend, Arnold wrote in his diary that he went to bed "dreaming about alligators lurking in the shadows awaiting such prey as came their way." Quoted in Thomas A. Julian, "The AAF and the Warsaw Uprising," *Air Power History,* Summer 1995, p 33.

30. "Revision of Policy with Relation to Russia," JCS 1313, Apr 16, 1945; "Arrangements with the Soviets," JCS 1301/2, Apr 5, 1945; "Military Position of the US in the Light of Russian Policy," JCS 1545, Oct 8, 1945, all in RG 218, CCS 092 USSR, 3–27–45, Box 209, NARA.

31. Memo for AC/AS–4 from AC/AS–3, subj: VLR Fighters, Aug 31, 1945, *Case History Supporting Documents.*

32. "Report on Heavy Bombardment by Heavy Bombardment Committee convened to Report to the USAF Aircraft and Weapons Board," Nov 7, 1947, RG 341, DCS/Dev, Box 60, NARA. For a brief discussion of the HBC, see Robert Frank Futrell, *Ideas, Concepts, Doctrine: Basic Thinking in the United States Air Force, 1907–1960* (Maxwell AFB, Ala: Air University, Dec 1989), pp 232–33. For the Aircraft and Weapons Board, see Futrell, *Ideas, Concepts, Doctrine*, p 213.

33. Minutes of the First Meeting of the AF Aircraft and Weapons Board, Aug 19–22, 1947, RG 18, 337, Meetings, Box 780, NARA.

34. *Ibid.*

35. Ltr, LeMay to Chairman of the Military Liaison Committee to the Atomic Energy Commission, subj: AAF RAND Project—Proposed Summer Session on Capabilities and Limitations of Nuclear Powered Aircraft, Jul 18, 1947, RG 18, 334 Boards, Box 620, NARA.

36. In fairness, it should be noted that Partridge's comment did not apply to the

envisioned "work-horse" bomber carrying conventional bombs that would need to be used on two-way missions if the AF was to be able to preserve a bombing capability with the relatively small number of aircraft that he and the other Board members believed postwar defense budgets would allow them to buy. USAFAWB Minutes.

37. HBC Rpt. The towing pairs offered as examples included the B-29 with the B-36, B-29 with the B-50, B-50 with the B-36, and the B-47 with the B-36. At various times "other means" included towed gliders filled with fuel which were to be cut loose when empty, a similar method to that prototyped by the *Luftwaffe* during the war with a towable airfoil. I am indebted to Dr. Richard Muller, historian of the *Luftwaffe* and currently a member of the USAF Command & Staff College, for this information, telephone conversation, Aug 12, 1994. As late as the early 1950s, the AF did have organized units whose sole mission was to penetrate Soviet airspace to pick up bomber crews after a one-way strike. A good friend of the author's assigned to one of these units was killed during a training accident.

38. The Air University appears to have been the source of this idea, and the Air Staff pushed for testing the new P-80 and P-84 jet fighters as "parasites" for the B-36. Memo, Sep 17, 1947, Maj Gen L. C. Craigie, Chief, R&D Div, ACAS/A-4, to CG, AMC. RG 341, entry 172, Box 51, NARA.

39. *Case History*, p 11; ltr, Brig Gen Thomas Power to CG, SAC, Nov 12, 1947, *Case History Supporting Documents*. McMullen was apparently heartily detested at SAC and also seems to have been a factor in Kenney's later relief as SAC Commander. However, in his account of the Pacific War, Kenney speaks highly of McMullen's wartime service under him in Fifth AF, Thomas M. Coffey, *Iron Eagle: The Turbulent Life of General Curtis LeMay* (New York: Crown, 1986), pp 273-74; *General Kenney Reports: A Personal History of the Pacific War* (Washington: Office of AF History, 1987), pp 535, 542.

40. Ltr, Brig Gen Thomas S. Power to CG SAC, subj: Air-to-Air Refueling Proj, Nov 12, 1947, RG 241, 172, Box 51, NARA.

41. Routing & Record Sheet to All Buyers from the Chief, AMC Procurement Div, subj: "GEM" Prog, Mar 29, 1948, *Case History Supporting Documents*.

42. Ltr, CG, AMC, to Chief of Staff, USAF, subj: In-flight Refueling Prog for Bombardment Aircraft, Mar 15, 1948, *Case History Supporting Documents*.

43. Memo for the Dir of Intel from Maj Gen E. M. Powers, Asst Dep Chief of Staff, Materiel, subj: Lancaster Refueling Equipment, Mar 9, 1948, RG 341, 172, Box 51, "Early Refueling Methods" folder; ltr, Lt Gen H. A. Craig, DCS/Materiel, to Undersecretary of the AF, subj: Foreign Procurement—In-Flight Refueling Equipment, Mar 15, 1948, *Case History Supporting Documents*.

44. Lt Col H. E. Warden, rpt, subj: Visit of AMC Personnel to Flight Refueling, Limited, Apr 8, 1948, RG 341, 172, Box 39, NARA. As noted above, the actual requirement had risen to forty sets by mid-Mar, and the reason for the discrepancy is not known. It may be that the unit price was high enough to compel reduction of the numbers to be bought to obtain the other British assistance that was desired.

45. Ltr, Col George E. Schaetzel, Chief, Aircraft & Missiles Sect, Procurement Div, AMC, to Boeing Airplane Co, subj: Refueling Equipment for Tankers and Bombers, Apr 8, 1948, *Case History Supporting Documents*.

46. Memo for the Secretary of the AF, from Nicholas S. Ludington, Mar 24, 1948, *Case History Supporting Documents*.

47. Memo for Col Mark Bradley, HQ AMC, from Capt Lyle C. Freed, USAF Plant Representative, Boeing, Wichita, Jun 6, 1948; Weekly Status Rpt on Dev of Air to Air Refueling under Contract AC 20413 from Vice President, Boeing Wichita, Jul 8, 1948, *Case History Supporting Documents*.

48. Memo for Boeing Airplane Co, Wichita, Kansas, from the Acting Chief, Aircraft and Missiles Sect, Procurement Div, AMC, subj: Contract No. AC-20413,

Air-to-Air Refueling "Flying Boom;" *Case History Supporting Documents.*

49. *Ibid*

50. "Supplement to Dept of the AF Presentation to Joint Chiefs of Staff Committee on the Dev of the AF During Fiscal Year 1950," Aug 25, 1948, RG 343, John McCone Papers, NARA. I am indebted to Park Temple for bringing this document to my attention and generously providing me with a copy. Casey and Baker, *Fuel Aloft,* p 16.

51. H. W. Hoyt, "Final Rpt on Functional Test (CHC) of Flying Boom Refueling System" [ca. Oct 1950], RG 341, 44, Office of the Surg Gen, Box 268, NARA.

52. *Case History,* p 16. See the set of Long Range Prediction Calculation Sheets with accompanying marked charts in Curtis E. LeMay Papers, Box 95, MD/LC.

53. HBC Rpt, p 22; Daily Activity Rpt of R&D Office, DCS/M, HQ, USAF, Oct 8, 1948, RG 341, Box 51, NARA, *Case History,* p 16.

54. The SAC tankers utilized included 29 KC–10s and 193 KC–135s deployed in the area of operations (AOR) at the peak of Desert Storm with another 17 KC–10s and 69 KC–10s operating in direct support from outside the AOR, *Gulf War Air Power Survey,* vol IV, *Weapons, Tactics, and Training* (Washington: GPO, 1993), p 361, fn 3. The problem was not lack of tankers but congested airspace that precluded establishing more tanker refueling orbits, *Gulf War Air Power Survey,* vol III, *Logistics and Support,* p 179.

55. *Air Force Magazine,* vol 77 (May 1994), pp 126–27.

56. Project CHECO Southeast Asia Rpt, *Aerial Refueling in Southeast Asia, 1964–1970* (HQ PACAF, Jun 17, 1971), p 39; CHECO Rpt, *USAF SAC Operations in Southeast Asia: Special Report* (HQS PACAF, Dec 17, 1969), p 47.

57. Knaack, *Bombers,* p 121.

Irving B. Holley, Jr., is professor emeritus of history at Duke University. Dr. Holley holds a B.A. from Amherst College and an M.A. and Ph.D. from Yale. He has lectured widely on military doctrine and the history of technology at the nation's military academies. Among his better known writings are *Ideas and Weapons* (Yale), *Buying Aircraft* (GPO), and *General John M. Palmer: Citizen Soldiers and the Army of Democracy* (Greenwood Press). For ten years he chaired the Secretary of the Air Force's Advisory Committee on History and has served on numerous advisory other boards in government and the history profession. He received Social Science Research Council Fellowships in 1955 and 1961 and was a Smithsonian Institution Fellow in 1968. An Associate Fellow of the American Institute of Aeronautics and Astronautics, he is the recipient of numerous awards, including the Army's Outstanding Civilian Service medal, and the USAF Exceptional Civilian Service medal, Distinguished Service Medal, and Legion of Merit. He was an aerial gunner in World War II and retired in 1981 as a Reserve major general.

100

Technology and Doctrine

I. B. Holley, Jr.

When Robert Fulton, the inventor of the steamboat, attempted to secure the financial support of Napoleon Bonaparte, the Emperor brushed him off:

> What, sir? Would you make a ship sail against the wind and currents by lighting a bonfire under her deck? I pray you excuse me. I have no time to listen to such nonsense![1]

I won't stop to speculate on what might have been the outcome if Napoleon had developed a steam navy to employ against the British. Here I want to touch upon a few of the factors which have inhibited the development of sound doctrine in the wake of highly promising technological innovations.

Lack of imagination, as Napoleon's response suggests, which is to say, a failure of vision, has been a repeated source of difficulty. Let me give you an air arm example. In 1936, the Air Corps sent Lt. John W. Sessums out to New Mexico to visit Robert H. Goddard's experimental rocket station. Lt. Sessums reported back that the rocket appeared to have "little military value." He did say that rockets might possibly be used as targets for antiaircraft gunners, but no less a person than the Chief of the Air Corps dismissed even this secondary application as impractical.[2]

Clearly, lack of imagination along with bureaucratic arrogance can be particularly harmful in the matter of developing doctrine for technological innovations. To give you a more recent example, let me quote Secretary of Defense McNamara writing at the time of the war in Vietnam:

> In the contest of modern aerial warfare, the idea of a fighter being equipped with a gun is as archaic as warfare with bow and arrow.

What our fighter pilots thought about this whiz kid pontification is evidenced in a Nellis *Fighter Weapons Review* cartoon lampooning the idea. It shows an Asiatic armed with a six-shooter and a USAF pilot armed with a rifle engaged in combat inside a phone booth.[3]

An F–4 in Southeast Asia with a centerline gun pod. Early model F–4s
had no internal gun. In Southeast Asia, however, they were needed.

A classic example of bureaucratic arrogance is to be found in the British Air
Ministry decision to go into production with the Bolton-Paul Defiant, a two-
place fighter with a turret in the rear cockpit. The rationale of the Ministry in
deciding to procure 450 of these planes, enough for nine squadrons, was that the
two-place Bristol was highly successful in World War I, so the Bolton-Paul with
a turret should be even more successful. This was in 1938, a decision made over
the objections of the RAF.

The Defiant was fatally flawed. It weighed half a ton more than the
Hurricane but was powered by the same engine, so its performance was miser-
ably inferior. What's more, it had no forward firing guns. German fighters soon
caught on to this fact and slaughtered them in droves. As one trained as an aerial
gunner, I am particularly appalled by another design defect of the Defiant; when
the power failed, there was no way to rotate the turret, so the gunner had no way
to escape.[4]

In our Sperry ball turret, one at least had a hand crank to rotate the ball.
That crank was an important psychological asset. There you are at 10,000 feet,
crouched in a fetal position with your knees near your ears and your back
against the door. You begin to think about the tiny latch which holds the door
shut and your hands squeeze harder on the control handles.

Bureaucratic arrogance, let us call it design hubris, is not confined to lethal
weaponry. Consider the case of the B–47. The original design made no provision for electronic countermeasures. That bomber was expected to fly so high
and so fast as to be beyond the reach of interceptors. Tests at Eglin Air Force
Base, even when using older model fighters, showed the bomber could be
intercepted. So, back to the drawing board.[5]

What looks like bureaucratic arrogance is often nothing more than a tragic disconnect between whose who design, engineers and manufacturers, and those who fly and fight. The B–52 offers a fine example of this. Opening the bomb bay doors on the bomb run over Hanoi gave a large radar signature that alerted the enemy that the bombers were beginning their final approach. By simple triangulation they could predict the release point and aim their missiles just prior to that point. This overcame the lag time since missiles accelerate slowly in the first 10,000 feet, but reach top speed thereafter.[6]

One of the worse examples of disconnect between designers and users is one I personally experienced in World War II. The B–25 twin-engine medium bomber was originally designed with a Bendix lower turret offering 360° rotation. It was a monstrosity. The gunner rode in a kneeling position with chest upon a padded support and his eye pressed into a padded eyepiece looking straight down into a periscope lens. Imagine trying to sustain that uncomfortable position for long periods during a mission.

Worse yet, when tracking an incoming fighter, the target seemed to tumble and turn upside down as the periscope mirror rotated. Imagine trying to keep oneself properly oriented while tracking smoothly and computing the proper lead! Mercifully, the Bendix turret was dropped from the B–25. It was salvaged later in the war when it appeared as a chin turret on the B–17. There is not much deflection when aiming the chin turret against in-coming nose attacks, but the periscopic sight was still impractical. This was quickly abandoned and a standard heads-up reflector sight substituted.[7]

One of the most important means of keeping doctrine aligned with technological realities is the use of trained intelligence officers as debriefers to interrogate returning crews. We have a nice example of this from RAF experience in World War II. Time and again RAF fighter pilots would report getting on the tail of a German fighter and ready for the kill, only to have the enemy escape destruction by executing a pushover, a sharp, plunging dive. When the Spitfire pilot tried to pursue, he suffered a sudden loss of power which left him far behind. Spits were equipped with float carburetors, so in a pushover, the float tumbled and induced fuel starvation. Debriefings revealed this flaw and immediate steps were taken to install Stromberg carburetors which functioned under negative g conditions.

Spitfire fighters began to rack up many more kills. Those who were killed did not get home to report that the tried and true *Luftwaffe* evasive action doctrine was no longer working. But inevitably, a few *Luftwaffe* pilots survived such encounters and escaped. Even so, many German fighters continued to use the pushover tactic. Why so?

There was an important organizational difference between the RAF and the German Air Force. In the RAF, every squadron had an intel officer, whereas the German Air Force only had intel officers at the group or *geschwader* level. So word of the need to revise evasive doctrine and escape tactics, spread through

103

German squadrons only informally and causally by word of mouth, pilot to pilot.[8]

This business of the critical role of intelligence officers in doctrinal modification is all too often overlooked. In the Gulf War, intel officers had a low priority on shipment into the theater. So, as one fighter pilot reported to me, none were available at his base during the first five days of operations against Baghdad. As we all know, the learning curve is steepest in the early days of combat. The absence of trained debriefers was not the only critical cost in lost opportunities for modifying doctrine; even more significant was the absence of appropriate intel types for ECM operations. Our Air Force lacked an adequate data base for enemy emissions to be used by Wild Weasels for suppression. So our fighters had to face hazards which might otherwise have been suppressed.[9]

Speaking of the F–4 Wild Weasels, I am told that only one F–4 out of the nine on base in Turkey was able to accompany the fighters into Iraq because of lack of maintenance. This was attributable to the lack of highly skilled maintenance people whose lower deployment priority delayed their arrival in theater. The electronics of the Wild Weasels require continual delicate adjustment to keep them fully effective.

Admittedly, the early days of combat are almost always going to be chaotic. Our fighters deploying from Germany to Turkey flew 10 hours to get there. Two hours later they flew a mission into Iraq. What further proof does anyone need that our pilots have got the right stuff?

One might readily argue that these early problems were resolved in time. That's true, but my point is that the steep learning curve in the first few days of combat is precisely the time when the relation of innovative technology and doctrine is crucially important. Doctrine formulated in peacetime is largely hypothesis. Only when a new weapon is realistically subjected to the test of combat is the hypothesis confirmed or denied.

One pilot operating out of Turkey against Baghdad told me that for want of sufficient video tapes, his unit lacked records of the first seven missions over Iraq. I do not know if this deficiency was related to the low priority of intel officers or resulted from a glitch in the supply system. But without such tapes, analysis of after action reporting was seriously degraded, with consequent loss to the evaluation of the doctrine employed.[10]

This brings me to a point I have been making for years with evident lack of impact. We have failed to institutionalize adequately the practice of after action reporting. After action reporting is the lifeblood of doctrinal revision. By no means do all commanders put sufficient emphasis upon it. The content, format, and procedures for securing and processing such reports remains somewhat tentative, if not casual.

From what I have already said, it should be clear that formulating doctrine to accommodate novel technology involves many factors, not least among them exhaustive testing before innovative technology is put into full production and

F–4 Wild Weasels refueling before a mission during the Gulf War.

issued to the troops. But testing is extremely difficult to accomplish in any realistic fashion, especially without the environment of actual wartime combat.

Let me illustrate this problem by recalling the tribulations of Watson-Watt in his 1935 experiments with radar. Using a breadboard apparatus, he got good tracking results on a plane coming in from some 15 miles away. Encouraged by this modest success, he arranged to put on a demonstration before the powerful Tizard Committee which controlled R&D funding by the Air Ministry. There were many bugs and limitations in this early radar which only time and money could hope to resolve.[11]

The demonstration was a disaster. The breadboard apparatus refused to repeat its earlier promising performance. The radar would pick up only intermittent snatches of the incoming plane at an entirely unsatisfactory range of no more than eight miles. This was clearly an entirely unacceptable performance. Only later did Watson-Watt learn that the problem was bad atmospherics which had disrupted radio transmissions all over the British Isles that day. This episode delayed the massive funding required to get radar off the ground for some months. Eventually the funds were made available, and as we know, radar played its pivotal role in the Battle of Britain.

One of my favorite examples of the difficulties encountered in testing concerns an RAF training aircraft, the Tiger Moth. This plane had an appalling tendency to spin. But why? The plane was sent to the Boscomb Downs Experimental Establishment where it was subjected to exhaustive trials by the most experienced test pilots. It behaved perfectly, never showing the slightest tendency to spin unbidden. So back too the training squadrons it went. Immediately there were more disastrous spins.[12]

Eventually the cause was discovered. It was a bomb rack that disrupted the air flow at certain critical speeds. But bomb racks were scarce in the training

Technology and the Air Force

squadrons, so when the Tiger Moth was sent to Boscomb Downs, the bomb rack
was removed for local use. Hence no spins when flown by test pilots. I am not
suggesting this particular case had an undesirable impact on doctrine. I use it
only to illustrate how difficult it is to conduct effective tests on weaponry.

I have suggested that after action reports are the lifeblood of doctrine. But
to be fully effective in shaping sound doctrine, such reports must be interpreted
with great care. Those who attempt to revise or update doctrine must go about
it with the same sort of skills employed by historians seeking to make objective
interpretations of the past. Doctrine writers must be at pains to analyze all the
factors impinging on a given situation.

Once again a World War II example affords us insights on this problem. In
September 1940, the *Luftwaffe* modified twenty-two Me 109s to carry 1,000-
pound bombs and sent them to London. They were detected by RAF radars. But,
being fighters, they were not immediately attacked, as Spitfires and Hurricanes
were vectored against bombers. As a consequence, the Me 109s got through and
dropped their bombs on London with devastating effect. They were so successful the German Air Force commander, General Kesselring, ordered the tactic to
be repeated. By this time, however, the RAF, with its superior procedures for
after action analysis, was well aware of the tactic.[13]

The heavily-laden Me 109s were unmaneuverable and fell easy prey even
to older model Hurricanes; this in spite of the more agile fighter escorts accompanying them. *Luftwaffe* fighter pilots were disgusted when their modified
aircraft were redesignated "light bombers" and expressed their contempt by
referring to their planes as "Light Kesselrings." Finally, after four months of
continual losses, the practice of loading heavy bombs on fighters was abandoned.

What can we learn from this episode? The surprise introduction of a
technological innovation may give a decided advantage. But that advantage may
be fleeting if one's opponent is agile in responding with appropriate countermeasures. Are we to conclude then that fighters should never be used as fighter-
bombers? Clearly, other factors are involved. If we have substantial and effective local air superiority, then fighter-bombers might operate with impunity. If
we have developed fighters with a sufficient margin of power, presumably they
then might maneuver successfully to cope with contenders. Or again, fighter-
bombers might survive without jettisoning their bombs if their escorting fighters
have perfected their shielding tactics sufficiently and are present in adequate
strength. Given all these variables, one can readily see how demanding the job
of the doctrine writers has become.

Now let me take you down another line of investigation to illustrate how
seemingly small and insignificant design changes can have a substantial impact
on doctrine. At the end of World War II, I had occasion to be concerned with
the Rolls-Royce Merlin engine contract. I remind you that the RAF licensed
Packard to manufacture the Merlin in the U.S. as the power plant for the P–51.

106

The original contract called for 9,000 engines, but by the end of the war, Packard had produced nearly 55,000.[14]

Of course, Rolls-Royce kept improving the engine all through the war. Where in 1940 the Merlin turned out 1,260 horsepower at an altitude of some 12,000 feet, by the end of the war its output was 1,800 horsepower, all this with relatively slight increases in weight. Just to give you a feel of what an engineering triumph this was, let me remind you that the Merlin had 11,000 parts, of which 4,500 were different. There were twelve different grades of steel in the engine, not to mention the various kinds of aluminum, bronze, brass, all to demandingly high specifications.

So design improvements jacked up combat performance. Where the Merlin got 1,260 horsepower at around 12,000 feet, it produced only 1,175 horsepower at 21,000 feet, with a significant degradation in performance at that altitude. However, with improved supercharging and a new carburetor, by 1942 the Merlin got maximum power at 21,000 feet.

Supercharging alone did not make the difference. Part of the gain in power derived from finding a more efficient coolant. This allowed a redesign cutting down the size of the radiator, which reduced drag. These gains in power and performance came, of course, at a price. To get maximum power at high altitudes, the engineers found they had to circulate hot engine oil around the carburetor to prevent icing. And erratic behavior of the magnetos at altitude required further design changes.

It should take little imagination to see how improved performance at altitude radically influenced tactical doctrine. Nonetheless, even with more power at altitude making it possible to get the jump on the enemy, the technology doctrine equation can be convoluted. When Merlin-equipped planes were used as night fighters, a new difficulty appeared. Probably most of you in this gathering have stood on the ramp before daylight and watched those plumes of blue and yellow flames from the exhaust manifolds light up like beacons. Obviously that would never do for a night fighter. So flame suppressors had to be added. Extending the manifolds increased drag and also reduced the extra thrust gained from exhaust afflux. So performance was degraded with a consequent impact on tactics.

All these seemingly minor technical details have implications for tactical doctrine. The point I am making should be evident. Only when alert officers observe the consequences of modified technology and write them up can we hope to keep doctrine fully abreast of developments.

Yet another dimension of the doctrinal problem is induced by the shifting character of the threat. Today, with the greater probability of brushfire wars in remote and undeveloped areas, it is appropriate to ask if we are giving enough attention to bare base operations. Will our weapons be able to operate from primitive runways or are we tied to ten thousand feet of reinforced concrete? Can we operate in the absence of ground handling equipment? I was fascinated to observe not long ago that the Soviet Air Force had hand crank and gear winching arrangements in the wingtip so that heavy ordinance could be mounted without the presence of an external hoist.[15]

Increased speed has long been a major objective of designers, whether one speaks of muzzle velocity in tube weapons, the speed of airplanes, or of cruise weapons. But as speed rises, there is a cost. The faster a cruise missile flies, the more fuel it consumes; the more fuel, the smaller the payload of explosives. If we keep increasing the speed to avoid interception, we end up with a weapon capable of doing very little damage.

Increasing speed in piloted vehicles has a high cost also. A World War I fighter at, let us say 100 knots had a 35°/second turning circle. A World War II fighter at, say 250 knots had about a 24°/second turn. A jet in Korea brought this down to 15° and in Vietnam even lower. As the speed goes up and the Gs mount we approach the limits of human capacity. But G forces are not the only ceiling on human performance. As all our wonderful sensors provide more kinds of information faster and more frequently, at some level the information overload reaches the point of no return. Just as the speed of light puts a ceiling on scientific investigation, so too the speed of thought is becoming the upper limit at every echelon of command.[16]

I have tried to offer here a few of the many technological factors which impinge upon doctrine. In each instance, the solution seems to lie with those perceptive individuals who observe what is taking place around them, those individuals who actually experience the problems. Gen. Giulio Douhet, the Italian theorist of air power, was often wrong in his predictions. But he was certainly right when he said, "Experience, the teacher of life, can teach a great deal to the man who knows how to interpret experience." Formulating doctrine means interpreting experience correctly.[17]

The problem, as I see it, is that we leave the task of formulating doctrine largely to the officially designated doctrine writers. Isn't it evident that they are utterly dependent upon all the rest of us, the rank and file in the operating Air Force, to take the initiative in bringing our experience to the attention of the doctrine writers? Doctrine should be everybody's business. But it is hard to generate much enthusiasm for this when a very senior Air Force general pontificates before an audience of officers by declaring "Doctrine is bullshit."

Let me conclude by quoting one of my former students, now a very successful lawyer, who became chairman of the Duke University Board of Trustees. He said "People do act wisely—after they have exhausted all other possibilities."[18]

Notes

1. Tidal W. McCoy, "Longbow to Laser," *Government Executive*, vol 19 (Oct 1987), p 26.

2. E. C. Goddard and G. E. Pendray, eds, *The Papers of Robert Goddard* (New York: 1970) vol II, p 1028; Willy Ley, *Rockets, Missiles, and Space Travel* (New York: 1958), p 205.

3. Capt Mark DeCesari, "Mission Support System", *USAF Fighter Weapons Review*, vol 37 (Fall 1989), p 32.

4. Vincent Orange, *Sir Keith Park* (London: 1984), p 69; F. K. Mason, *Battle Over Britain* (Garden City, New York: 1969) pp 102, 178.

5. Interview with Lt Col Daniel Kuehl, Apr 13, 1991.

6. Interview with Col Smith, Air War College, Jul 17, 1991.

7. Gen Ira Eaker, CG, Eighth AF, to Maj Gen Barney Giles, OCG/AAF, Jul 18, 1943, AF Assn, Murray Green Mss folder 3.

8. Richard Hough and Denis Richards, *The Battle of Britain* (New York: 1989), p 309.

9. Interview with Lt Col Michael Worden, Sep 13, 1991.

10. *Ibid.*

11. Robert Watson-Watt, *Three Steps to Victory* (London: 1957), p 127.

12. Allen Henry Wheeler, *that nothing fail them* (London: 1963), p 16.

13. Cajus Bekker, *The Luftwaffe War Diaries* (London: 1967), pp 178-79.

14. "The Rolls-Royce Merlin XX Aero Engine", *Engineering*, vol 154 (Mar 6, 1942), pp 154-56; Irving Brinton Holley, Jr, *Buying Aircraft: Matériel Procurement for the Army Air Forces* (Washington: Center of Military History, 1964), pp 367, 392, 581.

15. John W. B. Taylor, "Change on the Wing" *Air Force Magazine*, vol 73 (Jan 1990), pp 58-64.

16. Capt J. W. Bodnar, USNR, "The Military Technical Revolution" *Naval War College Review*, vol 46 (Summer 1993), pp 12, 19.

17. Quoted by Clark Reynolds, *Naval War College Review*, vol 38 (Nov 1985), p 117.

18. Neil Williams, chairman of the Duke University Board of Trustees at the Duke University Convocation, Dec 10, 1987.

Jacob Neufeld is a senior historian with the Air Force History Support Office. From 1992-1994 he was the Director of the Center for Air Force History. Currently he is also the editor of Air *Power History,* the quarterly journal of the Air Force Historical Foundation. Neufeld earned the B.A. and M.A. degrees in history at New York University and did doctoral work at the University of Massachusetts, Amherst, and at the University of Maryland. He is an adjunct professor in U.S. history at the latter institution. Neufeld has authored and edited numerous works in military history and in the history of technology, including *Development of Ballistic Missiles in the United States Air Force, 19451960; Reflections in Research* and Development in *the United States Air Force;* and *Makers of the United States Air Force.* In addition, he has published histories on World War II and the Vietnam War.

Ace in the Hole: The Air Force Ballistic Missiles Program*

Jacob Neufeld

The first generation of Air Force ballistic missiles — the Atlas, Thor, and Titan I — were developed, tested, and deployed under a national top priority crash program. Remarkably, in the space of only seven years, this program produced a formidable force of nuclear-tipped missiles. This achievement, however, did not come easily. It was the result of the leadership, know-how, and energy of three singular individuals: General Bernard Schriever, Dr. John von Neumann, and Assistant Air Force Secretary Trevor Gardner. Without their persistence and dedication, the program surely would have foundered. Above all, these three men recognized the importance of "technology push" — the need to foster scientific progress — lest the United States lag behind in the arms race with the USSR. In the process, the Air Force missile program overcame many obstacles, including technical challenges, intra- and interservice rivalry, and budgetary limitations. Ironically, even before the Air Force missile force had attained operational status, its successors — second generation Minuteman and Titan II missiles — was already underway.

Background

The Air Force's involvement with missiles followed two separate and distinct paths. The first concerned the development of remotely controlled and guided aircraft. These were variously called pilotless aircraft and airbreathing, or aerodynamic, or cruise missiles. The second path involved ballistic missiles, weapons that were launched by rocket engines; these were not airbreathing but carried their own supply of oxygen. Rockets had traditionally belonged under the purview of the Army Ordnance Corps. The United States first experimented with pilotless planes in 1917. Over the next two decades, work continued sporadically on controlled and guided military and civil aircraft. During World War II, a special weapons group in the Engineering Division at Wright Field,

---

*Based on the author's *The Development of Ballistic Missiles in the United States Air Force, 1945-1960* (Washington: Office of Air Force History, 1990).

111

Gen. Bernard Schriever (top left), Dr. John von Neuman (above), and Trevor Gardner (left).

Ohio, developed various remotely controlled bombs and aircraft fitted with explosives for wartime use. Ultimately, however, these special weapons played only a limited role in the war because of the effectiveness of the large numbers and types of conventional aircraft that were produced.

In June 1944, when Germany began launching the jet-propelled V–1 "buzz bombs" against England, the United States first evinced interest in these aerodynamic missiles. Air Materiel Command obtained parts of the V–1 and, within a couple of months, produced an American copy, called the JB–2. Although the United States built nearly 1,400 of these weapons, they were not used in the war because both the V–1s and JB–2s proved too slow, carried a small payload, had limited range, and were inaccurate.

Modern ballistic missile history began in the United States with the work of Dr. Robert H. Goddard. Early in the twentieth century, Goddard began to experiment with rocket engines and fuels. At the close of World War I, he had tried, but failed, to interest the Army in the utility of rockets. Nonetheless, Goddard persisted and, in March 1926, successfully fired the world's first

A JB–2 missile in flight.

liquid-fueled rocket. Unfortunately, the United States did not capitalize on his work because it failed to foresee its military significance. However, the Germans, took an interest in Goddard's work and built upon it. The reason for this was that the Versailles Treaty had forbidden German rearmament, but said nothing about rockets.

A second major American effort in rocketry involved the work of a group of students at the California Institute of Technology in the early 1930s. Encouraged by Dr. Theodore von Kármán, the students — including Frank Malina, Hsue Shen Tsien, Apollo Smith, John Parsons, Edward Foreman, and Weld Arnold — experimented with various rockets and fuels.

During World War II, both Goddard and the Caltech group separately experimented with jet-assisted takeoff units for the Army and Navy. The JATO units were actually propelled by liquid and solid rocket engines, which were designed to provide additional thrust to help heavily laden military aircraft become airborne. The scientists adopted the term JATO because the press had ridiculed rocketry as too "Buck Rogers," that is, futuristic. JATO development evolved from canisters that provided 200 pounds thrust for eight seconds to units that delivered 3,000 pounds for longer than one minute.

In these endeavors, von Kármán and Malina were obliged to establish their own company because they could not interest commercial firms to undertake the work. Their company, Aerojet Engineering, later became Aerojet General. In 1942 the Caltech group developed liquid propellants that burned red fuming nitric acid and gasoline. They also produced GALCIT–53, a castable solid propellant composed of an asphalt-potassium percolate. Goddard developed a very successful LOX-gasoline fuel that was perfected by Reaction Motors, Inc., later a division of Thiokol Chemical. However, Goddard refused to collaborate with the Caltech students, fearing that they would steal his work.

As noted above, the Germans — having exploited Goddard's work — made the greatest strides in employing rockets in warfare. Beginning development

Technology and the Air Force

before World War II, the German V–2 ballistic rocket was a fourteen-ton missile that stood forty-six feet high, measured five feet in diameter, and carried a 1,650-pound bomb (composed of ammonium nitrate and TNT) over a distance of about 200 miles. Although the V–2 proved wildly inaccurate, it was tremendously successful as a terror weapon and virtually unstoppable. On October 3, 1942, a V–2 flew for 120 miles, becoming the world's first ballistic missile. By the war's end Germany had launched about 3,800 V–2s against England and continental targets.

As World War II drew to a close, the United States and Soviets raced to round up as many German scientists as possible. Under Operation Paperclip, the U.S. Army recruited some 600 German scientists, including about 130 experts in rocketry. The best known of these, Dr. Wernher von Braun, led a team of former German researchers at the Redstone Arsenal, Huntsville, Alabama. In 1958, the von Braun team was transferred to the National Aeronautics and Space Administration.

Postwar Program

After the war, Dr. Theodore von Kármán led a group of scientists to study Germany's World War II weapons. They produced *Toward New Horizons*, with thirty-three volumes, that urged the Army Air Forces emphasize jet propulsion in the postwar period. With respect to missiles, however, the report writers recommended pursuing a sequential approach, featuring airbreathing missiles. They were less enthusiastic about ballistic missiles (or rockets), reasoning that such weapons were years away from practical use and that more pressing problems demanded attention.

Subsequently, the AAF mapped out a broad program of missile research, including all four basic types of missiles: air-to-air, air-to-surface, surface-to-air, and surface-to-surface. Further, the AAF divided its missile projects into two groups: those that could be developed almost immediately, especially missiles that complemented aircraft, and those that would take at least five years to complete. As expected, priority went to the missiles that could be developed the soonest.

Compared to the Army and Navy programs, however, the AAF missile program lagged far behind. One major reason was that the AAF simply had not invested as much in research and development funds as had the other services. Other impediments to speedy missile development were the weight and dimensions of atomic weapons. AAF leaders firmly believed that over the next ten years the manned bomber alone could deliver atomic weapons. It is hardly surprising, therefore, that surface-to-surface missiles were relegated to last priority.

In the spring of 1946, the AAF missile program included the study of a supersonic intercontinental ballistic missile. It contracted with Consolidated

114

Vultee, predecessor of Convair and later General Dynamics, to conduct a missile study project — called the MX–774B. The MX–774B prototype would lead to an ICBM capable of carrying a 5,000-pound warhead, over a 5,000-mile distance, and striking within one mile of its target. Convair's project manager, Dr. Karel "Charlie" Bossart started with the V–2 design as a model, but soon concluded that major alterations were necessary to meet the AAF's stated requirements. Bossart used Convair's $1.4 million contract to introduce these innovations:

> He removed the V–2's double wall arrangement and stored the propellants inside two enclosures.
> He designed a separating nose cone (which held the warhead) to reduce the friction of having the entire missile reenter the atmosphere.
> He removed the stiffeners that supported the airframe and instead used nitrogen gas to pressurize the airframe. These weight-saving measures reduced the ratio of the airframe weight to the propellant weight by a factor of three.
> He designed gimbaled (swiveling) engines to control the direction of flight. This modification replaced the movable vanes in the exhaust that the Germans had used. Bossart had discovered that the vanes reduced thrust by about 17 percent.

Convair chose Reaction Motors, Inc., to build the MX–774B engine, principally because that firm was already building a 1,500-pound-thrust engine for the Bell X–1 rocket plane. Convair contemplated using four rocket engines, each generating 2,000 pounds of thrust. The MX–774B guidance system was a simple device based on a gyrostabilized autopilot. This was succeeded by the Azusa precise-phase comparison system. Signals from the missile were received by two pairs of ground stations. Phase variations due to differences between the stations and the missile transponder were fed into a computer that enabled the missile's flight to be compared to an ideal trajectory up to the point of nose cone separation. Corrective signals were sent to the missile for guidance.

Despite these advances, the missile program fell victim to fiscal constraints. In December 1946, President Harry S. Truman drastically cut Fiscal Year 1947 funding, obliging the AAF to follow suit. Again, the AAF elected to stress near-term programs over long-term ones. Consequently, the AAF assigned top priority to two aerodynamic missiles — the Navaho and the Snark — which were expected to become operational within eight to ten years, while it was estimated that MX–774B would take more than ten to complete. Another consideration involved a series of difficult technical problems that had to be solved before an ICBM could materialize. These problems included reentry, range, accuracy, more efficient and more powerful motors, and higher specific impulse fuels than a LOX-alcohol combination.

Test of the MX–774 at White Sands Proving Grounds, New Mexico (left),
and of the Navaho at the Air Force Missile Test Center, Florida (right).

Nonetheless, Convair was permitted to continue flight testing the MX–774B; three tests were run between July and December 1948. Although rated as only partly successful, these tests validated the soundness of Bossart's design. At the end of 1948, the USAF tried to enter the MX–774B as a high-altitude research vehicle. However, it lost to the Navy's Viking program.

Subsequently, two major world events helped to restore funds for missile research. The first event was the Soviet A-bomb test in August 1949; the other was the outbreak of the Korean War in June 1950. Moreover, some promising solutions to technical problems emerged, and in December 1950, a RAND study supported the technical feasibility of building a long-range ballistic missile. Early in 1951, with money appropriated as a result of the outbreak of the Korean War, the USAF awarded Convair a study contract for Project MX–1593. This study was expected to recommend either a ballistic type or a glide type missile. The general operational requirements were for the ICBM to carry an 8,000-pound warhead to a range of 5,000 nautical miles and hit within 1,500 feet of the target.

In September 1951, Convair rendered its verdict, concluding that, from an operational perspective, the ballistic missile approach was superior to the glide missile approach. The company proposed building an ICBM — to be called the Atlas — that would be 160 feet long and 12 feet in diameter. Convair considered using a multiple-engine configuration based on either North American's LOX-alcohol engines or RMI's LOX-gasoline engines. The preferred combination would be four first-stage engines, rated at 133,000 pounds each, and one 123,000-pound central engine, generating a total of 656,000 pounds thrust. All of the engines on this "one and one-half" stage missile would be started on the ground because the designers were uncertain whether the rocket engines could be started while in the air. Guidance was provided via an on-board inertial autopilot transponder-receiver and a ground-based station which included a radar tracker and computer.

Meanwhile, U.S. intelligence reported that the Soviets were already busy building various ICBMs and that they supposedly had a rocket engine that could generate 265,000 pounds of thrust — or twice the power of any existing American engine.

A breakthrough in thinking occurred in December 1952, when the Atomic Energy Commission predicted that nuclear weapons weighing only 3,000 pounds would be developed. This prediction led the Air Force SAB to recommend relaxing some of the stringent ICBM requirements, including easing accuracy from 1,500 feet to one mile. In March 1953, an SAB panel, meeting at Maxwell AFB, Alabama, reviewed plans for three different ICBM versions: a one-engine (X–11), three-engine (X–12), and five-engine (XB–65) Atlas and recommended a phased, ten-year approach. The SAB plan was designed to complete research in 1956, development in 1961, and testing in 1963. However, the Air Staff opposed the SAB recommendations, fearing that any reductions in ICBM requirements would adversely affect the ongoing development of Navaho and Snark aerodynamic missiles.

By spring 1953, a compromise was reached, calling for a smaller — but still formidable — ICBM. This new ICBM would be 110 feet long, 12 feet in diameter, and weigh 440,000 pounds. It would carry a 3,000-pound warhead to 5,500 nautical miles, but its required accuracy remained at 1,500 feet. Five engines, generating a total thrust of 656,000 pounds, would be needed.

Another internal Air Force debate concerned whether to integrate missile and aircraft development, or to establish a separate missile program instead. Initially, the integration advocates won, arguing that their strategy would better demonstrate the close relationship between missiles and aircraft among industry contractors. For several years afterwards, the Air Force referred to all missiles as pilotless aircraft and dubbed missile fins as wings. The USAF even went so far as to assign aircraft designations to missiles. Thus, for example, the Atlas ICBM became the XB–65, for experimental bomber 65!

A Radical Reorganization

In June 1953, Secretary of Defense Charles E. Wilson, of the newly inaugurated Eisenhower administration, ordered the Air Force to conduct a thorough review of all U.S. guided missile programs. The Air Force's representative in this review was Trevor Gardner, at that time a special assistant to Air Force Secretary Harold Talbott. Gardner appointed two committees: one to study all guided missiles and a second committee to consider only the strategic missiles. The latter, called the Tea Pot Committee, was headed by Dr. John von Neumann, a former AEC commissioner and a world renowned mathematician.

Von Neumann was among a select group of the scientists who had predicted that more powerful nuclear weapons, weighing as little as 1,500 pounds, were feasible. If true, this meant that Atlas's 440,000-pound weight could be cut in half and that extreme accuracy would not be required. Subsequently, a new 240,000-pound Atlas design [XSM–65 (WS–1076A–1)] emerged, powered by two 135,000-pound thrust North American engines and one 60,000-pound sustainer engine, for a total 330,000 pounds of thrust.

Issued in February 1954, the Tea Pot Committee report, issued simultaneously with a companion RAND report, noted that the Soviets had tested the H-bomb in August 1953 and were making excellent progress in the ICBM field. The Tea Pot report urged that the United States undertake an immediate crash program to develop an ICBM. This effort would transcend Convair's capabilities and would succeed only if it was accompanied by a radical reorganization within the Air Force. Trevor Gardner, too, was convinced that the program could not succeed unless it was made a separate undertaking, so as to avoid Pentagon red tape.

Brig. Gen. Bernard A. Schriever was appointed to head the new missile development entity, called the Air Force Western Development Division, located in Inglewood, California. Schriever received extraordinary control and access to the USAF leadership. Thus, he reported directly to the Commander of Air Research and Development Command and was permitted to handpick his staff. Another innovation was WDD's employment of the Ramo-Wooldridge Corporation to perform systems engineering and technical direction — replacing thereby the Air Force's traditional single prime contractor approach. Schriever also campaigned for, and ultimately won, control over his own budget.

In terms of the scope and challenge it posed, the ICBM program resembled the World War II Manhattan Project, which built the A-bomb. Schriever adopted a method of parallel development, whereby separate contractors were selected to produce the major subsystems — airframe and assembly, propulsion, guidance, computers, and nose cones. Contractors were rated according to seven criteria, including managerial performance, manufacturing capability, financial condition, development capability, cost and delivery record, security, and vulnerability. Although some criticized this approach as uneconomical, Schriever argued that it saved time through competition, ensured that failure by any one contractor would not halt work, and that it permitted the pursuit of advanced designs without unduly jeopardizing the ICBM program. It also permitted making subsystems interchangeable and dramatically expanded the industrial base for missiles. Finally, the parallel development approach led to the development of a new, alternate ICBM, the two-stage Titan.

A Family of Missiles

In January 1955 — serving as advisors to both the Department of Defense and the Air Force — the Tea Pot Committee recommended the development of a tactical ballistic missile. Initially, General Schriever had opposed the recommendation, fearing that it would divert resources and talent from the ICBM program. Eventually, the TBM evolved into an intermediate-range strategic missile — the Thor — adding yet another set of contractors.

Missile development, however, ran headlong into fiscal conservatism. In February 1956, when the Air Force R&D budget underwent drastic cuts, Trevor Gardner resigned and made public his disagreement with the administration. Then, in July 1956, Air Force Secretary Donald Quarles — probably influenced by the Congressional airpower hearings — disapproved the FY 1957 missile budget and substituted for it his own "Poor Man's Approach." This action extended the ICBM's Initial Operational Capability date from March to December 1961 and reduced the number of strategic missiles to be deployed by one-third.

As a result, throughout 1957, ICBM and IRBM production rates were cut back sharply. The IOC dates were postponed for the Atlas to 1964, Titan to 1965, and Thor to 1956, respectively. Not even the Soviets' announcement that they had successfully test fired an ICBM in August 1957 dissuaded Secretary Wilson. Only after the October 4, 1957, launch of Sputnik — the world's first artificial satellite, placed into orbit by a Soviet ICBM — did the administration act.

By that time WDD (now renamed the Air Force Ballistic Missile Division), had created the nucleus of a missile force, albeit one consisting of support units rather than operational missiles. However, the Sputnik launch definitely reversed the conservative fiscal trend and sped development of operational missiles.

At this time, too, political considerations intruded with the Democrats charging that the United States was on the wrong end of a "missile gap." Although President Eisenhower vigorously denied that the nation was behind the USSR, two Presidential studies — the 1957 Gaither Report and the 1958 Killian Report — seemed to contradict the President and fuel the debate. Also, Eisenhower could not reveal intelligence information to the contrary — that in fact no missile gap existed — without compromising that intelligence.

Other factors influencing the numbers and mix of ICBMs to deploy was the variety of configurations that had evolved — each offering different capabilities and potential. Also, technical problems, including some highly publicized and

Launch of an Atlas from Vandenberg AFB, California (left), and
a Titan from the Air Force Missile Test Center, Florida (right).

spectacular flight test failures, further complicated matters. Congressional
scrutiny investigated not only the technical problems, but also probed the
ICBMs exorbitant costs. The hearings showed that building, testing, and
installing the missile force represented not only an enormous engineering
enterprise, but a considerable management challenge as well.

With the advent of the Kennedy administration, the ICBM program was
reevaluated once more. Meanwhile, the so-called missile gap faded as interest
shifted from the numbers of missiles available to their reliability and flexibility.
The Thor IRBM became operational in the United Kingdom between June 1959
and April 1960; Atlas D and E models went on alert between August 1960 and
November 1961; Titan I and Atlas F became operational during April to
December 1962; and Jupiters were installed in Italy in 1961 and in Turkey in
1962. In all, thirteen Atlas and six Titan I squadrons became operational. Even
as these missiles were put in place, important decisions were made with respect
to their successors — the solid-fueled missiles.

As early as 1955, solid-fueled missiles were candidates for tactical roles,
and technical progress in 1957 showed that the solids might be adapted to fly
over longer ranges. By 1958, the promise of solid-fueled missiles had gained
broad support. Even General Curtis E. LeMay, the Commander in Chief of
SAC, a skeptic with respect to missiles, supported the development of a solid-

fueled ICBM. The Navy was especially interested because of the impracticality of carrying liquid-fueled missiles aboard ships. While the Army, Navy, and the Defense Department opposed yet another ICBM crash program, General Schriever persisted. In March 1961, Defense Secretary Robert S. McNamara returned from a visit to BMD convinced of the necessity for building a solid-fueled ICBM, now called the Minuteman. The development of the Minuteman was so rapid and so successful that it accelerated by several years the phaseout of the first generation, liquid-fueled ICBMs. By December 1964, Atlas Ds came off alert, and by June 1965, Atlas E and F and Titan I were retired.

The first ten Minuteman I missiles came on alert in time for the Cuban Missile crisis in October 1962. Eventually a force of 1,000 Minuteman and 54 Titan II ICBMs were fielded.

Epilogue

Titan IIs remained in service for some twenty-five years, until they were retired in 1987, while the third generation ICBM, the MX (Peacekeeper) had become operational in late 1986.

Fielding the ICBM fleet was a monumental achievement. There were controversies over the feasibility, necessity, and control of ballistic missiles among the services and within the Air Force. The USAF ICBM program was a crash effort that incurred an estimated cost of $17 billion. Not all of the spending involved technological development for new airframes, propellants, and guidance systems. Large outlays were also required to build the missiles launchers and control facilities — the environment for the missile weapon system.

One lasting legacy of the ICBM program is that it elevated subcontracting to a grand scale. At the end of 1955, for example, 56 contractors worked on the Atlas. Two years later that number had climbed to 150, and by the end of the decade, it stood at an astounding 2,000. At the start, the missile program did not have enough trained personnel, not enough manufacturers, and too few test facilities. These would come in time. Also, the success of the missile program caused the Air Force to rethink its plans and doctrine. It became obvious that missiles of all types would play increasingly important roles in the USAF. Indeed, the Air Force fielded a mixed force of aircraft and missiles, and that trend has continued in the USAF budget as well.

David R. Mets is the Professor of Technology and Innovation at the School of Advanced Airpower Studies, Air Command and Staff College, Air University, Maxwell AFB. He completed a 30-year career in the Navy and Air Force as an electronic technician, midshipman, instructor navigator, instructor pilot, Air Force Academy professor, West Point professor, and editor of the professional journal of the Air Force. He flew two tours in Southeast Asia and became the commander of an AC–130 gunship squadron. He has written three books (including a biography of General Carl Spaatz sponsored by the Air Force Historical Foundation) and served as a civilian historian at Armament Division of Systems Command for six years. He left Eglin AFB for his present position in 1990. He holds a B.S. in engineering from Annapolis, a M.A. in history from Columbia University, and a Ph.D. from the University of Denver.

Stretching the Rubber Band:
Smart Weapons for Air-to-Ground Attack

David R. Mets

Precedents

Historians typically try to reach back as far as they can for precedents about the subject at hand. I am sorry I only made it back to the thirteenth century, when Genghis Khan had a Chinese city under siege in the middle of the Eurasian heartland. Time went by, and his troops were getting hungry, he was running out of water for the horses, and something had to give. So Genghis sent a truce team to communicate with the town fathers. They informed the mayor that Genghis did not intend to harm them.

The mayor believed that an arrangement could be made and that everybody could go away happy. All that the good Khan wanted was a little bit of self-fulfillment, as at the apex of Maslow's triangle.[1] He did not want to hurt them, just a little tribute from the defenders. The truce team said, "Well, gentlemen, the Khan has authorized us to inform you that if you will give him 1,000 pigeons and 1,000 cats, that will satisfy his self-esteem, and he will fold his tents and go away, and everybody will be happy."

The Chinese could not believe their good fortune. They had wall-to-wall cats and wall-to-wall pigeons. They were reproducing all the time–it was a piece of cake. They sent the pigeons and the cats out to Genghis's camp. He tied a piece of cotton to the leg of every one of them. He then dipped it in incendiary material, set it on fire, and you had your first autonomous seekers, your first guided munitions. The birds all flew back into the attics of the town and the cats went into the cellars, with unerring accuracy — a beautiful example of the virtues of precision-guided munitions.

So, you see, the desire to be able to attack an enemy from a long distance, too far for him to retaliate, has existed for a long time. This urge most likely began well before Genghis Khan in the thirteenth century.

I was having some trouble documenting that tale because I figured there might be some historians in this audience who would want a footnote for it. So I asked all of my colleagues about this anecdote. None of them had ever heard it. I used the Air University Library and looked up all the biographies on

123

Technology and the Air Force

Genghis Khan. I could not find anything, and I was really worried that somebody, like Jack Neufeld, would say, "Okay. Give me the footnote." I was speaking to my good wife at the supper table Saturday night, telling her how worried I was because I could not find the documentation. She said, "David, I just saw that on the Learning Channel, and it's true."

Well, I have better documentation for a World War II experiment. Vannevar Bush published a book known as *Modern Arms and Free Men* in 1949, in which he informed us that ICBMs will never happen, or will not happen in our lifetime, anyhow. Another thing he spoke of in that book, when I was a plebe at Annapolis, was that ordnance people are very stodgy.[2] You need to watch them. Many folks call them flaky, and flaky they are.

One of Vannevar Bush's colleagues, though–and this can be documented without the Learning Channel–had the idea that if you lower the temperature of bats, they will go into hibernation. Consequently, there are no problems in handling. FDR got wind of the notion, and it thus was assigned to the Air Proving Ground at Eglin and became a matter of some interest. They got a bunch of bats, they refrigerated them, took them up in a B–17. They went out over the Mojave Desert, with an incendiary device attached to the leg of each one of these bats. They opened the hatch, threw them overboard. The theory was that the bats would thaw out on the way to the ground, and glide into the nearest belfry (the obvious candidates being Japanese cities). The incendiary devices would go off and burn down the towns. Unhappily, the test bats did not thaw out, and when they hit the ground, they all died. The test was not considered a success.[3] That brings me to my next point, the rubber-band theory of military history and doctrine.

Smart Weapons Precedents

The development of guided munitions has been ongoing since the Kettering Bug in World War I. All of our guidance systems, except the laser-guided bomb, were conceived during World War II. The Navy even had a launch-and-leave missile or glide bomb called the "Bat." It had a radar sensor on it, an active radar sensor, and it actually got some ship kills in the Pacific.[4]

The Army Air Forces were working on television guidance at Eglin Field during World War II. But General Arnold cut that off that in favor of a gyroscopic-controlled glide bomb with a guidance system similar to the German V–1. The Eglin Proving Ground also explored infrared and television sensors during World War II. However, those were cutting-edge technologies. The Army was uncomfortable with the idea that it may not be possible to bring them to fruition in time to make any difference in the war. So it concentrated on a visually guided, radio-directed guidance system for a bomb known first as the "Azon," which was controlled in azimuth only. The Proving Ground was also developing "Razon," which was controlled in both range and azimuth, and it

The JB–2, a copy of the German V–1, during a static test at Eglin Field.

provided an early case of the axiom that "The perfect is the enemy of the good enough". The scientists themselves were not pushing Azon as hard as they might have. Doubtless they thought that something better, Razon, was just around the corner.[5]

Among the things that made me think of a rubber band were the initial tests on the Azons. They included a remarkable development that nobody had ever thought about before. When one drops a bomb, a dumb bomb, from an airplane, it is a kind of a self-correcting device. One never gets perfect fins on a bomb, and the casing is never perfectly symmetrical. When you drop it from an airplane, it may veer to the right for a while. However, a bomb rotates on its way down because of the imperfections. When it turns over, it starts going left; and when it turns back again, it starts going right again. So that if you drop a volley of bombs out of a bomb bay, you get a rather small pattern on the ground.

However, when they tried this with Azons, they had to stabilize the bomb so it would not roll and the bombardier could control it with his joystick. So a gyro control unit was installed to stop this rotation. Much to their amazement, the bomb he was controlling out of the volley would come much closer to the target than was the case of the dumb bombs. However, with all the rest, the removal of the rotation spread the pattern, and they missed by more than would otherwise have been the case.

So the testers at Eglin decided it was a "piece of cake." The "fix" was easy — merely drop a volley of four bombs out of a B–29, roped together to yield a small pattern. What to use? A steel cable. Well, they underestimated the stresses by a wide margin; test and evaluation is much more sophisticated these days than it was in the late-1940s. They released these bombs, which immediately charged off in different directions. When they got to the end of their tethers, the cables snapped with impressive vigor.

Well, as many folks know, Eglin AFB is in the middle of a big playground. Naturally, there were some sailors in the research, development, and testing effort, and doubtless one of them realized the obvious. One never uses a rigid line for anchors. You do use something like rayon for your halyards and things like that, but for an anchor you need spring in it. So they decided to use nylon. They tied up the next set of bombs with nylon line, and then when they came out of the bomb bay, it worked. They did not snap their tethers. Rather, they went out to the end really stretching, and then, springing back: wham, they collided right underneath the airplane. Fortunately the air flow had not armed them, so they did not explode, and the guys lived to describe what the problem was. The answer was to develop enough channels in the radio link so that you could control each weapon individually from different bombers.

Breaking the World War II Rubber Band: Air-to-Air

That brings me back to an idea I call a "rubber-band theory of war," which could be applicable today. You could choose any number of examples, I guess. It is hard to see what is going on in the enemy's mind. Dr. Holley referred a little while ago to the experience with escorts during World War II. He noted that we very soon found out that without them the bomber would not always get through with acceptable losses, and we had to do something about that.[6]

I do not agree that the Air Corps was as rigid as some folks think, but that misperception was confirmed when we got into war. The air arm was to do the initial combat while the infantry trained. It had to self-deploy overseas with the Eighth Air Force which brought along its P–39s as far as Maine, but had to leave them there. The fighter groups continued to England, and then were re-equipped with Spitfires. However, they could accompany the bombers only so far toward Germany. The Germans were no dummies, so why tangle with the Spitfires? They would just wait until the escorts had to turn and go home, and then they would jump on the bombers. The escorts seemed to have little or no effect.

There was a great gun on those bombers, the M–2 .50 caliber, but the Germans had the solution for that. As there were no American fighters around, they could just put bigger guns on their airplanes. The Me 109 came with a 20-mm gun firing through the hub. Gradually as the war went on, the Germans would further load it with larger caliber guns and rockets, and wait for the Spitfires to go home. Then they would come up behind the B–17 — no change in the azimuth, no change in range. They could just sit outside the range of the

126

Republic P–47 Thunderbolt escorts with drop tanks.

.50 caliber and pop away until they got some kills or cripples and caused the disintegration of the American formation.[7]

Well, we had to have a solution for that. We brought on the "Jug" (Republic P–47), with longer range than the Spitfire. The Jug went out a little farther. The Germans backed off a little more, and still we could not do anything to engage them in that last sanctuary at the far end. Therefore, all fighters had to turn back. Well, we could fix that too. We put drop tanks on the Jug, and the Germans backed off a little more.[8] Still, the effect of the escorts was not apparent.

The USAAF brought in some P–38s with drop tanks. (It was an old idea, incidentally. Billy Mitchell mentioned in his 1925 book that the Air Service had used drop tanks in World War I.)[9] The 8th Fighter Command put drop tanks on the Jug and the escorts could go a little farther, that stretched that rubber band a little more. Still, the Germans would delay their attack a little more. Bring in P–38s with drop tanks, and the *Luftwaffe* backed off a little farther.

Finally, the P–51 came along with laminar flow wings and (fortuitously) an efficient Rolls-Royce engine in addition to drop tanks. It could make the entire trip to Berlin and back, closing that last sanctuary.

If you look at the casualty figures among the Germans, P–47s killed more enemy fighter pilots than P–51s did.[10] There were more of those young, untrained *Luftwaffe* pilots who died in approach and landing accidents than were

North American P–51 Mustang escorts with drop tanks.

killed by P–51s. However, the Mustangs had a disproportionate effect because they ended the capability of the *Luftwaffe* to back off yet one more time.

Although the P–51 pilots were not killing all that many fighter pilots, there was no place for the Germans to hide anymore. The *Luftwaffe* pilots found themselves with these huge guns and rockets aboard. They had lost their agility, and although the P–51s had a much smaller gun, a .50 caliber, the German pilot could not shoot his 20-mm or 30-mm at a P–51 that was behind him. His big guns did not matter. If you look at the casualty figures on our side and on their side in the spring of 1944, you will see that ours take a dive around April or May, a rather steep dive, suggesting that the rubber band had snapped, that the last sanctuary had closed.[11]

I will suggest here that Arnold and the rest back in America were near the limits of their frustration. They were gravely disappointed with those enthusiastic prewar predictions of Douhet and Mitchell, the Air Corps Tactical School, and the others. We just could not see how the enemy could stand up against our superior air power. We had been fighting with air superiority ever since the middle of 1943, and we just never could understand how it was possible for the Germans to survive this aerial onslaught. It was clear enough that our bombers were having a rough ride; but it was not so clear that the *Luftwaffe's* rubber band was being stretched tighter and tighter. In the air battle, it was a long time coming, but when it did snap, our loss figures went down in a hurry. Closing the last sanctuary in the air-to-ground battle was to be a longer process.

Stretching the World War II Rubber Band: Air-to-Ground

During Generalfeldmasrchall Gerd von Rundstedt's postwar interrogation, he said it was like fighting with one hand tied behind his back.[12] He had lost the capability to move in the daytime. He still had a night sanctuary, but in the long days of June in northern Europe, that was not much. In World War I, what Billy Mitchell called strategical bombing was bombing of a rail center maybe thirty clicks beyond the front. There really was no strategic bombing at all (except for the Zeppelins and Gothas over London). There really could not be much deeper penetration than a few kilometers with the aircraft of those days — nor could they hit much of anything. The Kaiser's ground people could always back off, and thus enjoyed the sanctuary of distance.[13]

Well, airpower pretty well removed that during World War II. We made serious inroads on the distance sanctuary, but von Rundstedt still had the asylum of darkness. Five years later in Korea, with only the primitive guided bombs we were using (the Razon and the Azon in very small numbers)[14] we nonetheless did cause real mayhem on the enemy lines of communications. Nothing much had changed and standard World War II fighter-bombers achieved that. However, the problem was that the Chinese learned quickly. They also had their sanctuary of night. Stretched as they were as the first winter came on, they had the good fortune of the increasing sanctuary of both darkness and bad weather. However, when Vietnam came we were to seriously undermine the former refuge.

Various things helped achieve that. One was the AC–130 gunship that two of my former students in the audience will affirm was the answer to all sorts of problems. It carried low-light-level TV and infrared sensors to which it could slave a laser designator (in effect, its cannons). Gunship tactics included one for targets that were too stout even for its 105-mm gun. At that point, the back seater in a fighter or someone on the ground had to have visual contact with the target in order to employ laser-guided bombs. The enemy asylum of darkness still handcuffed our fighters. However, employment of the gunship low-light television or infrared sensors in combination with the fighter LGBs penetrated that sanctuary. The AC–130 would call in a fighter with some Mark 82s or 84s.[15] The gunship would designate the target for it while orbiting at a safe altitude (safe in the absence of SAMs, that is). The F–4 would come along, drop the laser-guided bomb into the basket, and blast the targets in the middle of the night. This generated the byproduct of great deal of psychological difficulty for the enemy truck drivers. The story was that the North Vietnamese were chaining the truck drivers to their steering wheels. That prevented their teamsters from quitting their mounts for the security of the jungle at the mere sound of a passing C–130 or jet — as did many tank drivers in Kuwait much later.

So we made a serious inroad on that sanctuary of darkness during Vietnam. Another thing I would like to emphasize here is that PGMs did not come upon

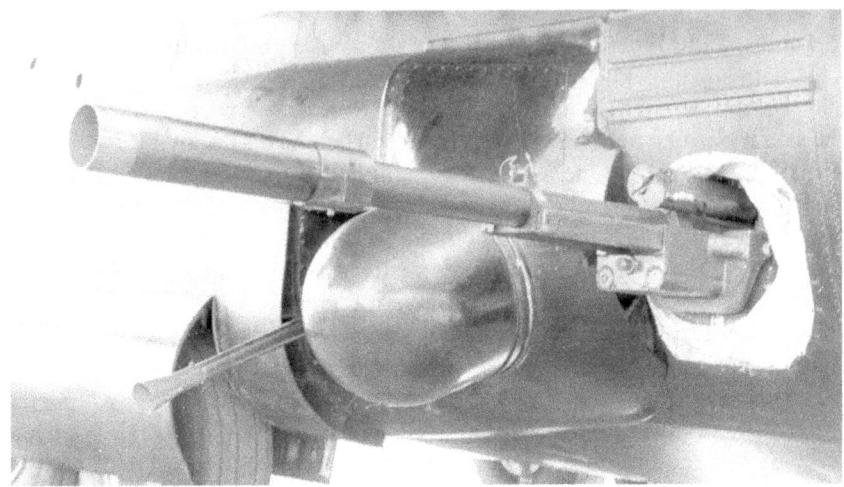

The left side of an AC–130 with a 105-mm howitzer and a 40-mm gun.
In between is the tracking radar that directs the two guns.

the scene all of a sudden in Desert Storm. We had another important experience with them during the North Vietnamese Easter Offensive of 1972. During Linebacker I that spring, the laser-guided bomb reached maturity as a result of a very remarkable research and development effort. The scientific principles underlying laser light (monochromatic, single-frequency light) and their possible application to weapons guidance were not revealed until 1958.[16] In 1967, an Eglin AFB test unit was in Vietnam with laser-guided bombs, ready to use them in combat, and they were so tested. The reason that they did not get the publicity is that just about the time the USAF started dropping them, President Johnson called a bombing halt. Bombing in the jungles of South Vietnam did not generate the kind of media attention that PGMs later got from Desert Storm. The Eglin unit employed the test items extensively in South Vietnam in 1968 while the bombing halt was operative up north, and the results were highly encouraging.[17]

In any event, in the spring of 1972, for Linebacker I, the most famous PGM case was the dropping of the Thanh Hoa bridge. It had been a target for five or six years, and we had never been able to destroy it. Well, Thanh Hoa went down to PGMs in a trice.[18] Further, especially in Military Region I up around Quang Tri in northern South Vietnam, there was much additional mayhem caused by the redeployment of American air power. This time it came with laser-guided bombs, and created terrible choke points at all the river crossings. So much so that destroying the stacked up traffic became an easy matter even for non-precision air weapons. South Vietnamese ground forces supported only by American airpower stopped the North Vietnamese in their tracks.[19]

Well, that did not make it into the media very much, but it is an important part of our sample. Airpower at least partially shut down the night asylum

F-4s with guided bombs brought down the Thanh Hoa bridge.

during the Vietnam War, but still there was the weather sanctuary. We got into Desert Storm, and one heard a lot of gnashing of teeth about laser-guided bombs only getting a direct hit about half of the time. Looking back to the B-17, less than one percent of the bombs were hitting, so 50-percent direct hits is pretty good — perhaps bordering on the revolutionary.[20] By then, some of our fighters, like the F-117, had their own infrared apparatus with which they could do their own designation at night. Further, imaging infrared seekers on some bombs and missiles worked as well in the darkness as television did in the light.

Toward Breaking the Air-to-Ground Rubber Band

Still, we know from Tom Keaney's and Eliot Cohen's book on the Gulf War and many other sources, the weather, even in that dry climate, was still somewhat of an impediment for PGMs, thus a weather sanctuary.[21] Now there are several programs afoot in the Navy and the Air Force that aim to overcome that. The one I am most familiar with is the Joint Direct Attack Munitions program down at Eglin. JDAM's first phase aims to help us clamp the weather sanctuary shut so that there will be no place to hide in the air-to-ground battle.[22]

JDAM's first phase plans to do this with the inventory conventional bomb without a seeker. The PGMs are a bargain. They have a pretty high unit cost, but they are a bargain. We are likely to use them a lot, as we know from Desert Storm. Weapons are still the lesser part of the total cost of a sortie, even with their high unit costs. They are a still smaller fraction of the cost of the entire strike package. So, any way in which you can reduce the requirement for either sorties or large strike packages is likely to pay for itself. The unit cost of the rocket-assisted stand-off television weapon, the AGM-130, runs to several hundred thousand dollars.[23] Even that of the LGB amounts to several tens of thousands. Further economies are conceivable — especially if the target does not demand the last increment of precision. (We used very few AGM–130s in the Gulf War, but indeed consumed many LGBs, notwithstanding that the latter are not launch-and-leave bombs.)[24]

So, for the long haul, the Air Force is equipping the standard bombs in our inventory with a combination of guidance that will not yield quite the same accuracy as laser, television, or infrared guidance. However, they will deliver something better than twenty-five meters of Circular Error Probable.[25] If you get a 2,000-pound bomb within twenty-five meters of you, it will ruin your afternoon for sure.

The plan is to use a combination of inertial guidance updated with Global Positioning System data. That will yield the desired accuracy for just any old dumb bomb, which the aircrew can drop and immediately forget. Both inertial and GPS guidance are immune to weather interference. The bomber can launch the weapon at some distance from the target and still have some standoff protection from the most dangerous ground defenses. The crew can then begin its escape earlier than if it were using a laser weapon for which it would have to stay and designate the target until impact.[26] James Canan, writing in the April 1995 issue of *Seapower*, cites a standoff distance of 12 miles.[27]

The developers argue that they have a gnawing problem with all this under control. Any system requiring external inputs is theoretically subject to jamming. When the Germans fielded the radio-controlled Fritz to sink the Italian battleship *Roma* in 1943, they had already anticipated that problem. With blazing speed, before the landings at Anzio early in 1944, the Allies had already prepared a countermeasures ship to jam the Fritz's radio link. However, the Germans had anticipated that and developed a wire-guided version invulnerable to jamming.[28] JDAM's inertial system is completely self-contained and therefore unjammable. The requirements mandate that it be able to get within 30 meters even in the absence of GPS, and that is one hedge. Still, there are those who still worry that the antijam measures being built in will not be enough.[29] One hedge, albeit expensive, would be to add the radar seeker originally planned for Phase III.

The first phase of JDAM is to get the GPS/inertial kits installed on the standard bombs as soon as possible. That will diminish the weather sanctuary with accurate fire through the clouds and smoke and whatever. The next phase is to add a programmable fuse, and originally the plan was for a third phase to add a radar seeker for the last increment of precision by terminal guidance. Yet economics may stand in the way of this in that the seeker would multiply the unit cost of the weapon — and fifteen meters is more than close enough for most work.[30]

The Air Force Development Test Center has recently opened a state of the art Guided Weapons Evaluation Facility to experiment with synthetic-aperture radar, millimeter-wave radar, laser radar, as well as other kinds of seekers. In order to develop apparatus that would serve the purpose of the last phase of

JDAMs, the weapon would have to contain a computer that could store the enormous numbers of templates (algorithms) to cover all possible targets from all possible angles. Hopefully, the GWEF experiments plus computer flight dynamics analysis will reduce the need for actual flight testing enough to make the desired seeker-processor and the algorithms affordable. It would have to be robust enough to endure carrier catapult launches. It would have to be small enough to fit inside a weapon that we could load on fighter pylons or bomb racks. The radar-computer combination would have to take a target return and compare it to the algorithms to distinguish between, say, a school bus and a tank — and to do it infallibly.

However, you can imagine what a horrendous problem it is to develop the parameters for this seeker to distinguish between a school bus and a beer truck. It must not only to tell the difference from the side view or the top view. It also must work from all distances and from every perspective throughout the whole hemisphere above that target. This has become possible only since Vietnam or even since Desert Storm. The continued miniaturization of electronics and increasing processing capacity make feasible very small computers. Hopefully Materiel Command can develop one to fit inside a seeker, that will fit inside a weapon, that can be launched from inside an aircraft's weapons bay.

We could go on and on in our exploration of the possibilities. There are numerous programs afoot that aim to reduce or eliminate the remaining sanctuary — to give the rubber band the final stretch that will snap it. One is the Joint Standoff Weapon in which the Navy is the lead service. Its guidance principle is very similar to that of JDAM's inertial/GPS combination. However, it has a set of folding wings that will make it into a glide bomb with considerably more standoff range than with the JDAM — safer for the aircrew, but more expensive. The important point is, though, that it also would deliver accurate strikes in all weather conditions, though in its early phases it will not have the precision of an LGB or the AGM–130. Two of its variants are to be area-type weapons in any event — containing submunitions. One would deliver 145 Combined Effects submunitions into the immediate vicinity of mobile targets. Another spreads a set of Sensor Fused Weapons that themselves can recognize a hot moving target and fire a potent slug at it. Yet another development is the Joint Air-To-Surface Standoff Missile in which the Air Force is the lead service. This will be expensive in unit costs compared to both JDAM and JSOW, but it will reach out farther than either. It will yield both long standoff and extreme precision. In part, the purpose is to take down the adversary air defenses so that the other, cheaper, weapons can move in to complete the work with a lessened threat of lost airplanes. It, too, will be an all-weather system, but its initial operational capability is doubtless farther into the future than with either JDAM or JSOW.[31] Smart weapons are only one part of the technical equation that might yield a revolution in military affairs. They do form a synergy with many other technologies like low observables in the F–117, B–2, and F–22, which carries its advanced air-to-air missile. Others include information superiority arising from intelligence, command and control technologies, and organization, including the Airborne Warning and Control System and Joint Surveillance Target Attack Radar System aircraft.

Toward a Broken Rubber Band Theory and Doctrine

I would like to suggest here that we may be in a position similar to that facing the whole world after the Battle of Cambrai in 1917. Then humanity had just suffered a long, long, long agony from 1914 to 1917. It had been a terrible war of attrition, maybe the worst in history. We were stretching the rubber band and stretching the rubber band and stretching the rubber band, and time after time after time, we were telling our sons to go over the top into the face of murderous machine gun fire. Time after time after time, they were slaughtered and driven back into the trenches.

Then all of a sudden, bingo, here comes the tank at Cambrai (and other places). It broke through the line, but it only got a few miles before it collapsed. The exhaustion, the process of slaughtering our young, went on for yet another year and a half before that agony finally came to the end. Why?

Now we are looking at a military technical revolution, we say that is only one part of the process, that once you get the technology to close that last sanctuary, you have to recognize it as having been closed. Once you have the technological revolution, you must make the doctrinal adjustments, as Dr. Holley has pointed out. You must look at the meaning of this, and only after you have determined what the doctrinal significance of the new technology is can you make the required organizational changes that will result in a whole new kind of military power.[32]

In the sample I chose, the French, by 1940, had more tanks, and according to some arguments, better tanks than the Germans. But in 1917, the tank was an infantry-support weapon. What do you do with an infantry-support weapon? You farm them out to every battalion in the army. And the French, most of them, did not make the doctrinal revolution to go along with the technical revolution. They did not make the organizational revolution to go along with the doctrinal and technical changes involved with armored warfare.

And on the other side of the line for various reasons the Germans did. One of them was that they had lost the last war and they were angry about it. So they came up with new doctrinal concepts, and they organized their army so that the tank would be an independent weapon, one for going through the hole and getting past the hard front lines and tearing up the rear areas instead of supporting the infantry. Finally, in 1940, that rubber band broke — the long stretch of the stalemate of the trenches was over.

Are we in a similar situation right now? We overcame the distance sanctuary between the wars with long-range bombers — and afterwards

An AC–130 fires its 105-mm howitzer at night.

extended that to the fighter world with tankers. Starting with Vietnam, I think, we overcame the night-time sanctuary — nowadays it is the flyers and not the truck drivers who long for the sunset. It seems clear enough to me that we are on the verge of eliminating the last refuge, the weather sanctuary. But are we making the doctrinal changes that these things demand? Are we treating our air power as it is just another auxiliary weapon to be used in the interdiction battle commanded by a green suit CINC or corps commander to do what artillery does?

Are we going to miss the opportunity that may exist out there by saying that interdiction was a disappointment in Italy. Interdiction was a disappointment in Korea. Interdiction was a sad disappointment in Vietnam. But just because it is alleged to have failed all those times, it does not necessarily follow that it would fail again. Arguably, the equation is complete, the last sanctuary is extinguished, and the rubber band has finally snapped. I would urge the reader to keep track of these things; and if, indeed, you become convinced that is the case, maybe the words that we find written by Drs. Tom Keaney and Eliot Cohen are appropriate, and it is time to turn to this thing from a military technical revolution into a revolution in military affairs.

Here are the words they used:

To be sure, given the mitigating effects of political circumstances, training, technology, geography, and force ratios that heavily favored the Coalition, some caution is indicated. We may require a sterner test against a more capable adversary to come to a conclusive judgment. But if air power again exerts similar dominance over opposing ground forces, the conclusion will be inescapable that some threshold in the relationship between air and ground forces was first crossed in Desert Storm.[33]

Well, I think it was first crossed at the time of Genghis Khan or certainly in Vietnam, but it gives us all some food for some very serious thought. Perhaps we are now on the other side of the threshold and it is time to think about armies in support of air forces — and about the appointment of CINCs who are airmen.[34]

Notes

1. Abraham H. Maslow, *Motivation and Personality* (New York: Harper and Row, 1987), explains that the primary motive for normal folks is survival, and once that is assured, then creature comforts, status and finally "self-actualization" become important. Often this is presented in the form of a triangle with survival at the base and self-fulfillment at the apex—the famous Maslow's Triangle.

2. Vannevar Bush, *Modern Arms and Free Men* (New York: Simon & Schuster, 1949), p 25.

3. Robert Frank Futrell, "Science and Air Warfare," in William S. Coker, ed, *The Military Presence on the Gulf Coast* (Pensacola, Fla: Gulf Coast History and Humanities Conference, 1978), p 136.

4. Norman Polmar and Thomas B. Allen, *World War II: America at War, 1941–1945* (New York: Random House, 1991), p 136; J. D. Gerrard-Gough and Albert B. Christman, *History of the Naval Weapons Center*, vol 2, *The Grand Experiment at Inyokern* (Washington: Naval Hist Div, 1978), p 278.

5. "Eglin and the Bridges of North Korea," *Eglin Eagle*, Jun 7, 1985, p 18.

6. See Irving B. Holley, Jr., "Of Saber Charges, Escort Fighters, and Spacecraft: The Search for Doctrine," *Air University Review*, vol 34 (Sep–Oct 1983), pp 2–11.

7. *United States Strategic Bombing Survey*, vol 59, *Defeat of the German Air Force* (Washington: GPO, 1947), pp 12–18.

8. See William R. Emerson, "Operation POINTBLANK: A Tale of Bombers and Fighters," in Harry R. Borowski, ed, *The Harmon Memorial Lectures in Military History, 1959–1987* (Washington: Office of AF History, 1988), p 441–72.

9. William Mitchell, *Winged Defense: The Development and Possibilities of Modern Air Power, Economic and Military* (New York: Dover, 1925, 1988), p 183.

10. Emerson, "Operation POINTBLANK," p 464.

11. Williamson Murray, *Strategy for Defeat* (Maxwell AFB, Ala: Air University, 1983), pp 235, 345.

12. Generalfeldmarschall Gerd von Rundstedt, "Interrogation," Sep 2, 1945, copy in Spaatz Papers, Manuscripts Div, Library of Congress, Washington, DC.

13. Alan J. Levine, *The Strategic Bombing of Germany, 1940–1945* (Westport, Conn: Praeger, 1992), pp 2–5; Lee Kennett, *The First Air War, 1914–1918* (New York: Free Press, 1991), pp 49–62.

14. Both required that visual contact with both bomb and target be maintained by the bombardier until impact—an exception being the case of the "Droop Snoot" P–38s modified with a second seat and bomb sight for a bombardier. When the B–29s released the Azons, they were free to break away while the Droop Snoots took over control from a higher altitude.

15. The projectile for the gunship's 105-mm cannon is sixty-odd pounds, about half of which is the propellant. The Mk–82 bomb weighs 500 pounds, and the Mk–84 weighs 2,000. Less than half the weight of general purpose bombs like these is in explosive material, but they are nonetheless far more potent than the 105-mm projectile if delivered accurately. (Blast effect diminishes with the cube of the miss distance.)

16. During World War II, Gen Henry Arnold convened a group of scientists under the leadership of Dr. Theodore von Kármán. It produced a multi-volume work known as *Toward New Horizons* that has gained fame as the USAF roadmap into the space age. Early in the 1950s, the USAF asked von Kármán to write another edition of his report, but he declined, saying that it was now beyond the capability of one person or his group. The USAF turned to an older group, the National Academy of Sciences, for guidance. The result was the Woods Hole Summer Study Group that met in 1957 and 1958 on the "Old Whitney estate." The scientists brought their

families, and it was a low pressure setting with many recreational activities. This was thought to be conducive to innovation. The *Toward New Horizons* was reviewed at the outset. There was some feeling that it had been remarkable for its conservatism. The original report had included a recommendation for the launching of a space satellite. However, the USAF, according to von Kármán had discouraged that during the intervening years because of the congressional feeling that it was pie-in-the-sky thinking and a waste of money. Then, in the fall of 1957, the U.S.S.R. launched such a satellite. This so disturbed the United States, and her scientists, that it stimulated the next summer's meeting to some radical thinking on the possible scientific advances—one of the thoughts (arising within the Limited War Panel) being that LASER light, just becoming known, might be utilized in weapon guidance, Michael H. Gorn, *The Universal Man: Theodore von Kármán's Life in Aeronautics* (Washington: Smithsonian, 1992), pp 113–22, 138–43.

17. "Bombs—Renaissance by Terminal Guidance," *NATO"s Fifteen Nations*, vol 25 (Oct–Nov 1980), pp 72–8; James A. Puffer and Vernon L. Reirson, "Development and Flight Test Evaluation of the M–117 Laser Guided Bomb," AFATL–TR–69–152 (Eglin AFB, Fla: Armament Lab, AF Systems Command, Dec 1969), copy at Tech Library, Eglin AFB.

18. Though LGBs have received most of the glory for the victory, some of the weapons employed were early versions of television-guided bombs.

19. Eduard Mark, *Aerial Interdiction in Three Wars* (Washington: Center for AF History, 1994), pp 386–87. John E. Doglione, *et al, Airpower and the Spring Invasion* (new imprint, Washington: Office of AF History, 1985), pp 105–08.

20. Thomas A. Keaney and Eliot A. Cohen, *Revolution in Warfare? Air Power in the Persian Gulf* (Annapolis: Naval Institute, 1995), pp 191–93, 206; John A. Warden, III, "The enemy as a System," *Airpower Journal*, vol IX (Spring 1995), pp 54–55, in which Colonel Warden argues that the quantum jump in accuracy has made possible "parallel war" in place of "serial war" in the past.

21. Keaney and Cohen, *Revolution in Warfare*, p 144.

22. *The Nation's Air Force 1996 Issues Book* (Washington: HQ USAF, 1996), p 30; "Semi-Annual Historical Reports, Joint Direct Attack Munitions Joint System Program Office input to ASC/HO" (Wright-Patterson AFB, Ohio: AF Materiel Command, Aeronautical Systems Center, Oct 27, 1995), copy on file at ASC History Office.

23. This weapon also comes in an imaging infrared variant (IIR), but that is equally expensive, though it works in darkness as well as daylight.

24. Keaney and Cohen, *Revolution in Warfare*, p 90.

25. Bruce Rolfsen, "The Boss commends JDAM team's work," *Northwest Florida Daily News* (Mar 15, 1996), p 1B, in which it is claimed that the unit cost for the JDAM kit has been brought down to $14,000 which is a fraction of that of the LGB; Roy Braybrook and Eric Biass, "Not-too-Close-Encounters of the Air-to-Ground Kind," *Armada International* (Feb/Mar 1996), p 34, cites the accuracy at 15 meters. CEP is the standard measure of bomb or artillery accuracy. It is the radius of a circle within which 50 percent of the rounds fired or dropped can be expected to fall; "AF Fact Sheet/Joint Direct Attack Munition," undated, states the accuracy that must be met by the contractors is 13 meters with GPS and 30 meters without.

26. The act of designation is when a sensor operator holds the beam of laser light on the target by means of a joy stick or some similar apparatus so that the light will be reflected back to the seeker on the bomb—perhaps similar to using a computer mouse to identify a word in a manuscript for deletion.

27. James W. Canan, "Smart and Smarter, JSOW and JDAM: The 'Most Significant' New Weapons," *Seapower* (Mar 1995), p 96.

28. Charles H. Bogart, "German

Remotely Piloted Bombs," *US Naval Institute Proceedings*, vol 102 (Nov 1976) pp 62–8.

29. Stanley B. Alterman, "GPS Dependence: A Fragile Vision for US Battlefield Dominance," *Journal of Electronic Defense* (Sep 1995), pp 52–5; "US Reviews GPS Policy," *Military Technology* (May 1996), pp 8–9; Stephan M. Hardy, "Will the GPS Lose its Way?" *Journal of Electronic Defense* (Sep 1995), pp 56–61.

30. Braybrook and Biass, "Not-So-Close," p 34; Canan, "Smart and Smarter," pp 93–6.

31. "Eglin/Hurlburt Industry Days," briefing slides, AF Materiel Command, AF Dev Test Center, Apr 23–25, 1996; "Report of the Secretary of Defense to the President and Congress," (Washington: DOD, Mar 1996), pp 184–85.

32. Irving B. Holley, *Ideas and Weapons* (Hamden, Conn: Archon, 1971), pp 3–22; James R. FitzSimonds and Jan M. van Tol, "Revolutions in Military Affairs," *Joint Forces Quarterly* (Spring 1994), pp 24–31.

33. Keaney and Cohen, *Revolution in Warfare*, p 209.

34. Admittedly, Gen Lauris Norstad was in charge in Europe in the late 1950s, and also some of the PACOM CINCs have worn Navy wings and AF Officers have commanded Alaskan Command. I leave it to the reader to decide whether those things tend to deny or confirm my point.

Kenneth P. Werrell graduated from the United States Air Force Academy and piloted WB–50 aircraft during his brief Air Force career. Dr. Werrell did his graduate work at Duke University and, since 1970, has been teaching American and military history at Radford University with one year at the Army's Command and General Staff College and two years at the Air University. His books include a bibliography of the Eighth Air Force, a history of the cruise missile, a history of flak, a B–17 unit in the Mediterranean, and a study of the bombing of Japan. Currently, he is beginning a study of air power in the Korean War.

The USAF and the Cruise Missile:
Opportunity or Threat?

Kenneth P. Werrell

The cruise missile has presented the Air Force with a seemingly terrible dilemma. While many airmen pursued it for decades in a quest for an accurate device to hit vital targets without risking aircrew, when it became a reality, it seemed to threaten the airplane, the thing most dear to airmen. As a result, the cruise missile has met with a mixed reception in the Air Force, certainly not with the enthusiasm of its supporters. In 1982, a selected sample of Air War College students were surveyed about the cruise missile. Although these officers did not believe the cruise missile would replace the manned penetrating bomber, they did believe that it would have a major or considerable impact on the USAF. Their ranking of the various versions of the cruise missiles indicates that they were thinking of the cruise missile primarily as a strategic nuclear weapon, as they ranked the nuclear version above the conventional version.[1] It would seem that those officers were correct on some aspects and probably accurately reflected Air Force views, but were off the mark on others. It is correct that the cruise missile did not replace the bomber, and that in the Air Force view, the nuclear version is most important. On the other hand, the cruise missile has not had a major impact on the USAF, and the downgrading of the conventional role of the missile has proven short sighted. The how and why of all this is the crux of this paper.

It was not until 1991 that the cruise missile was to see action, and then not in a nuclear war against the Soviets, but as a conventional weapon (non-nuclear) in a regional war. While it won public acclaim and posted a distinguished record in the Gulf Conflict, today the cruise missile's role in the Air Force is in a state of flux. Why? Is this because the end of the Cold War has made it superfluous, that other technologies, such as the stealth aircraft, have surpassed it? Or has the USAF never fully appreciated its potential, or perhaps seen the cruise missile not as a complement to its inventory, but as a threat to its very existence, because it could be seen as a substitute for the manned bomber? The answer to these questions will soon be apparent in how this weapon is treated in an era of force and budget reductions. In this forum and format, I can do little more than present the outlines of the history, raise some of the questions and issues, and

141

only hope that some of you will be stimulated (or provoked) to investigate and answer the more important of the relevant questions. The most important is that of the relationship of this weapon, a long-range, precision standoff weapon, to the manned bomber. In brief, is it a threat or an opportunity?

Let me add two quick points before I proceed. First, this paper is based on the open literature. Second, I have defined the cruise missile as a long-range (over 100 nautical miles), precision, unmanned, air breathing, winged missile for the attack of land targets. Some might instead use the term long-range, powered, "standoff" weapon.

History: an Idea Before its Time, Pre-World War I to the 1970s

Even before World War I, there was talk of unmanned "flying bombs" or "aerial torpedoes." The United States built and tested two different versions of such weapons during that war, experiments that extended after the war's end. Each was essentially a small aircraft controlled by a preset device using gyroscope technology connected with the inventor Elmer Sperry. The first was a U.S. Navy effort in 1917–18 that also involved the aviation great Glenn Curtiss. A little later, a U.S. Army project that lasted from 1918 to 1919 was headed by Charles Kettering and included Orville Wright. Both devices proved impractical: only 8 of 36 attempted tests were successful because of launch and mechanical problems. The Army and Sperry experiments continued in the 1920s, with the focus shifting from preset to radio control guidance, but again without marked success. In the 1930s, the Navy worked with the idea, as it looked for a flying antiaircraft target. The Navy had an ambitious program during World War II for such weapons and was able to demonstrate its potential in combat tests with 46 drones in 1944. Meanwhile, the Army had renewed its interest in the late 1930s, but it wasn't until Kettering got back into the picture that the Army began to put its money into the program. He convinced his old friend Hap Arnold of the program's potential, and as a result, a prop-powered device was tested stateside in 1941–43 with only limited success.

The AAF also flew a number of remotely controlled aircraft. The only combat use was with "war weary" B–17 and B–24 (Aphrodite) bombers that were crammed full of explosives and guided by radio control sent against German targets with little success. (It is best, or worse, remembered for the parallel Navy project in which Lt. Joseph P. Kennedy, Jr., was killed.) As Dave Mets has so well explained earlier, the AAF experimented with controlled bombs during the war, the Azon (azimuth only) bomb and a glide bomb that carried a television seeker.

The Germans got the first cruise missile into combat with its pulse-jet-powered V–1, firing over 10,000 against Britain and over 7,000 against targets on the continent. They carried a one-ton payload at speeds between 340 mph and 400 mph (increasing as fuel burned off) and at an average altitude of 2,300 feet over its 150-mile range. It was not an effective military weapon as the preset gyro guidance achieved accuracies averaging about ten miles. Nevertheless, the buzz bombs had a marked psychological impact on both the British public and decisionmakers; it was an effective terror weapon. Allied defenses rapidly improved and consisted of attacks on German factories and launch sites, fighter patrols, antiaircraft belts, the first use of the proximity fuze in the European theater, and a belt of barrage balloons. During the course of the campaign against Britain, the defenders downed about 38 percent of the number launched, whereas mechanical problems accounted for another 20 percent destroyed. Just over 2,700 Britons and another 4,700 on the continent were killed by the attacks. It should be noted that although the Germans launched the majority of these missiles from ground positions, about 15 percent were launched from aircraft.

Two aspects of the V–1 campaign should be noted. First, the V–1 although not militarily effective, was cost effective, as the missile attacks diverted a vast Allied effort. A wartime British study concluded that the defenders spent 3.8 times as many resources defending against it than the Germans did in developing, building, and operating the V–1s. Second, the weapon emphasized the importance of intelligence in an air campaign, especially one without man directly observing the impact area. The British fed the Germans information that the V–1s were impacting beyond their aiming point through their network of captured agents (Operation Double Cross), encouraging the Germans to shorten the range of the missiles that already were landing short of the center of London. Belatedly the British realized that the Germans could determine the location of impact through obituary notices, therefore in short order they censored this information. The lack of German aerial reconnaissance allowed these deceptions to go on.[2]

The United States quickly (within three weeks) reverse engineered the device and tested it under the designation JB–2. The AAF planned wide-scale use against Germany and Japan, but neither effort worked out. (There were plans to produce 5,000 of the missiles a month.) The missile was extensively tested, with launches from both ground and air, and with both a preset and radio controlled guidance. In tests after the war, the AAF improved accuracy to 5 miles at a range of 150 miles with preset guidance and 1/4 mile at 100 miles with radio controls. The Navy used the same device (called Loon) and fired them from both a research ship and submarines on the surface. In all 1,385 JB–2s were built.

At the same time, the AAF developed its own version of the cruise missile. In July 1944, a month after the V–1s began to slam into Britain, they ordered the JB–1 from Northrop. In the Northrop tradition it was a tailless missile, in this case powered by turbojets. A host of problems dogged the program, and tests were much less than successful. In February 1945, the missile was redesigned to accommodate a modified V–1 pulse jet engine, a version that was redesignated JB–10. But Northrop built the device to aircraft specifications, an act of gross overengineering, and as a consequence its unit cost was approximately $55,400 compared to the JB–2's $8,600. Less than a year after the war's end, it was cancelled.[3]

The USAF continued its quest of cruise missiles with a bit more success in the 1950s and 1960s. It briefly deployed the intercontinental range Northrop Snark in the period 1959–61. It was guided by a one-ton inertial system updated by stellar navigation. But, as with its predecessors, it was expensive, technically flawed, and in the end, unsuccessful. There were numerous aerodynamic problems, and test failures were so frequent that some wit dubbed the waters off of the test site at Cape Canaveral as "Snark-infested waters." (One missile, however, went too far. It was last seen by the USAF after its launch in 1956; in 1982 a Brazilian farmer in the Amazon basin found it!) Its designated follow-on missile was no better, as the North American Navaho is probably best remembered for the rhyme, "Never go, Navaho." The USAF did best with the Martin Matador/Mace missile that was operational between 1955 until 1969 in both Europe and East Asia. It was about the size of a fighter and used a number of different guidance systems: radio control (Shanicle), radar map comparison method (ATRAN), and inertial. But like its big brother the Snark, the Matador/Mace's record was hindered by troublesome engines, guidance, as well as low reliability and accuracy. The Navy had about the same luck (or lack of luck) with its Chance Vought Regulus, a missile that was very much like the Matador in appearance and performance. It did give the Navy a nuclear punch and was liked by some naval officers.

Test launch of a Snark at the Air Force Missile Test Center, Florida.

But the greatest success the United States achieved with this type of weapon in the three decades following the war was with the air-to-ground Hound Dog. It emerged from a March 1956 request for an air-to-surface missile for the B–52. Powered by a jet engine and guided by an inertial system, it could carry a 1,742-pound (4-megaton) warhead between 340 and 562 nautical miles at supersonic speeds (depending on altitude and speed). Two of the missiles could be carried by the Boeing bomber. It was operational between 1961 and 1976, until it was replaced by the smaller, faster, but much shorter ranged (25–100 nautical miles), SRAM, a ballistic missile.

In brief then, American airmen attempted a number of times almost from the outset of manned powered flight, with a number of companies, to develop an unmanned, explosive carrying, winged missile. But the technologies of the day were inadequate to the task. The two principal failings were reliability and accuracy. In essence, American industry fielded a missile that looked like and was about the size of an airplane (a small bomber in the case of the Snark and a fighter in the case of the Matador/Regulus), was powered by an aircraft engine, and had essentially the same flying performance; but was less reliable, accurate, and versatile; and on the basis of a one-way mission, was more expensive than an aircraft. Therefore the cruise missile of the day could not compete with the manned aircraft. But what killed the cruise missile was the introduction of the ballistic missile that was faster, more accurate, invulnerable to enemy defenses, and easier and cheaper to maintain. Thus the United States went to the Triad concept built around manned bombers, ICBMs, and SLBMs.[4]

A Technological Revolution

Several technological breakthroughs transformed the cruise missile from a disappointing failure into a potent weapon. The first was the development of a new guidance system that could get the weapon within tens of meters of its aiming point. By 1970, aircraft inertial guidance systems had improved and accuracy was degraded by only 1/3 nautical mile per hour of flight, as compared to its previous inaccuracy (drift) six times that figure. Nevertheless, to be a precise navigational system, inertial guidance required accurate updates over the hours of flight connected with the cruise missile's intercontinental range (unlike the minutes of flight of the much more expensive inertial systems in ballistic missiles). A navigational system that could update the inertial system was available, called TERCOM (terrain contour matching). About the same time, what can be easiest described as a multisource data-crunching method, Kalman filtering, had been developed that allowed even small computer processors (16K) to be used. This system permitted accuracies stated in the open literature of between 100 to 600 feet over intercontinental distances. Later this was mated with terminal guidance that reduced the inaccuracy to less than 10 meters, according to some published reports. Thus, it came close to approaching the goal

145

A B–52 carrying a Hound Dog missile under each wing.

of one top civilian DOD official who wanted a missile that could hit a "gnat's ass."[5]

Concurrent developments in the space program made TERCOM practical. Despite the doubts and reservations of many, if not most, within the military, the Defense Mapping Agency was able to supply maps of sufficient accuracy and in adequate numbers to support the cruise missile program. Later, space-based satellites would enable even greater accuracy with GPS (Global Positioning System).

Miniaturization also was important to the acceptance of the cruise missile. The hardware for the inertial system had been reduced from 300 pounds in 1960 to 29 pounds ten years later. Meanwhile, nuclear warheads were greatly reduced in size at the same time their power was massively increased. The impact was to permit a notable reduction in the size of the missile. Another crucial invention that made the smaller cruise missile possible was the appearance of a small jet engine. Developed by the Sam Williams, a brilliant inventor and entrepreneur, it evolved from an engine for drones and the strange, futuristic "jet belt," into a critical element for the "new" cruise missile. In the 1960s, the Williams Research Company built and demonstrated an engine that had good fuel economy, adequate power, and most significant, was one-tenth the size of the next larger engine. The small missile that resulted could use existing launch platforms (bombers and submarines), which meant that the potentially great costs of new platforms were avoided. It also meant that numbers of these missiles could be carried. For example, in contrast to the Soviet development of large cruise missiles, two to an aircraft (similar to the Hound Dog), the United States was able to fit 20 missiles aboard B–52s. Reduced costs were critical during this period as the military underwent the drawdowns following the Vietnam War.

The reduced size also gave the missile operational advantages. Precise navigation not only meant getting a warhead closer to its target, but also that the vehicle could fly lower. Low flight coupled with the device's small size made it very difficult to detect and down.[6] The potential of a small low-flying device was proven in the very hostile skies over North Vietnam. Against the most dense defenses in the world of the time, in 1971 and 1972, only 81 reconnaissance drones were downed on 743 sorties.[7]

These new technologies produced the modern cruise missile, but not without further complications. The direct line of development for the Air Force missile was the desire to protect the bomber fleet. The USAF wanted a replacement for the Quail decoy missile, and the new technologies indicated that a much longer range, yet smaller, device was now possible. In January 1968, Headquarters Air Force issued a requirement for a subsonic, armed cruise missile. A week later SAC issued a requirement for an improved unarmed decoy. The difference between the two was significant, and the issue of whether or not to arm the new device was to remain and haunt the USAF. While SAC wanted a short-range decoy missile, Air Force Systems Command pushed for a long-range armed missile. The USAF compromised, seeking an unarmed decoy with a later arming option, concepts summarized in the project's name, Subsonic Cruise Armed Decoy (SCAD). The service's insistence that the new missile fit into the B–52's SRAM rotary launcher, severely constricting its size and shape, was seen by some to indicate that the USAF was neglecting the missile's potential, if not deliberately ignoring it. While the Air Force's position on retaining the expensive SRAM launcher might be understandable, its slowness, interpreted as reluctance, to incorporate a more accurate guidance system and the arming option that DOD proposed is more difficult to justify, except that the USAF favored a more "orderly" sequential development. Whatever the reason, the Air Force's actions and inactions delayed the missile.[8] *Air Force Magazine* summed up the USAF position that it "makes no sense to substitute a small, subsonic, relatively inaccurate missile for the ballistic missile" and that SCAD was too small to be a standoff weapon.[9]

It didn't escape the leaders in DOD or in Congress that the Air Force was responsible for the SCAD's limited range and accuracy capabilities. In June 1971, Senator William Proxmire charged that the Air Force leaders were obstructing the missile in an attempt to protect the B–1 bomber. A battle between the Air Force and the Deputy Secretary of Defense over the device led to the project's cancellation in June 1973. The Senate Armed Service Committee noted, "The Air Force has proceeded with this program solely as a decoy, not

Four reconnaissance drones at a base in Southeast Asia,
with the C–130 that carried them in the background.

withstanding the direction of the Congress. It is generally recognized that the Air Force has resisted pursuing SCAD with an armed warhead because of its possible use as a standoff launch missile. This application could jeopardize the B–1 program because it would not be necessary to have bomber penetration if a standoff missile were available as a cheaper and more viable alternative."[10]

That would have been the end of the story except that the Navy was developing a similar device, albeit from a different sequence of events and for a different platform. In short, the Navy, prodded by the success of the Soviet Styx missile that sank an Israeli destroyer in October 1967, was seeking a long-range ship killer. From this came the potent Harpoon, a tactical antiship missile. paper. What is pertinent here is that a spin-off program from the Harpoon project was working on a longer range strategic missile. This effort was both different and technically difficult because the Navy sought a missile that could be launched from a submerged submarine, in essence a flying torpedo. There are a number of interesting twists in the story, including plans to refit old Polaris submarines, Admiral Rickover, and arms limitations talks.

In August 1973, Undersecretary of Defense William Clements authorized a Navy program to proceed with cruise missiles from all possible platforms (sea, land, and air) that encouraged the Air Force to follow suit. Clearly the Air Force did not want to have "a torpedo [Navy missile] rammed up its bomb bay!"[11] In

January 1977, DOD established a joint office and ordered the services to cooperate, the former to share its engine and high-energy fuels and the latter to share its TERCOM guidance system. This allowed engineering and manufacturing development of the Boeing ALCM and General Dynamics Tomahawk missiles.[12]

These missiles survived a joint program office, technical problems, testing difficulties, and international arms agreements. But the USAF's B–1 bomber did not, at least for the moment. The ALCM was a factor in President Carter's June 1977 decision to cancel the bomber. (Stealth technology was another.) This also led to renewed interest in the ALCM as the Air Force was left with the admittedly old B–52 and without a follow-on penetrating bomber. Studies of a variety of carriers for the cruise missile were conducted, from USAF transport aircraft to commercial aircraft. Unsurprisingly, the Air Force determined that the best cruise missile carrier would be a B–1. In 1981, President Reagan restored the Air Force bomber as the B–1B.[13]

After a fierce competition with the General Dynamics Tomahawk, the Boeing-built AGM–86B armed with a nuclear warhead went on to arm the Air Force's bomber fleet. The Air Force fielded another cruise missile for a brief time. To counter the newly deployed three-headed, Soviet SS–20 intermediate-range ballistic missiles in Europe, in 1977 the United States pushed two tactical nuclear delivery systems, a ground-launched cruise missile (GLCM) along with Pershing II ballistic missile. The cruise missile chosen for this duty was the Tomahawk, mounted four in a canister atop a truck. This missile also encountered military resistance as neither the Army nor Air Force wanted it. The Army refused to send a representative to the office developing the missile, and one Air Force general stated that the USAF "didn't want GLCM, we were issued it."[14]

The Air Force further resisted the idea of using the Navy missile by referring to it as the BGM–109G and eventually naming it the Gryphon, in other words, calling it anything but Tomahawk! After initial deployment in December 1983, it was traded for the removal of Soviet missiles from Europe, in the INF agreement of December 1987, with the first leaving Europe in April 1988. It had achieved its goals.

An effort to field a conventionally armed cruise missile, MRASM, was not as successful. This program was tentatively authorized by Clements' 1973 memo but did not catch fire until then Under Secretary of Defense William Perry wrote in March 1980, "It is a matter of national importance that a joint tactical medium range air-to-surface missile (MRASM) be added to our strike warfare systems as soon as possible."[15]

A joint office guided the missile's development, but it did not enjoy high-level support in either the Air Force or Navy. The Tomahawk was used as the vehicle, while work was done in the areas of terminal guidance and submunitions. In May 1978, a modified Tomahawk guided by TERCOM and SMAC (Scene-Matching Area Correlator) dropped 11 of its 12 bomblets squarely on its runway target after flying 403 miles from launch. More advanced guidance (DSMAC, Digital Scene Matching Area Correlator) and a number of other sub-munitions were tested. Despite this technical progress, the lack of Air Force and Navy enthusiasm for the program as well as escalating costs led to its cancellation in 1984. The Navy's effort to develop a cheap version, using a low-cost guidance (ring laser gyro) and IIR (Imaging Infrared Radar) guidance was cancelled. This seeker was later put on a Harpoon missile and fielded as the SLAM (Short-ranged Land Attack Missile).[16]

The Navy fielded four versions of the Tomahawk. Like the Air Force, the Navy deployed the missile as a nuclear-armed strategic weapon, and likewise there was resistance in the sea service against the cruise missile. Attack submariners saw the missile competing with torpedoes for limited space aboard their boats, while carrier aviators saw a challenge to manned aircraft. To make the point, one future CNO backed a naval officer into a corner after a cruise missile briefing, shook his finger under the officer's nose and emphatically stated: "We already have a cruise missile, it's an A–7. We don't need your cruise missile!"[17] The Secretary of the Navy expressed the same sentiment, testifying to a congressional panel that "our carrier aircraft are essentially 'manned cruise missiles.'"[18]

Nevertheless, it entered service as the SLCM (Sea Launched Cruise Missile) in 1983, first in the B version, the Tomahawk Antiship Missile (TASM), that I have defined out of our discussion. The next year, the Tomahawk A, Tomahawk Land Attack Missile-Nuclear (TLAM–N), entered service, but as part of a United States-Soviet agreement was withdrawn into "ready storage" in 1991. It had a 1,500-mile range. A third version is the TLAM–C armed with a 1,000-pound warhead that was deployed in 1983 on the U.S.S. *New Jersey*. The fourth version is the Tomahawk D variant, essentially the same except for its warhead, in this case a device that carries 166 bomblets, each weighing 3.4 pounds. These can be dispensed in batches to hit multiple targets. It entered service in 1988.[19]

ACM: Advanced Cruise Missile

In 1982, the USAF began a program to replace its ALCM. Featuring stealth technology (with such distinctive features as a swept forward wing and downward facing vertical stabilizer), a new engine, and new guidance equipment, the AGM–129 Advance Cruise Missile has greater range, accuracy, and survivability than its predecessor. It went into production in 1983 and the first was delivered for tests in 1985. But technical, testing, quality control, scheduling, and financial problems dogged the program. Senator Les Aspin called the ACM a "procurement disaster" with the worst problems of any of the eight strategic weapons programs his committee had studied. Design and quality difficulties were of such a magnitude to cause the government to stop deliveries

150

in 1989 and 1991. But other factors were to intrude. Originally the Air Force planned to buy 1,461 of the AGM–129, but in January 1992, the president decided to halt the program with a program buy of 460. This action was prompted by the demise of the Soviet Union and the program's costs and problems.[20]

The Air Force also pushed the development of a variant of the ACM for targets against which the ACM was considered "ineffective." SAC presented the requirement in 1985, and the USAF proposed to modify 120 ACMs for this task. But in 1991, Congress denied the request and told the Air Force to terminate the program. The next year, DOD ordered the Air Force to draw up plans to restart the program, an effort opposed by the GAO. The critics pointed to the end of the Cold War, the costs, technical problems, and scheduling risks as reasons not to build the missile.[21] There is nothing more about this mysterious missile in the open literature.

Combat

Although the cruise missile began to be deployed in the early 1980s, it was not until a decade later that it first was used in combat. Meanwhile, two incidents during this decade emphasized its utility. In Lebanon in 1983, two American aircraft were downed in a punitive strike, costing the lives of two airmen and a costly captive ordeal for a third. The 1986 strike against Libya was not only a failure in physical and propaganda terms, it cost one U.S. aircraft and two lives. It is unclear why aircraft and not cruise missiles were employed in these two cases.[22]

In any event, it was not used until 1991 when it proved to be one of the star performers of the Gulf War, along with the F–117 stealth bomber. In fact, these two weapons were the major thrust of the initial assault that helped to quickly and cheaply gain air superiority. Doubts about the cruise missile's reliability and accuracy led the planners to target multiple missiles against the same targets. Problems with battle damage assessment and strikes by bombs complicated the assessment problem, confusing planners then, and evaluators since, as to the weapon's effectiveness. The Navy fired Tomahawks at air defense, command and control, and key points such as electrical facilities, especially against targets in the well defended Baghdad area where only it and the F–117 were employed. Here, the Air Force said, the air defenses were seven times that around Hanoi at the height of the Vietnam War and more heavily defended than any East European city during the Cold War. Not only did the lights in Iraqi capital go out within minutes of the assault, but they stayed out until after the cease fire. By the end of the third day of the war, Iraq had lost 85 percent of its electrical power. The Navy fired 288 Tomahawks (116 in the first 24 hours of the war) and claimed that 85 percent were successful.[23] The cruise missile's contribution has been understated. Not only did it have remarkable accuracy, but it was much less affected by the weather that seriously inhibited the F–117s. In the first two days, the stealth bomber was able to release ordnance on only half of its sorties due to low clouds, and over the campaign was unable to release bombs on 19 percent of its sorties due to weather.[24] This underscores the dependence of laser-guided bombs on good visibility.

The bulk of the Tomahawks employed (TLAM–C) carried a unitary warhead, although apparently 27 (TLAM–D) carried submunitions.[25] Certainly the visuals of Tomahawks flying down the main streets of Baghdad in broad daylight had a profound impact on all observers. The use of the cruise missile in daylight made them easier to down, but had the positive value of applying psychological pressure around the clock, as the F–117 operated only at night. The tradeoff was that daylight employment of the cruise missile along with the lack of surface features and mapping forced the planners to use the same routes and "stream" tactics that aided the defenders. Thus, it is significant that, despite these advantages, the Iraqis were able to destroy few cruise missiles.

The USAF use of cruise missiles was considerably less in numbers, but not in drama. A year after the war, it was revealed that seven B–52Gs had flown 7,000 miles from Barksdale AFB to launch missiles and then returned to recover at their home field a day and half after takeoff and four air-to-air refuelings, clearly the longest combat strike of all time. They fired 35 missiles, 31 of which hit eight targets in northern Iraq (power plants) and southern Iraq (telephone exchange). These weapons were available because in June 1986 the USAF began to modify some of its nuclear-armed ALCMs (AGM–86B) for a conventional role by fitting them with a 1,000-pound warhead, GPS guidance, and the designation AGM–86C, variously called CALCMs or ALCM-Cs. These missiles were more sophisticated than the Tomahawks used in the Gulf War as their new guidance offered greater accuracy and simplified planning and permitted much greater flexibility in choosing attack routes. This missile went into service with SAC in January 1988. It was a demonstration of what the Air Force could do with forces based stateside.[26]

One last aspect of the cruise missile operations during the Gulf War deserves mention. Electrical power was one of the key targets of the air attack, and eight power facilities were targeted the first night. Three targets were hit by the CALCMs, two targets with conventional bombs dropped by aircraft, and the remaining three targets by an exotic device carried by Tomahawks. Carbon fiber wire was dropped over electrical transmission lines to short out the electrical system. The lines were not only more visible, vulnerable, and more difficult to defend than electrical generators, such attacks would deprive the Iraqis only briefly of electrical power by not destroying the costly generators.[27]

A quick, relevant aside. There is a report that the United States is fitting hundreds of its older air-launched cruise missiles with stealth technology and new types of nonlethal warheads. One type of warhead detailed is a non-nuclear electromagnetic pulse generator that has the potential to disable most electronic

A B–52 carrying twelve AGM–86 cruise missiles.

devices within hundreds of meters of the explosion: computers, radars, solid-state ignitions on vehicles, and aircraft electronics.[28]

Since the 1991 Gulf War, Tomahawks have had three further combat missions, two in Iraq. There were a number of Iraqi provocations just days before the 1993 presidential inauguration of Bill Clinton that prompted a number of air strikes by American aircraft. These were unsuccessful, hitting only half of the aiming points. Then on January 17, 1993, cruise missiles were launched against the Zaafaraniyah nuclear facility 13 kilometers southeast of Baghdad. All seven of the targeted buildings were hit (four were totally destroyed and two others were severely damaged) by 37 of the 45 missiles. Of the remainder, one went into the ocean, three fell short, three impacted on the grounds of the complex, but caused no damage, and one, apparently hit by flak, hit a hotel and killed two civilians, although its warhead did not explode.[29]

Six months later (June 1993), the Tomahawks went into action again, this time in retaliation for an alleged Iraqi assassination plot against former president George Bush. What makes this operation notable is that there were no American aircraft carriers on station, thus the attack was launched from forces available in the area, two ships, one sailing in the Red Sea, the other in the Persian Gulf. They fired twenty-three missiles, sixteen that hit and heavily damaged their aiming point, the Iraqi intelligence headquarters in downtown Baghdad. (Another four hit within the complex grounds.) Three others, however, landed 100 to 550 yards from the target and destroyed three houses and killed eight civilians and wounded a dozen others.[30]

Technology and the Air Force

Follow-ons to the Gulf War Cruise Missiles

There was an inevitable push to get better performance out of this marvelous new weapon. While some noted guidance problems, the major difficulty resulted from the intense mapping support for the TERCOM and the long time required for mission planning. The small payload was often mentioned. Finally, the lack of a man in the loop, the inability to retarget while in flight, was criticized. But probably the biggest problem was the cost.[31]

The Navy used Block II birds in the Gulf War. An improved version was in the works, and this Block III version began to see service in April 1993. Its principal advantages are improved penetration, a feature to permit controlled time of arrival, and greatly reduced planning time. The latter, a result of the Afloat Planning System, allows planning at the carrier battle group and theater command level. Block III missiles are equipped with GPS and TERCOM, giving the planners more flexibility for targets and routes. The missile's homing seeker (DSMAC) is also improved. Thus, inaccuracy as been cut to as little as ten meters. A new warhead has been fitted that reduces the weight by 200 pounds, permitting extra fuel to be carried, extending range almost 300 nautical miles to about 1,000 nautical miles. An improved powerplant produces 10–20 percent more thrust for additional power margin.[32]

Bosnia

The most recent use of the cruise missile came on September 10, 1995, when thirteen Block III Tomahawks were launched at targets in Serbian-held Bosnia near Banja Luka. First reports claimed the targets were "critical sites" defined as SAM positions and communications centers, while reports the next day stated the strike hit ten radio relay stations, antennas, and communication centers. The missile strikes were followed by air attacks. Military results have not been officially released, but a NATO officer stated that the missiles caused "severe damage" to their targets and that there was no evidence of any major collateral damage or injury to civilians. He went on to say that "the Tomahawk missile is a particularly accurate system and is used . . . because of that aspect of its operational capability."[33]

In any case, the Bosnian Serb leadership was "clearly shocked" by the attacks, and soon agreed to a cease fire, the overall objective of the NATO air campaign. Four aspects should be noted about this cruise missile strike. First, the airmen were somewhat concerned about air operations in this area, as in June, an American F–16 (piloted by Capt. Scott O'Grady) was downed. Second, the Tomahawks were used because the Italians, for political reasons, refused to permit basing of F–117 stealth bombers on their soil. Third, only a fraction of the numbers used in the 1993 strikes were used in the Bosnian operation, thirteen instead of the forty-five of the January 1993 strike and 23 of the June

154

1993 attack. Fourth, an important consideration was to inflict no damage or death on civilians. The use of the cruise missile in Bosnia indicates the military is gaining confidence in both its reliability and accuracy. This confidence was justified, as 13 turned out to be the right number to get the job done.[34]

TSSAM: Triservice Stand-off Attack Missile

In 1985 Congress pushed the armed forces toward developing a common, conventionally armed cruise missile. It was to use stealth technology, GPS navigation, and a new infrared homing seeker, all cutting-edge technologies, and was to be a low-cost missile to deliver conventional warheads at a range of over 100 nautical miles. But more tricky, and probably its most ambitious aspect, was that TSSAM was to come in various versions to be fired from eight different platforms (Air Force F–16 fighters and B–1, B–2, and B–52 bombers; Navy F–18, A–6, and A–12 aircraft, and the Army's Multiple Launch Rocket System) and carry five different warheads for the three services. In 1986, Northrop won the fixed price contract, and the Air Force was named lead service as it would get the bulk of the buy. (In April 1986, the Air Force was scheduled to get 5,000 TSSAMs, the Army 1,800, and the Navy 2,250.) The program bogged down with technical and manufacturing problems, time was lost, and prices escalated, which led to a souring of relations between the manufacturer and the Pentagon. There were cries of micromanagement against the military and mismanagement against the company. In January 1989, the GAO sounded the alarm that the project "was experiencing more than the normal amount of difficulty during development."[35]

By this time there were massive delays and cost overruns. Northrop attempted to get an additional $1.5 billion, but settled for a "mere" $.7 billion in 1990 to keep the program going. In February 1994, the Army withdrew, and as the numbers on order fell from 9,050 to less than half that (the Air Force's buy fell to 3,631 and the Navy's to 525), the cost per copy almost tripled (rocketing from $728,000 to $2.1 million). There were other problems as well. The early tests were successful, but there were seven failures in the final nine tests between mid-1993 and late 1994. With declining budgets, the military had to make some hard decisions. A March 1994 DOD study concluded that TSSAM was the most cost-effective weapon among several alternatives, primarily because of its performance in high-threat situations. The end of the Cold War meant that the most likely employment would not be against an technically advanced country, but instead in a third world conflict, so that such promised performance was unnecessary. Therefore, Secretary of Defense William Perry, a key player in the creation of both the F–117 stealth bomber and the cruise missile, pulled the plug on the AGM–137 in December 1994.[36]

This leaves the Air Force with its nuclear-armed AGM–86s and AGM–129s and its conventionally armed CALCMs. The Air Force is studying ways to improve the CALCM, but thus far has not reported funding any of these efforts.[37] Meanwhile, the Heavy Bomber Force Study mandated by Congress and carried out by the Pentagon and the Institute for Defense Analyses recommended a replacement for TSSAM with about the same characteristics: a stealthy missile with high accuracy able to deliver a 2,000-pound warhead over several hundred kilometers. Air Force Chief of Staff Gen. Ronald Fogleman requested $50 million from Congress in fiscal year 1996 to begin this program. The Air Force and the Navy are working together on a joint requirements document for what some have called "Son of TSSAM," but what is officially dubbed Joint Air-to-Surface Standoff Missile (JASSM).[38]

A number of possible replacements have been pushed for this requirement. The Tomahawk has been proposed as a reduced-range version and called "Air Hawk" by Hughes, its sole manufacturer. Texas Instruments has pushed its JSOW (Joint Standoff Weapon), an unpowered, bomblet-dispersing 75-kilometer-range glide bomb, with a later powered version. The Navy's interim solution is a derivative of SLAM (Stand-off Land Attack Missile) that was unsuccessfully used in the Gulf War (only 1 of 7 was successful). This version would have a larger warhead and extended range and be known (naturally) as SLAM-ER. The Army's ATACMS built by Loral Vought may be also in the running for this slot. The missile would be converted for air launch, its range doubled to 140 kilometers, and the bomblet warhead of antitank submuntions planned for the TSSAM would be used. Other aerospace companies have also expressed an interest in the project.[39]

Meanwhile, the Navy's Tomahawk Block IV version is scheduled to come into service at the end of the century. A forward looking imaging infrared sensor and GPS will replace both the TERCOM and DSMAC. It will also be fitted with a data link permitting real time damage assessment and man-in-the-loop control. These will employ such relays as satellites, unmanned aerial vehicles, or aircraft to monitor and control (and even retarget) the missiles in flight. In addition, it will have improved engines and warhead, presenting 60 percent greater accuracy, even greater penetration ability, and further reduce planning time so as to approximate that required for aircraft strikes.[40]

Finally, to end the story, there are reports that the Navy is considering building a special ship to carry as many as 500 missiles. In contrast to an aircraft carrier that cost $4.5 billion to build and $.5 billion a year to operate, this barge-like "arsenal ship" would cost about $.5 billion to build and tens of millions a year to operate.[41]

Conclusions

Army and Navy airmen have engaged in the development of cruise missiles from the dawn of manned flight. Until very recently, the Navy, despite its early reluctance, has shown more enthusiasm for the cruise missile than the USAF.

Only in the last few years has the Air Force begun to change its position. Until this point, the Air Force maintained a guarded attitude toward this weapon. The USAF slowly accepted the cruise missile as a nuclear delivery system that allowed the venerable B–52 to be an effective strategic bomber, and then pushed the development of nuclear armed cruise missiles. It has done less with conventional versions of the missile than has the Navy.

The cruise missile has shown its capabilities on a number of occasions, in a variety of circumstances, with different warheads. They demonstrated reliable, accurate, survivable, and lethal operations in both a full-scale conventional war and in a punitive, demonstration attack. In the former, they were valuable in neutralizing potent air defense systems and in repeatedly hitting crucial targets. The missiles showed that they can operate twenty-four hours a day in relatively featureless terrain, and despite repetitive daylight routing, can get through to the most heavily defended, high value targets, with minimum attrition. The cruise missile with its "bloodless" and "infinitely fearless" pilot and bombardier, can put ordnance squarely on target with minimal collateral damage. This characteristic is especially valuable in punitive attacks, for its great accuracy makes a statement by hitting precise targets and limiting collateral damage, while its lack of aircrew means that there can be no embarrassing and costly hostages.

Cruise missiles can be available and put on target in short order from U.S.-based B–52s, surface ships, or submarines. There is no need for Allied concurrence. There is also no requirement for a large support package required by aircraft. This item and the previous item should give pause to bad guys, and thus increase United States deterrent power.

At the same time, there are negatives, particularly in a wide-spread war on a scale beyond that of the Gulf War. Cost remains the biggest inhibitor, but there are also problems of mapping, and an enemy who increasingly gets smart. GPS and night operations help considerably.

The cruise missile story is one of great technological promise that until the last two decades has enticed, but mainly alluded American airmen. Until then, the problems were mostly technical, centering on reliability and accuracy. When these problems were solved, the military (both Air Force and Navy) resisted the weapon. Although it was initially developed and deployed as a nuclear-delivery system, fortunately, it has not been used in that role. It has been employed in two conventional roles: in a large-scale regional war and in three demonstrations or punitive attacks. As the most likely future conflicts for the United States over the next decades will probably be more of the same, the cruise missile is important.

What then for the Air Force in the future? Clearly the USAF must capitalize on this weapon. I'm happy to say that seems to be more the case today than in years past. The central question for the Air Force is the relationship between the cruise missile and the bomber. By that, I mean that today we only have a few aircraft (B–2s and F–117s) that can successfully attack heavily defended targets, and they are limited to night operations. We need a means to get more ordnance on target at less risk. We can do that by firing cruise missiles from other aircraft in the inventory.

There is a need to field an entire range of conventional warheads for the missile that will be effective against targets ranging from armor to hard targets. The USAF should continue to coordinate with the Navy to incorporate the latest technology updates at the lowest cost into Air Force missiles. The military does not like joint programs, but certainly in the case of the cruise missile, they have proven effective.[42] But most of all, the Air Force must promulgate a doctrine that effectively includes cruise missiles in future operations. Air Force decision-makers, staff officers, and planners are aware that cruise missiles can put a variety of ordnance accurately on target, day or night, in good or bad weather, and that their use can spare expensive aircraft and priceless airmen from having to penetrate lethal enemy defenses. They must incorporate this realization into their planning. There is a continuing but challenging role for manned aircraft, but this should be leveraged and enhanced by cruise missiles. I believe that there will probably always be a role for the penetrating aircraft, but this should be reserved for those missions that are truly critical and necessary. Despite its proud traditions, most of all BECAUSE of its history, there is no need to repeat Schweinfurt or Ploesti, nor Lebanon or Libya. There is a role, an important role, for the cruise missile in the Air Force. So let the concept of "The Two Headed Monster" of my title be left to journalists, not to the USAF.

My thanks to Rear Admiral Walter Locke, USN, Ret., for his comments and critique of an earlier draft and to Bud Bennett, Radford University's interlibrary loan specialist, who lent his invaluable skill and help.

Notes

1. Kenneth Werrell, *The Evolution of the Cruise Missile* (Maxwell AFB, Ala: Air University, 1985), pp 237-39, 244-45.

2. *Ibid*, chap. III.

3. *Ibid*, pp 68-70.

4. This is a gross condensation of the history detailed in *ibid*, chaps II–IV. The case for the bomber was that it was recallable, could be used as a show of force, and could carry more powerful warheads.

5. Kenneth Werrell, "The Weapon the Military Did Not Want," *Journal of Military History*, vol 53 (Oct 1989), p 422; John Toomay, "Technical Characteristics," in Richard Betts, ed, *Cruise Missiles: Technology, Strategy, Politics* (Washington: Brookings, 1981), p 37.

6. Merril Skolnik, *Introduction to Radar Systems* (New York: McGraw-Hill, 1980), p 44, notes that the Tomahawk's radar cross section is one-twentieth that of a small fighter.

7. William Wagner, *Lightning Bugs and Other Reconnaissance Drones* (Fallbrook, Calif: Aero, 1982), pp 199-200, 213.

8. Werrell, *Evolution of the Cruise Missile*, pp 144-49.

9. Edgar Ulsamer, "SCAD-Electronics Stand In for the B-52," *Air Force Magazine* (Nov 1972), p 47.

10. US Senate, Armed Service Committee, "Report Together with Separate and Individual Views," rpt 93-385, Sep 6, 1973, p 3383.

11. The quote is from the Boeing project manager, Ray Utterstrom, interview with author, Mar 1, 1982.

12. Werrell, *Evolution of the Cruise Missile*, pp 171-72.

13. *Ibid*, pp 177-78, 187-89.

14. Werrell, "The Weapon the Military Did Not Want," p 431.

15. William Perry, Memo for the Sec Navy, Sec AF, and Dir Joint Cruise Missiles Proj Office, Mar 27, 1980.

16. Werrell, *Evolution of the Cruise Missile*, pp 205-8.

17. Werrell, "The Weapon the Military Did Not Want," p 426.

18. *Ibid*, p 426n.

19. See Nigel Macknight, *Tomahawk Cruise Missile* (Osceola, Wisc: Motorbooks, 1995).

20. "Strategic Missiles: ACM, Opportunity for Additional Savings" (Washington: GAO, May 1992), pp 2-5; "Strategic Missiles: Issue Regarding Advanced Cruise Missile Program Restructuring," (Washington: GAO, May 1994), pp 1-2, 24; R. Jeffrey Smith, "Cruise Missile Reported Late, Over Budget," *Washington Post* (Apr 25, 1995), p A37; "D-5 Gets Top Rating Among 8 Weapons—Aspin," *Defense Daily* (Apr 25, 1988), p 317.

21. ACM Variant Program (Washington: GAO, Jul 1992), pp 1-2.

22. A number of possibilities exist. Perhaps the decisionmakers did not have confidence in the cruise missile or lacked adequate TERCOM maps.

23. James Winfield, *et al*, *A League of Airmen* (Santa Monica, Calif: RAND, 1994), p 122; Eric Schmitt, "The Day's Weapons of Choice, the Cruise Missile, is Valued for Its Accuracy," *New York Times*, Jan 18, 1993, p 8. This figure was to be greatly disputed, obviously depending on the measure of success. Barton Gellman, "Gulf Weapons' Accuracy Downgraded," *Washington Post*, Apr 10, 1992), pp 1, 37, writes that just over 50 percent hit their target. Eric Arnett in "Truth and Tomahawk," *Bulletin of the Atomic Scientist* (Jul-Aug 1992), pp 3-4, claims that less than 60 percent hit their target and that between 1 and 23 were shot down. It should be noted that not only was the task of damage assessment difficult, but also

that the AF planners responsible for targeting in many cases assigned an excess of cruise missiles against a target.

24. GAO, *Cruise Missiles: Proven Capability Should Affect Aircraft and Force Structures*, Apr 1995, pp 3, 26-27; Thomas Keaney and Eliot Cohen, *Gulf War Air Power Survey* (Washington: GPO, 1993), p 225.

25. Macknight, *Tomahawk, Cruise Missile*, pp 70, 78-79.

26. Paul Rogers, "Scramble for Supremacy," *New Statesman and Society*, Sep 9, 1994; Richard Hallion, *Storm Over Iraq* (Washington: Smithsonian, 1992), p 297; James Coyne, *Airpower in the Gulf* (Arlington, Vir: Aerospace Educational Foundation, 1992), p 79. The CALCMs have accuracy half as good as that of the Block III Tomahawks, GAO, *Cruise Missiles*, pp 17, 30. Hallion says that the cruise missile (presumably the Tomahawk Block II) has an accuracy only one tenth that of the F-117's laser guided bombs, *Storm Over Iraq*, p 250.

27. David Fulghum, "Secret Carbon-Fiber Warheads Blinded Iraqi Air Defenses," *Aviation Week and Space Technology*, Apr 27, 1992, pp 18-20.

28. David Fulghum, "ALCMs Given Nonlethal Role," *Aviation Week and Space Technology*, Feb 22, 1993, pp 20-22.

29. David Fulghum, "Pentagon Criticizes Air Strike on Iraq," *Aviation Week and Space Technology*, Jan 25, 1993, p 47; David Fulghum, "Clashes with Iraq Continue After Week of Heavy Air Strikes," *Aviation Week and Space Technology*, Jan 25, 1993, pp 38, 42. The damage to the hotel should make clear that the Tomahawk's bombload is much more than just its warhead and includes the airframe and remaining fuel.

30. John Broder, "Tomahawks Least Risky Option," *Los Angeles Times*, Jun 27, 1993, p A8; Norman Kempster and Melissa Healy, "Strike a Success; Some Civilians Hit," *Los Angeles Times*, Jun 28, 1993, p A1; John Lancaster and Barton Gellman, "U.S. Calls Baghdad Raid a Qualified Success," *Washington Post*, Jun 28, 1993, p A1; David Fulghum, "Low Tomahawk Kill Rate Under Study," *Aviation Week and Space Technology*, Jul 5, 1993, p 25.

31. Keaney and Cohen, *Gulf War Air Power Survey*, p 225; Winfield, *League of Airmen*, p 249.

32. WWW, CNN Interactive World News, "NATO Uses Cruise Missiles on Serb Targets; Serbs Say They Had A Deal for a Delay," Sep 10, 1995; Macknight, *Tomahawk*, pp 87-93.

33. WWW, CNN Interactive World News, "NATO Says Tomahawks Effective against Serb Targets," Sep 11, 1995.

34. Eric Schmitt, "Wider NATO Raids on Serbs Expose Rifts in Alliance," *New York Times*, Sep 12, 1995, p A1; Eric Schmitt, "Allied forces Hit Serb Air Defenses in Western Bosnia," *New York Times*, Sep 11, 1995, p A1; CNN, "NATO Uses Cruise Missiles." That the planners considered use of the F-117 and actually used the Tomahawks was either because of heavy air defenses in the area or, more likely, as a designed escalation of force. Now the most potent of America's weapons, those that had performed so well in Iraq were being used.

35. Bradley Graham, "Missile Project Became a $3.9 Billion Misfire," *Washington Post*, Apr 3, 1995, p A1; GAO, *Missile Development: Status and Issues at the Time of the TSSAM Termination Decision*, Jan 1995, pp 1-2, 8.

36. Graham, "Missile Project," p A1; GAO, *Missile Development: Status and Issues*, pp 3, 8, 10.

37. GAO, *Cruise Missiles*, p 30.

38. John Tirpak, "The Pentagon Declines More B-2s," *Air Force Magazine*, Jul 1995, pp 13-14; "Funds Sought for TSSAM Replacement," *Air Force Magazine*, Jul 1995, p 20.

39. Graham Warwick, "Choose Your Weapons," *Flight International*, Dec 21, 1994, p 34.

40. GAO, *Cruise Missiles*, pp 5, 14, 30, 55; Stanley Kandebo, "Cruise Missile Updates Pending," *Aviation Week and Space Technology*, Jan 17, 1994, pp 56-58.

41. Eric Schmitt, "Aircraft Carrier May Give Way to Missile Ship," *New York Times*, Sep 3, 1995, pp 1, 18. The AF is

inhibited from building a parallel carrier, cruise missile carrier, because international agreements do not distinguish between nuclear and conventionally armed cruise missiles. Therefore any aircraft carrying cruise missiles counts against the strategic bomber limit. But in view of the lessening need for nuclear armed bombers and the growing possibility of future (conventional) cruise missile use, the decision to arm aircraft with a maximum load of cruise missiles probably should be revisited.

42. The development of the AGM-86 and Tomahawk in a joint office was not only effective in terms of performance, cost, and schedule, it showed the advantages of competition and dual sourcing.

James E. Tomayko is a principal lecturer in computer science at Carnegie Mellon University and a senior technical staff member of the Software Engineering Institute. Dr. Tomayko directs the Software Development Studio that has produced a variety of software products. Prior to returning to Carnegie Mellon in 1989, he founded the software engineering graduate program at Wichita State University. He has worked in industry with NCR, NASA, and Boeing, among others. Dr. Tomayko has taught and lectured on software fault tolerance, development management, and process improvement at over 150 universities and companies in the U.S. and overseas. He has had a parallel career in the history of technology, specializing in computing in aerospace. He has written extensively on spacecraft computers and software and has researched the history of fly-by-wire technology. Dr. Tomayko is on the editorial staff of the *IEEE Annals of the History of Computing*.

162

Blind Faith: The United States Air Force and the Development of Fly-By-Wire Technology

James E. Tomayko

The challenge was ours. For fly-by-wire to become real and to be accepted by the "world," the experiment had to be performed. Fly-by-wire had to work as good as any mechanical control system from the safety, performance, and maintenance standpoints. We could not afford any mistakes. (James Morris, Air Force Flight Dynamics Laboratory, on the Survivable Flight Control System Project that led to an implementation of a fly-by-wire control system.)

These words come from a man convinced his view of the world is the right one, and planning a clear, irrefutable demonstration of his belief. As the 1960s drew to a close, Jim Morris and other engineers at the Flight Dynamics Laboratory, Wright-Patterson AFB, had finished enough preliminary research to know that the concept of fly-by-wire flight controls was viable. The half-century-old paradigm of a statically stable airframe controlled by cables was about to shift dramatically to one of an unstable, control-configured vehicle with control inputs carried via electrical signals. Within ten years, the paradigm shift was so complete that every new aircraft obtained by the Air Force would use fly-by-wire technology as the basis of "active" flight control systems. Ten years more, many civilian transport aircraft also incorporated fly-by-wire technology. This revolution in flight control is a manifestation of the faith of Air Force researchers who were able to form effective industrial partnerships in bringing the technology to full flower. However, the positive aspects of the fly-by-wire success story are balanced by some criticism that the faith of the engineers was blind to fundamental changes in aircraft design and piloting that may have unexpected negative impact.

Incorporating fly-by-wire controls enables an aircraft designer to meld together what used to be separate subsystems in an airplane. A cartoon making the rounds in aircraft manufacturers depicts a series of views of an airplane by different subsystem engineers: the propulsion team sees the plane as huge engine nacelles and tiny wings and fuselage; the aerodynamicists see big wings and stabilizing surfaces, and so on. Fly-by-wire technology makes it possible to use the control system to overcome instability, to incorporate engine controls in the flight controls, and, in general, blur the walls between the subsystems so they can be more fully integrated. Probably the simplest example of this is how the F–16's flight control system automatically compensates for recoil when the aircraft's cannon is fired. The airframe designers could place the cannon off the center of the longitudinal axis (where, presumably, it is in a more convenient location for them and for the other subsystems) and not have to worry the pilot about keeping the nose pointed by coordinating rudder pressure. This is also the source of some of the criticism. An F–16 pilot is unaware of many things a pilot in an older technology aircraft must explicitly control. The F–16's throttle, weapons systems, and control surfaces are so closely coupled that thrust settings, flaperon deployment, and slat angles are automatically set depending on certain inputs. The sidestick controller uses a different translational paradigm: instead of movement of the hand controller indicating proportional movement of control surfaces, it is pressure on the hand controller that is the basis for the signal. There have been some crashes due to pilots not internalizing all the new ways a fly-by-wire system can "bite." However, some of the aircraft involved would not be flying at all with conventional controls. For instance, it could be argued that the F–117, B–2, and later stealthy aircraft would be impossible to fly without active control, since the physical requirements of stealth do not lend themselves to aerodynamically stable designs.

For better or worse, this rapid change in the control system paradigm is possible due to one of the most remarkable stories of government-led technology development and transition. For 20 years, the Air Force Flight Dynamics Laboratory conducted a step-by-step research program in concert with industrial partners to make fly-by-wire possible. Beginning in 1956, the engineers at the Laboratory sponsored and participated in a graduated series of basic and applied research projects that culminated in the adoption of active flight control on the F–16 in the mid-1970s.

The F–16 was the first operational fly-by-wire aircraft designed as such. The total direct investment in Air Force fly-by-wire research prior to its design in then-year dollars is slightly under $20 million, inexpensive considering the pervasive results. The speed of this revolution in flight control is a direct function of the persistence of a team of U.S. Air Force scientists and engineers, and a loosely related group of NASA researchers, working closely with industrial contractors.[1]

How it Works

Since fly-by-wire technology enables active control of aircraft, they can be unstable in one or more axes. There are resultant advantages in maneuverability and reduction of the weight of control surfaces — advantages for both military and civilian aircraft. There are additional advantages for military aircraft in terms of survivability and weapons delivery. At the simplest level, the mech-

The Northrop B–2 Spirit.

anical cables leading from control devices such as stick and rudder pedals are eliminated and replaced with sensors at the base of a control column and other sensors to keep track of aircraft attitude and acceleration. Inputs from the sensors are sent to a computer which then calculates the appropriate commands to actuators that will accomplish the pilot's desires. Since all control signals are carried by wires rather than steel cables, the technology came to be called fly-by-wire.[2]

There is actually a range of possible fly-by-wire implementations. The simplest may be called the "electric airplane." The first example of this was the Mistel combination aircraft developed by Junkers in World War II Germany. A Mistel consisted of a Ju 88 bomber airframe which was due for a complete overhaul. The crew section was removed and eventually replaced with a structure-piercing warhead. An Me 109 or a Fw 190 fighter would be mounted atop the twin-engined bomber to enable the pilot to fly the device to the target, aim it, and then disengage to return to base. The Ju 88 had no hydraulic boost on its control system, and the control surfaces were moved by solid rods, rather than the cable-and-pulley systems common in Allied aircraft. Therefore, it was relatively simple to link potentiometers attached to the fighter's controls to ones providing signals to electric motors that moved the control rods. Thus, the fighter's remote control of the bomber is a fly-by-wire system using only pilot inputs carried by electrical signals.[3] This sort of system was also used in the Mercury spacecraft, except that the electrical signals turned on and off the attitude control jets rather than starting and stopping electric motors.

The next level of difficulty is coupling pilot inputs with sensor data such as that provided by inertial measurement units and pitot-static systems. This sort of system can handle the feedback necessary to overcome nonstatically stable

Technology and the Air Force

The Mistel, a Ju 88 with an Me 109 attached on top.

designs, and also to provide robust autopilots. The A–4 (V–2) rocket,[4] the Gemini and Apollo spacecraft,[5] as well as early fly-by-wire fighters such as the F–18 are of this type.

The fullest implementation of fly-by-wire is in control-configured vehicles. The hardware is much the same as on the F–18, but the design from the start is made to take full advantage of active control (the F–18 is statically stable). The B–2 is such a beast. It is a long way from Germany in the 1940s to the B–2. The path was shortened by the Air Force through direct research programs and contractor encouragement. We pick up the story in the middle 1950s.

The Air Force Flight Control Research Organization

The Air Force center of basic flight research is Wright-Patterson Air Base, fittingly located in the hometown of the Wright Brothers: Dayton, Ohio. In the early 1950s, the Aircraft Laboratory there housed most of the engineers and staff working on flight control. During 1954, a reorganization assembled several varied subgroups under the title "Flight Control Activity." As this conglomeration proved successful, the Air Force decided to create a new Flight Control Laboratory that would have responsibility for working on all elements of control systems. This decision had considerable internal support. Outside consultants such as Charles Stark Draper, the guidance and control genius who later led the design of the Polaris missile guidance system and that of the Apollo spacecraft, also recommended the reorganization.

Col. John Martin took over as chief of the "Activity" in June of 1954, and began to collect the pieces of what became the "Laboratory." These included the

166

Stability and Control/Flying Qualities group from the Aerodynamics Branch, Manual Controls from the Mechanical Branch, Automatic Flight Control Systems from the Armament Laboratory, and Instruments and Displays from the Equipment Laboratory. When the Laboratory officially opened on January 16, 1955, about 170 persons were working in it.

The First Ten Years of Fly-by-Wire Research, 1955–1965

As the United States entered the 1990s, a single fighter aircraft program, the F–22, a single bomber program, the B–2, and a single transport aircraft, the C–17, appeared destined to be the only new Air Force planes to enter service by the turn of the century (or, in the case of the F–22, possibly slip into the next). In contrast, during the 1950s, the new fighters under development or in operational use included the F–100, F–101, F–102, F–104, F–105, F–106, and F–107; bombers constructed were the B–52, B–58, B–66, B–57, and XB–70; and transports and tankers were the C–124, C–130, C–133, and KC–135. Some of these planes were improvements on existing technology, such as the F–100/F–107 and F–102/F–106 pairs, but others had widely differing designs and technological demands: the F–101 was the first aircraft with a high horizontal stabilizer, the F–107 (never built in quantity) had its engine intake above the cockpit, the F–102/F–106 were the first Air Force operational delta wing aircraft with no horizontal stabilizer, the B–58 was the first supersonic bomber, and the XB–70's high-altitude/high-speed requirements inspired extensive innovations. These programs most often had their new control needs "fixed" by incremental extensions to existing systems. Most were stability augmentation systems of limited to extensive authority. At any rate, there was much pressure on the Flight Control Laboratory to "do something" to improve the solutions to the individual problems the Air Force encountered on each aircraft program.

During the 1956–60 period, an awareness existed of the potential of fly-by-wire to enhance performance and to make flight control systems more generic. There was some initial planning of possible approaches to development, with private companies rather than the government leading the way. Convair studied fly-by-wire for future interceptors during 1956–57, Honeywell explored application to supersonic aircraft, North American planned a system for the XB–70, and General Dynamics wanted to apply the technology to the F–111. All four companies concluded that fly-by-wire offered significant weight and volume savings and could also solve some of the problems encountered by mechanical approaches to controlling modern aircraft; but in each case, uncertainty about reliability prevented complete implementation. In the end, only the F–111 spoilers were built as fly-by-wire and then because there was no simpler solution to the problem of running control cables in a large swing-wing aircraft.[6]

As the nature of the enabling technologies became clearer and the specific problems in applying fly-by-wire more apparent, the Air Force let study contracts concentrating on exploring those technologies. From 1960 to 1965, the laboratory expended roughly $400,000 to do so. The projects included a wide range of explorations in analog and digital technology, redundancy management techniques, and planning of flight test programs to bridge the gap between laboratory and actual service. An example of the type of work done during this period is the "Research and Feasibility Study to Achieve Reliability in Automatic Flight Control Systems."[7] With Wright-Patterson's James Morris as project manager, a team of General Electric engineers concentrated on increasing the reliability of digital systems through "code redundancy." The recommendation of the team was that "majority logic" (in which there are replicated systems using voting to decide who is right) offered the simplest and most efficient form of redundancy management. Duplicated components and voting circuits are the heart of such a system. The concept is owed to the legendary mathematician John von Neumann, who described it in a paper in 1955.[8] All fly-by-wire systems to date use this method of increasing reliability.

There are two examples of how this concept is implemented. The first is to take the output of multiple flight computers (A, B, & C) and use a comparator/voter to examine them and send the mid-range value on to the actuators. This architecture is common when analog computers are used, as it also compensates somewhat for signal drift. Both the F–16 models using analog computers and the F–117 use this scheme. Its only real problem is that it contains a single point of failure, the voter. The second architecture is more closely that of "code redundancy:" it avoids single point failures, and is the method used on the Space Shuttle orbiter. Four processors (P1 through P4) are loaded with identical software. Every time there is an output, an input, or a context switch (a change from executing one software module to executing another), the processors pause and send a three-bit signal on an intercomputer communication bus. These messages are compared, and a processor which detects that another processor has either sent the wrong message or has not sent any message in the previous four milliseconds is voted "bad," and the detecting processor no longer "listens" to it. The idea is that the processors will most frequently fail one at a time, and if three processors are voting one way while a single processor votes the other, it has failed. In case of massive failure or sufficient confusion, the pilots of the Shuttle orbiter can switch to a lone backup processor, which was code limited to returning the spacecraft to earth. This scheme was first implemented during Phase II of the NASA fly-by-wire research program.[9]

The GE project also explored the use of neural networks. The researchers thought they would provide safety in the sense that the loss of a few of the many neurons in the net would not have a great effect. This work is nearly 30 years prior to the much expanded research on neural nets among computer scientists and an indication of the wide-ranging nature of these early Air Force projects.

The team at Wright-Patterson also received a windfall in the early 1960s in that Boeing installed the hardware for the fly-by-wire X–20 Dynasoar space-

An F–111 landing, with wings swept forward and spoilers extended.

craft in a simulator in the Flight Dynamics Laboratory. Vernon R. Schmitt led several years of work and experimentation, while the Pentagon debated the worth and fate of the X–20, a reusable, winged lifting body similar to the Space Shuttle but much smaller [it was cancelled before it could fly].[10] Ironically, all of the manned spacecraft built in the 1960s in the United States had fly-by-wire control systems. In general, these spacecraft were wingless, either blunt reentry vehicles made with ablative materials like missile warheads, or those designed solely for use in a vacuum, such as the Apollo lunar landing vehicle. Thus, the flight controls in the cockpit connected to reaction control jets, simply turning them on or off. This is a very direct use of electrical connections for flight control. The concept of continuous modeling of the control system by an analog or digital computer was not fully realized in all these vehicles. At any rate, concentration on the race to the moon diverted NASA's attention during the time when the Air Force increased its fly-by-wire research.

The coalescing project that laid the groundwork for inhouse practical fly-by-wire development was number 8225, "Study and Research on Fly-by-Wire Techniques." Led by Vernon R. Schmitt and Flight Leader J. P. Sutherland, a Canadian officer, the project was in the Control Elements Branch of the Flight Control Division. The tasks of the project were to survey other work on fly-by-wire done outside the Flight Dynamics Laboratory and to set down and evaluate various approaches to the enabling technologies. Sperry's Phoenix-based operation got the subtask award that led to a highly useful report by F. L. Miller and J. E. Emfinger published in July 1967.[11] It contained information gleaned from visits to engineers across the country that gave the Flight Dynamics staff a good picture of the current state of the art. It also compared existing limited-authority applications of fly-by-wire, such as the F–111, CH–46 helicopter, and the

B–52H. But Miller and Emfinger spent most effort on an indepth system description, including some limited breadboard and simulation work, that could lead to a full-authority three-axis control system.

Miller and Emfinger hoped to build a real version of such a system as a follow-on contract to install in an existing aircraft. One of their recommendations strongly states their case: "To overcome the lack of confidence [in fly-by-wire], an existing aircraft, particularly one with known control system problems, should be converted to fly-by-wire control and flown to demonstrate its feasibility."[12] The Flight Dynamics Laboratory was already working on identifying a suitable airplane for its first flying experiments.

The Fly-by-Wire B–47

By 1966, the Flight Dynamics Laboratory was in the fly-by-wire business as a participant, rather than simply an observer. The years of monitoring other people's work were over.[13] In order to increase active participation in research, Schmitt traveled to the site of a B–47 Stratojet crash near Plattsburgh, N.Y. and returned with the largely intact tail section. Thus, the engineers at Wright-Patterson had a convenient testbed with which to try out fly-by-wire in the pitch axis.

Eventually a flyable B–47 became available. It had at least one deficiency in flying qualities that ought to be helped by fly-by-wire, if the engineers were right. The Stratojet is a six-engine medium bomber with swept wings. Boeing built nearly 2,500 of the bombers in the 1950s, and all were being rapidly phased out by the middle of the 1960s. During the initial test program, it was discovered that the plane needed a yaw damper. Otherwise, it was neither a simple nor difficult plane to handle, but it did have a tendency to be somewhat slow in pitch response, which naturally became more apparent during low-level, high-speed flight. This occurred because of stretch of the very long cables that ran to the tail. Since there is no stretch in electrical signals, this problem could provide a limited demonstration of an improvement in flying qualities. This problem also fit the Flight Control Laboratory decision to progress in a graduated and deliberate manner: at first the aircraft would have a fly-by-wire system for the pitch axis only. It would retain the mechanical reversion capability. Only later would the system be expanded to other axes.

According to Schmitt, another reason for choosing the B–47 was that it had two pilots. One could use the test system while the other acted as safety pilot using the mechanical controls. A fatal crash occurred in an F–4 carrying only a single pilot the year before, and Schmitt "wasn't about to let this happen on any R&D program I had."[14]

On December 14, 1967, the modified bomber departed Wright-Patterson's active runway for the first time with the fly-by-wire system installed. It carried two pilots and a controls engineer as the third crew member. The engineer occupied the former navigator/bombardier station in the nose of the aircraft. This station was converted into an airborne electronics laboratory in which the engineer could actually adjust the gains of the signals to and from different parts of the system. The pilot had a simple "on-off" toggle switch installed on his main instrument panel next to the attitude indicator. When he flipped the switch, control was transferred from the cable system to the fly-by-wire system. During the first flights, control signals to the pitch axis fly-by-wire actuator were generated by a sensor attached to the control column. These signals routed through an analog computer located at the engineer's station. The computer had rotary switches that the engineer used to fine tune the commands to the actuators. The hydraulic system could also be controlled from this station, and a special "failure injection panel" could be used to cause one or another of the redundant channels to be cut off.

For Gavin Jenney and his colleagues at Hydraulic Research, Inc., that supplied the actuators and associated hydraulic systems, the fly-by-wire experiment centered on the electrical connections and feedback from sensors. The toggle switch that activated the system not only resulted in electrical signals, but hydraulic pressure switched from the mechanically activated system to the electrical one.[15] The game centered on whether the electrical signal resulted in sufficient hydraulic pressure to be applied to the control surface actuators. Jenney or another engineer could actually adjust the gains and monitor the signals on an oscilloscope while the airplane flew.[16] These electrically controlled actuators are a key element of all follow-on fly-by-wire systems.

Sensors also required some imaginative engineering. The experimenters placed the accelerometer at the pilot's seat for a "more realistic" feel. The phrase "fly by the seat of your pants" actually has truth in it — the sensors placed at the seat of the pilot's pants worked better as part of the feedback loop of the system.[17]

For forty-five hours, spread over an eight-month period, the B–47 flew test missions of the single-axis system. Many times the airplane would travel south to Kentucky and follow power lines over the hills there, a good test of responsiveness in high-speed, low-level flight.[18] The chief project pilot, Maj. Barron Fredericks III, reported the flying qualities in that regime much improved. In a technical report summarizing the results of Phase I of the test program, he said, "The handling qualities of the fly-by-wire system in the area of precision and dynamic response were better than the normal elevator control system."[19] Thus encouraged, the Flight Control Laboratory expanded the program.

Phase II of the B–47 fly-by-wire experiment added roll control and a side stick. This is exactly the capability and equipment of the Airbus A320 commercial aircraft that is the first civilian example of the technology (neither the A320 or the B–47 ever had active control in the yaw axis). The engineers also improved the sensor suite by adding pitch rate gyros and an accelerator in the nose.[20] The "home workshop" atmosphere of the project is characterized by the

171

The test engineer's station in the nose of the B–47 modified
for fly-by-wire testing. (Photo courtesy Gavin Jenney)

fact that the side stick controller was a $25 retreaded radar antenna controller.
The base of the stick contained potentiometers to measure pilot input instead of
the now prevalent linear variable differential transformers because the engineers
thought that potentiometers gave better resolution.[21] Unbeknownst to them, their
predecessors in Germany agreed: the Mistel flying bomb also used potentiometers. Most later systems used linear variable differential transformers, which
measured pressure rather than displacement.

Although Sperry built laboratory models of a three-axis system based on the
B–47 data, it never flew.[22] Instead, the emphasis in Phase III of the B–47 test
program remained on the actuators and hydraulics, and the engineers introduced
quad redundancy for this final phase.[23] By the end of the program, over 40 pilots
tried the fly-by-wire system in the B–47.[24] Their glowing reports encouraged
further exploration. Even before the project ended, the Laboratory tried to get
the word out by sponsoring a meeting of industry and government to exchange
information on the state of research in active flight control. This meeting turned
out to be particularly effective in technology transition.

The Fly-by-Wire Flight Control System Conference, 1968

On Monday and Tuesday, December 16–17, 1968 [the 65th anniversary of
powered flight in an unstable airplane: the Wright Flyer], the Flight Dynamics
Laboratory hosted a meeting of 141 people engaged in fly-by-wire research or
vitally interested in its future. The conference was a showcase for the year-old
B–47 test program and the laboratory prototypes built by Sperry Flight Systems

Division and Douglas Aircraft Company. It also gave attendees an opportunity to speculate about the nature of fly-by-wire systems in future aircraft.

The conference papers largely reported on work in progress. The hidden agenda was to create a demand for fly-by-wire so great that further research would be sponsored by the Flight Dynamics Laboratory and its industrial partners. The early results, though promising, still did not fully convince the money controllers in Washington. If the Laboratory personnel and contractors could sell the industry attendees on the idea, then pressure would be applied to the government for further support. As Col. Charles A. Scolatti, Chief of the Flight Control Division, said in the conclusion to his welcoming remarks, "I hope that this conference will provide you with reinforcement on the potential, soundness, and maturity of fly-by-wire flight control systems and open the doors which will permit you to consider fly-by-wire for flight control system tradeoff studies for our future aircraft and aerospace vehicles."[25] In short, the people at Wright-Patterson were sold, and now it was time to sell the others.

Maj. J. P. Sutherland began the technical part of the conference with an overview of what fly-by-wire really meant. He made the point that fly-by-wire is a significant paradigm change for both pilots and flight control system designers: from control of surfaces to controlling vehicle motion directly. Acknowledging the reluctance of both parties to abandon mechanical systems (he showed a cartoon of Snoopy shooting at the Red Baron while thinking "Security is a mechanical flight control system!"), he nevertheless went on to show that the B–58 and F–111 needed their electrically based stability augmentation systems to successfully accomplish their missions. The retention of the mechanical systems in those aircraft resulted in "many of the disadvantages of a mechanical system" in high-performance aircraft that would benefit by a change to fly-by-wire. In order to further assure the audience that the time had come for fly-by-wire, Sutherland wrapped up his presentation with a discussion of redundancy, showing how fly-by-wire systems could actually be made more dependable than mechanical systems with relatively few penalties.

The reliability question came up many times at the conference, as it was the chief stumbling block in many minds to the full adoption of the technology. It is no coincidence that much of the first decade of Air Force fly-by-wire research aimed at developing techniques for increasing reliability, rather than the basic hardware. The next paper in the conference concentrated on that theme.

The Sperry group presented following Sutherland. Jack Emfinger reviewed the progress of the three-axis laboratory model, including a detailed discussion of the methods of equalizing redundant signals and voting. His team used DC amplifiers in the computer, and Weston Hydraulics built a triple-redundant actuator (three push rods side-by-side) to demonstrate that part of the technology. Actuators are among the critical components of fly-by-wire, as they translate the control signals into control surface deflection. Sperry also had a control simulator specific to the B–47 that pilots could use to familiarize themselves with the system. The most important part of the presentation was the detailed data on reliability. Sperry, Weston, and Rome Air Development Center did calculations on the major components of each channel as a single unit, then created a failure equation for a triply redundant system. Ironically, some of the highest failure rates were on such well understood components as hydraulic power sources (750×10^{-6} probability) and rate gyros (56×10^{-6}), while the computational electronics fared better (48.2×10^{-6}). The triply redundant system raised the reliability rate to 2.698×10^{-9}. An "ultrareliable" system is often measured in values to 10^{-9}, so the Sperry team was getting quite close to achieving high-confidence reliability.

V. C. Sethre of Douglas Aircraft presented the results of yet another proto-type program to the conference. In some ways, the Air Force contract with Douglas seemed like a "back-up" to the Sperry contract, but it was not. Schmitt wrote the contract to encourage the "marriage" of an airframe manufacturer, such as Douglas, with a flight control/electronics supplier — a necessary prereq-uisite to technology transfer.[26] Beginning a year before Sperry, the Douglas team made better use of analog computers to simulate the entire system, and also achieved "particularly significant" results in demonstrating redundancy techniques. Douglas split the pilot's flight command signal into three, then transmitted them separately over three channels, and reconverted them by continuous voting into a single channel for the actuator.

If the attendees were not convinced by the two experimental approaches of Sperry and Douglas, the third paper attacked the doubters directly. Gavin Jenney presented the results of actual flight tests of the B–47, which had been in the flying phase for a year at that point. He detailed specific responses of the system in various flight regimes, and left little room for speculation that the fly-by-wire dream could not be realized.

Following Jenney was a paper by a McDonnell engineer named Fred M. Krachmalnick. He was the chief guidance and control mechanics engineer at the St. Louis plant that developed the F–4 Phantom II aircraft. McDonnell also built the fully fly-by-wire Gemini spacecraft there, thus some internal technology transfer from spacecraft to aircraft could happen (although there are no direct references to this). Krachmalnick reported on experiences developing control augmentation and stability augmentation systems for the F–4, both of which were flight tested. He also led the development of two fly-by-wire configu-rations, one with dual electrical supplies and the other with three independent electrical supplies. Otherwise they were identical, with quad-redundant com-puters, sensors and servos with triple-redundant hydraulic systems. This study directly led to a project that would serve as the focus of fly-by-wire activity at Wright Patterson for the next six years and would result in the first flight of a fully fly-by-wire aircraft.

The remainder of the conference had papers on a broad spectrum of topics: control of helicopters, control of large strategic aircraft, redundancy, a compar-

174

ison of analog and digital computers as the heart of a control system, and others. The conference ended with a panel discussion open to participants. Some of the key points that emerged from the interplay are that the true benefits of fly-by-wire could not be realized until an airplane would be designed from the beginning to use the technology. Otherwise, the savings in weight and size from limiting the need for large control surfaces would not happen. Also, rumors of pilot opposition to fly-by-wire were largely discounted. The attendees felt that the step-by-step research programs would increase pilot confidence to the point where there would be few questions.

In general, the conference was an eye-opening experience for all the participants. The enthusiastic participation of such a wide range of government and industry researchers demonstrated to the Flight Dynamics managers that any progress they could make in demonstrating the efficacy of fly-by-wire would be warmly welcomed by the community. As Major Sutherland's last slide said: "The era of Fly-By-Wire has come!" The Flight Dynamics Laboratory determined to make certain the era was a successful one.

The Survivable Flight Control System Project

During Major Sutherland's talk at the beginning of the fly-by-wire conference, he showed a slide with a cutaway diagram of an F–4C Phantom II tactical fighter revealing the installation of a two-fail-operational active control system. He publically announced that the next step in the Laboratory's program to demonstrate fly-by-wire would begin in mid-1969 with this aircraft modification. The impetus for the project came from two directions. On one hand, the laboratory sponsored the Tactical Weapons Delivery research program (known as TWeaD) to improve bombing accuracy. Using a jet fighter as a bombing platform is always problematic. High approach speeds and low-level flight work against the pilot having the leisure to set up properly for manual bombing. The TWeaD program is one of a series of research efforts that studied ways of applying avionics to assist the pilot. Part of TWeaD was determining the control laws of the F–4 in order to design some stability and control augmentation, these could then be directly applicable to design of the fly-by-wire modification.

The other immediate source of support for fly-by-wire was a study of the surprisingly heavy combat losses of the F–4 and F–105 tactical fighter-bombers. Nearly one-half of the more than 700 F–105s built were shot down in Vietnam. An Air Force study of fighters that survived damage revealed that the aircraft that did actually return to base had few or no serious hits to areas where control system cabling converged. The truth became obvious: those that did not return probably had serious damage to those areas. The simple fact is that even redundant control cables have inevitable points of vulnerability due to the fact that they must have proper clearances in the cramped internal fuselage and wings. For example, the loss of a Boeing 747 in Japan occurred because a hand grenade exploding in an aft toilet severed all three redundant control cabling leading to the tail control surfaces. Unlike steel cables, the electrical wiring needed for fly-by-wire could snake in even the narrowest runs in the aircraft. Thus, a fly-by-wire system could be designed to have no single points of failure by dispersing the electrical runs and even the control computers and sensors.

The Flight Dynamics Laboratory engineers used the combat survival argument to convince funders that the work on fly-by-wire could be useful. They thus avoided the direct approach of having to convince the Pentagon that fly-by-wire is better for reasons such as handling qualities or other such vague and subjective concepts. As part of the sales pitch, the engineers used a cartoon of Snoopy as the World War I flyer with three of four electrical cables shot off by the Red Baron, captioned "This fly-by-wire is great!" The message apparently took, as the "Survivable Flight Control System" project received precious funding. James Morris, a lead engineer on the project in the Flight Dynamics Laboratory, makes the point that this practical approach is the primary reason why the funding came through.[27] Gen. George Brown essentially "ordered" that the project succeed. Therefore, there was no room for anything that appeared even remotely "experimental." Morris thinks that this is the reason why the Air Force did not cooperate with a very similar NASA project underway at the same time on West Coast, even though the NASA program had basic objectives identical to the Air Force project. He acknowledges that the "sense of competitiveness" helped prod the Air Force engineers.[28]

Survivable Flight Control System Technology

Converting an F–4 to fly-by-wire while retaining the mechanical control capability required some innovation, but the study contracts previously awarded to McDonnell Douglas and reported in the conference laid the basis for the actual flight configuration. The aircraft modified in the program was the 266th F–4 built, originally intended to be a Navy F–4B, but renumbered as a YRF–4C, an Air Force prototype of the reconnaissance version of the F–4. It continued its career with the unique designation YF–4E, the testbed for the internal cannon and leading edge slats. After that it went on to be the only aircraft in the fly-by-wire programs. The first physical change in the F–4 most easily identifiable by the untrained eye was the addition of a side stick in the front and rear cockpits. The requirement to keep a partial mechanical system intact made retention of the center stick in the front cockpit necessary. However, there are other advantages to side sticks that were worth proving: they allow the pilot a better view of the instrument panel, promote a reclining posture that provides better distribution of G forces in high-speed maneuvers, and they let the pilot's arm rest.[29]

Out of sight inside the aircraft were the Sperry-built quad-redundant analog computers and sensor suite. McDonnell Douglas tested the control laws for the

F–4 using a CDC 6600 mainframe computer to simulate the sensor and pilot feedback.[30] The system compared each analog computer output to the response measured by rate gyros and accelerometers. The difference between the two was the control surface deflection.[31] It was possible to provide automatic trim, in which the pilot did not have to continually put pressure on the stick to maintain straight and level flight. In fact, even with a total failure of the stick and its transducers, the pilot could fly the plane using direct input to the trim system via thumbwheels used for manual settings.[32] This is not as unusual as it sounds. The reason why large German aircraft in World War II could make do without hydraulic boost at the control surfaces is that ingenious placement of trim tabs enabled the slipstream to act as "power steering."[33]

Retaining the mechanical control system, a safety measure, became a point of contention. Roll control was all fly-by-wire, but the pitch and yaw cabling was left in. On April 29, 1972, the modified F–4 took off from Lambert Field in St. Louis using the mechanical system, then switched to all fly-by-wire once the landing gear retracted. About a month later, the NASA test aircraft took off under all fly-by-wire. NASA felt that it was important for the technology to be tested in complete form to be convincing.[34] After the 27th flight, engineers permanently disabled the F–4's mechanical system, and the remaining 57 flights were all fly-by-wire.[35] A large part of the reason why the mechanical system could be taken out of the loop was increasing pilot confidence and pleasure with the fly-by-wire system. Charles P. Garrison, the McDonnell Douglas test pilot, reported that the control of the F–4 "noticeably improved" even on its first flight. The engineers tweaked the control laws to alter maneuver rates in the three axes in order to make the aircraft even more responsive. One immediate positive result was the near elimination of a pitch transient that used to occur during rapid deceleration from supersonic to subsonic flight.[36] These positive results whetted the appetite of both the Air Force and McDonnell Douglas to see if more gains could be had.

Introducing Instability

Both the Air Force's F–4 and the F–8 aircraft modified by NASA were statically stable. The greatest rewards for using fly-by-wire could only be gained by placing the control system in an unstable aircraft. The aircraft would be more maneuverable in at least one axis, and the control system would prove its true worth, since the unstable airframe would not be able to be flown by a human pilot otherwise. ARPA's X–29 (forward-swept wing) program demonstrated such results in a aircraft that was designed unstable. McDonnell Douglas merely added a set of canards to the F–4 as part of the "Precision Aircraft Control Technology" (PACT) program and made it possible to do so. The canards moved the neutral point forward and caused the longitudinal axis of the aircraft to be unstable when flying subsonically.[37]

RF–4 number 12200, after conversion to three-axis fly-by-wire (top), and the same aircraft in its PACT configuration, unstable in the pitch axis (bottom).

The canards, coupled with some other changes such as leading edge slats on the wings, made some significant performance gains possible. The four G maneuvering ceiling rose roughly 4,000–5,000 feet to 50,000 feet. The turn radius of the F–4 also improved.[38] Most spectacularly, the aircraft could point the fuselage (and thus its guns) without changing the direction of flight.[39] The stabilators and the canard were electronically "geared" to minimize drag, and could essentially act as additional lifting surfaces, making it possible to fly the aircraft at very high angles of attack. The utility of this in a fighter aircraft is obvious.

At the completion of the PACT program, the modified F–4 remained in St. Louis. Pilots from the Air Force, the Navy, the Marines, and NASA flew it during its over 100 missions. The one remaining service, the Army, finally got its chance. In late 1978, Morris heard from friends at McDonnell Douglas that the F–4 was about to be scrapped. In an effort to save a piece of aviation history, he was able to get the Air Force Museum to accept the plane as a donation. The problem was getting it from St. Louis to Dayton. He phoned the 272nd Transportation Company, Fort Sill, Oklahoma, and explained his problem. The commanding officer at the other end thought that moving the plane would be a great training exercise, so he dispatched a CH–54B Sikorsky Skycrane with a couple of other helicopters as escorts. On January 10, 1979, the Skycrane's pilot flew the F–4 to a soft landing on the ramp by the Museum, where it currently resides in the Annex, serving as a visual reminder of the challenge of building a fly-by-wire system met and conquered.

Fly-by-Wire in "Big Iron" Planes

A project working in parallel to the tactical aircraft effort expanded the data set to airplanes more like commercial airliners. Flying large transport aircraft is significantly different than flying lighter, higher power-to-weight ratio aircraft. In fact, the larger planes stood to benefit as much or more than fighters in using active control systems, with great implications for commercial transport aircraft. The B–47 project gave a taste of these improvements, but modifications to a Lockheed C–141 Starlifter more fully exploited the technology.

The C–141 is a durable, high-capacity, long-range transport. Nearly thirty years after its introduction, it was the mainstay of the U.S. portion of the buildup during Desert Shield and of resupply in Desert Storm, flying at high frequency and full loading throughout the operations in 1990–91. However, the manual control system on the C–141 has about a 20-degree "slop" in its controls along the roll axis.[40] Honeywell and the Air Force Flight Dynamics Laboratory combined to produce a fly-by-wire system in the pitch and roll axes of a C–141, a system very similar to that simultaneously being tested on the modified F–4. The control device was a side stick at the copilot's station.[41] During 22 flights in the August to October 1973 period, pilots found remarkable improvements in the big plane's handling. One pilot report read, "After two and one-half hours of flying (four approaches, all three fly-by-wire modes) I do not feel like I have been flying. The workload's that good."[42] The best effect of this is that pilots with limited large aircraft experience could compensate more quickly for the C–141's quirks, improving transition time and safety. These two factors are very important to the commercial air transport industry.

The Light Weight Fighter Program and the Legacy of the YF–4E

As the fly-by-wire test programs wound down, the Air Force began the Light Weight Fighter competition that led to the adoption of the General Dynamics F–16 Falcon by that service and the McDonnell Douglas/Northrop

F–18 by the Navy. As the first aircraft program begun after the success of the fly-by-wire technology demonstration projects, there was an understandable desire on the part of the Flight Dynamics Laboratory team to have the winning fighter adopt active flight control. A study by McDonnell of the YF–4E database concluded that the concept of a control configured vehicle must be designed in from the beginning in order to take advantage of minimum size and weight plus maximum performance.[43] This was the first opportunity to do so.

General Dynamics appeared to move the most quickly. Its Convair division had studied fly-by-wire as early as 1957 while grappling with the supersonic B–58 Hustler delta wing design.[44] In 1964, again faced with a leading edge design problem, this time the variable geometry F–111 Aardvark, the company used a stability augmentation system that was electrical in nature.[45] Thus they were ready nearly a decade later to plunge in more deeply. McDonnell Douglas had all the data it needed from the YF–4E program. In fact, one of the lead engineers on the YF–4E, Bob Kisslinger, next worked on his company's entry in the Light Weight Fighter Program. Jim Dabold of the Aeronautical Systems Division at Wright-Patterson made certain General Dynamics had the same information to encourage both competitors to adopt full fly-by-wire.[46] When General Dynamics chose to use relaxed static stability in the F–16 design, the die was cast in favor of fly-by-wire, and the aircraft won the competition.[47] Ironically, McDonnell Douglas' entry, the YF–17, a statically stable aircraft, was later modified to use a digital fly-by-wire system in its reincarnation as the F–18.

Transfer to the Civilian Sector

From the standpoint of a commercial user of fly-by-wire technology, the Air Force's emphasis on safety and reliability helped accelerate adoption. NASA's airplane first flew with a single digital computer and a triple-redundant analog backup. The NASA engineers knew that single-string systems would hardly be considered for civilian, or even military, operational use. The Phase II of the NASA program used three digital computers, mirroring the Air Force's triple analog system.

The technology was sufficiently mature for Boeing to use it in the YC–14 STOL cargo aircraft prototype in the mid-1970s. Boeing was able to demonstrate all the advantages of fly-by-wire in a medium sized transport. Why the company let Airbus be the first to use the technology in the commercial sector is unknown. Either the 757 or 767 aircraft could have been fly-by-wire. One speculation is that they wanted Airbus to go through the certification agony first. Nevertheless, Boeing is now a major player in active flight control, following Airbus as the European company applied its system to ever larger aircraft. The contribution of the Air Force's research program is indirect: by encouraging multiple companies to enter the fly-by-wire arena, sufficient experience built up to make conservative commercial aircraft builders amenable

The Boeing YC–14.

to using the technology. However, there has been some criticism of the implementations made to date.

Changing the Piloting Paradigm

Negative experiences with fly-by-wire appear to be centered on the changing control paradigm made possible by the technology. The Airbus Industrie A320 has suffered multiple crashes since it entered service. All of the crashes to date have been evaluated as pilot error, not due to malfunctioning of the fly-by-wire system, but rather failures of the pilots to fully internalize the interplay of integrated systems.

The Air Force has had at least two crashes that point out significant differences in how fly-by-wire aircraft can cause trouble relative to conventionally controlled planes. On April 20, 1982, Lockheed test pilot Bob Ridenauer sat in the cockpit of the first production F–117A stealth fighter on the runway at Groom Lake in Nevada. He had done his usual preflight checks, waggling the control surfaces, and working his way down the checklists. Cleared for takeoff, he applied power to the plane's twin engines and accelerated. When the F–117 reached rotational speed, Ridenauer eased the stick back, expecting the nose to come up. Instead, there was an immediate and rapid yawing motion, followed by an equally violent pitch up that flipped the aircraft onto its back, sliding tail first down the runway.

Ridenauer, though seriously injured, survived a crash caused by an error difficult to detect in fly-by-wire systems. Inputs from the pitch and yaw rate gyros were connected in reverse. When he signalled the flight control system to pitch up by moving his control stick back, the system interpreted the pitch motion as yaw, and tried to compensate by issuing a yaw command in the opposite direction. The resulting rapid yawing was sensed as a pitch action, and the control system thus sent the pitch compensation command that flipped the airplane onto its back.[48]

Why had this problem not shown up in the preflight checks? A fly-by-wire aircraft sitting on a runway is generating no sensor feedback. Moving the control surfaces actually has no result on the attitude or direction of the aircraft, as it is sitting still. Once in motion, the sensors have something to work with. Observers of the first taxi tests of Northrop's YF–23 advanced tactical fighter prototype noted rapid control surface deflections even when the aircraft hit bumps in the runway. These can be damped with software changes, but they point out the potential sensitivity of the control system.

In another example of an immature control system, the second YF–22 had a spectacular encounter with the ground in 1992. Despite nearly 20 years of digital flight control experiences, the YF–22 prototype aircraft showed how difficult it is to get everything right the first time in designing such a system. Just before the test aircraft was to be put to work in some ground tests, it was making a few farewell "photo runs." Video photographers from multiple angles captured the aircraft doing touch-and-go landings on a desert airstrip.

The video does not even have to be in slow motion to clearly show the horizontal stabilators "flapping" through large arcs as the YF–22 approaches the runway in a "dirty" (flaps extended, gear down) configuration. What appear to be extreme control surface movements are a result of the software switching the control system into a mode used for low-speed, landing configuration flight. When the gear and flaps are raised, and the aircraft increases speed, the software makes a transition to another mode that deflects the control surfaces in a more limited manner. This makes sense: higher speeds means that less deflection is needed to gain the same forces. The pilot is unaware of these movements and transitions, which is food for the critics of active control systems.

On the tape, it appears that the afterburner is being applied while the gear and flaps are still fully extended. On any airplane, when you add full power in this configuration, there is usually some immediate pitch response for which the pilot has to compensate. The active flight control system is supposed to be helping. As the pilot tried to adjust the angle of the nose, the software was still in a state that made the pitch response so great that the flight computers tried to limit it. The pilot responded with an opposite control input, and the airplane "porpoised" down the runway in ever increasing angles of pitch, until it finally hit the concrete with the gear up. The aircraft slid for hundreds of feet, and the friction of scraping the runway ignited some of the metal in the tail.

This incident demonstrates how difficult it is to adequately identify the requirements for the control system. The particular interaction of pilot and airplane in a certain configuration is near impossible for a software designer or systems analyst to imagine while sitting in a cubicle in front of a terminal. The nearly irreproducible set of circumstances is almost impossible to properly anticipate during the requirements phase of any real-time software development.

182

The YF–22.

The lesson is to use simulations and prototyping even more extensively than is done now.

Successful Promotion of Technology Transition

Even with the unfinished paradigm shift, fly-by-wire technology is a case of successful technology transition. Beginning with the new fighters, the middle of the 1970s were marked by the appearance of first generation operational fly-by-wire aircraft. The Air Force facilitated this revolution with an expenditure of less than $20 million: $400,000 for various exploratory projects in 1960–65, $900,000 for the B–47, $1.5 million for TWeaD, and $17 million for the YF–4E SFCS and PACT.[49] Many more millions had been spent in other parallel projects, such as NASA's, and even more would be spent in implementation, but rarely has a government-sponsored research program yielded such spectacular and cost effective results. It can be said that "breakthroughs" and "revolutions" can hardly be planned, but the consistent and seemingly well-coordinated approach taken by the Flight Dynamics Laboratory leads to the opposite conclusion.

The current pervasiveness of fly-by-wire in new designs is a justification of the faith of the Wright-Patterson flight control community. Their vision and persistence made possible a new era in flight. Airplanes have become more integrated as complete systems than ever before.

A Note on the Sources

The administrative archives of both Wright-Patterson and Edwards AFBs are, unfortunately, devoid of useful documentation on the various programs described in this paper. The story has been derived from technical reports and interviews of participants in Wright-Patterson research programs in the late 1960s and early 1970s. The paper has been reviewed for technical and general accuracy by James Morris and Vernon Schmitt of the Flight Control organization and by Gavin Jenney, now an independent contractor.

Based on the author's personal experience, the use of technical reports as primary sources results in very good accuracy relative to ideas, implementations, and even dates. However, the individual stories of the participants tend to be undeveloped using only those sources. I apologize to the participants if I did not do justice to their personalities. All those I encountered who are associated with the Air Force's fly-by-wire research were unfailingly helpful. They are intelligent men with great vision, and they served the Air Force and the United States well.

Notes

1. This article concentrates on the Air Force role in the development of the technology. For more on NASA's influence, see James E. Tomayko, "Digital Fly-By-Wire: A Case of Bidirectional Technology Transfer," *Aerospace Historian*, vol 33, no 1 (Mar 1986), pp 10–18.

2. A more complete overview of the technical characteristics of fly-by-wire is in James E. Tomayko, "The Airplane as Computer Peripheral," *American Heritage of Invention and Technology*, vol 7, no 3 (Winter 1992), pp 19–25.

3. Interview with Dr. Fritz Haber, Junkers Project Engineer for the Mistel, Bridgeport, Conn, Aug 17, 1995.

4. James E. Tomayko, "Helmut Hoelzer's Fully Electronic Analog Computer," *Annals of the History of Computing*, vol 7, no 3 (Jul 1985), pp 227–40.

5. James E. Tomayko, *Computers in Space Flight: the NASA Experience* (New York: Marcel Dekker) 1987. Also published as NASA Contractor Report 182505, Mar

1988. A simpler, revised explanation is in *Computers in Space: Journeys With NASA*, Alpha Books, 1994.

6. F. L. Miller and J. E. Emfinger, "Fly-By-Wire Techniques," AFFDL–TR–67–53, AF Flight Dynamics Lab, Research and Technology Div, Wright-Patterson AFB, Jul 1967, pp 44–45.

7. J. J. Fleck, *et al*, "Research and Feasibility Study to Achieve Reliability in Automatic Flight Control Systems," WADD Tech Rpt 61–264 (Mar 1961).

8. John von Neumann, "Fixed and Random Logical Patterns and the Problem of Reliability," American Psychiatric Association, Atlantic City, New Jersey, 1955.

9. Tomayko, "Digital Fly-By-Wire."

10. Gavin D. Jenney, Hydraulic Research, Inc, interview with the author, Sep 20, 1990.

11. Miller and Emfinger, "Fly-by-Wire Techniques."

12. *Ibid*, p 240.

13. James Morris, AF Flight Dynamics Lab, interview with the author, May 18,

1990.

14. Vernon Schmitt, AF Flight Dynamics Lab, letter to the author, Apr 12, 1993.

15. Gavin D. Jenney, "JB–47–E Fly-By-Wire Flight Test Program (Phase I)," AFFDL–TR–69–40, AF Flight Dynamics Lab, Wright-Patterson AFB, Sep 1969, p 92.

16. Jenney interview.

17. *Ibid.*

18. *Ibid.*

19. Jenney, "Fly-By-Wire Flight Test Program"

20. *Ibid.*

21. Jenney interview.

22. J. E. Emfinger, "A Prototype Fly-By-Wire Flight Control System," AFFDL–TR–69–9, AF Flight Dynamics Lab, Wright-Patterson AFB, Aug 1969, p 21.

23. Jenney, "Fly-By-Wire Flight Test Program," p 59.

24. Jenney interview.

25. This quotation, and the other information in this sect is taken from J. Sutherland, ed, "Proceedings of the Fly-By-Wire Flight Control System Conference," AFFDL–TR–69–58, AF Flight Dynamics Lab, Wright-Patterson AFB, Aug 1969.

26. Schmitt letter to the author.

27. Morris interview.

28. Tomayko, "Digital Fly-By-Wire."

29. G. H. Hunt, "The Evolution of Fly-By-Wire Control Techniques in the UK," in *11th International Council of the Aeronautical Sciences Congress*, Sep 10–16, 1978, p 70.

30. F. M. Krachmalnick, *et al*, "Survivable Flight Control System Active Control Development, Flight Test, and Application," presented at NATO Conference on the Impact of Active Control Technology in Airplane Design, Paris, Oct 14–17, 1974, p 8.

31. David S. Hooker, *et al*, "Survivable Flight Control System Interim Report #1," Tech Rpt AFFDL–TR–71–20 Air Force Flight Dynamics Lab, Wright-Patterson AFB, May 1971, p 84.

32. John H. Watson, "Fly By Wire Flight Control System Design Consideration for the F–16 Fighter Aircraft," in AIAA Guidance and Control Conference, San Diego, Calif, Aug 16–18, 1976, p 3.

33. Haber interview.

34. Interview with Dwain Deets, NASA Dryden Flight Research Center, Jul 5, 1990.

35. Kractmalnick, *et al*, "Survivable Flight Control System," p 4.

36. *Ibid*, p 10.

37. *Ibid*, p 16; Morris interview.

38. Kractmalnick, *et al*, "Survivable Flight Control System," pp 21–23.

39. G. J. Vetsch, R. J. Landy, and D. B. Schaefer, "Digital Multimode Fly-By-Wire Flight Control System Design and Simulation Evaluation," in AIAA Digital Avionics Systems Conference, Nov 2–4, 1977, p 203.

40. Jenney interview.

41. H. B. Larson, *et al*, "Military Transport (C–141) Fly-By-Wire Program, vol I: Control Law Development, System Design and Piloted Simulation Evaluation," AFFDL–TR–74–52, AF Flight Dynamics Lab, Wright-Patterson AFB, Apr 1974, p 1.

42. *Ibid*, p 126.

43. R. B. Jenney, F. M. Krachmalnick, and S. A. La Favor, "Air Superiority with Control Configured Fighters," *AIAA Journal of Aircraft*, vol 9, no 5 (May 1972), p 370.

44. Miller and Emfinger, "Fly-By-Wire Techniques," p 44.

45. B. R. A. Burns, "Fundamentals of Design VII: Flight Control Systems," in Air International, vol 18 (May 1980), p 232.

46. Morris interview.

47. Carl S. Droste and James E. Walker, "The General Dynamics Case Study on the F–16 Fly-By-Wire Flight Control System," in the AIAA Professional Study Series, p 18.

48. Bill Sweetman and James Goodall, *Lockheed F–117A: Operation and Development of the Stealth Fighter* (Osceola, Wisc: Motorbooks, 1990).

49. James W. Morris, "Background Fly-By-Wire Historical Information," AF Flight Dynamics Lab memo, Aug 11, 1989.

Frederick L. Frostic is the Deputy Assistant Secretary of Defense, Requirements and Plans. A 1963 graduate of the U.S. Air Force Academy, he earned a Master of Science and Engineering degree from the University of Michigan. He was an honor graduate at the Army Command and General Staff College, and a research fellow at the National War College. He served as an F–4C fighter pilot in the Air Force, flying 225 combat missions in Vietnam. Before assuming his present position, he was RAND's Associate Program Director for Force Structure. While on active duty, Mr. Frostic was Commander of the Northeast Air Defense Sector, Vice Commander and Deputy Commander for Operations of the 50th Fighter Wing, and a member and Deputy Director of the USAF Chief of Staff's Special Study Group. Earlier, he was Deputy Division Chief and Study Director, USAF Center for Studies and Analyses, and Assistant Professor of Engineering Sciences, USAF Academy. He has written several studies on the uses of tactical air power and technology for RAND and in defense and technical journals.

The New Calculus Revisited

Fredrick L. Frostic

This is an occasion I have looked forward to, for it is a night to talk about air power and technology. It is also an opportunity to reflect on "The New Calculus" and to give a historical perspective about what we did then and what we should be looking at four years later.

At the time, we concluded that the results of our analysis indicate that the calculus has changed and air power's ability to contribute to the joint battle has increased. This conclusion should be even strengthened today. Tonight we will investigate the history that was and the history that might be.

It is a fitting time, too. For I am convinced that all ideas need revisiting as capabilities, challenges, and conditions change. Now, like the time when we produced the New Calculus, the Earth is beginning to shift. I will talk about the shift later, but first we should look back at the conditions that launched our original enterprise and some of the background behind it all.

The conceiver of our enterprise was Gen. Michael P. C. Carns. Every enterprise should be blessed with a thoughtful sponsor like General Carns. We started in August 1991. The Base Force had been reintroduced and was defined as "the minimum force necessary for the nation's enduring needs." However, even as the Base Force was established as the basis for our strategy and force structure, General Carns was concerned that budget pressures would cause us to go below the Base Force level, and therefore, he wanted an analytical foundation for future forces.

Beyond General Carns' instincts about future budgets, it was also an appropriate time to look beyond the Base Force. The Cold War was over. Desert Storm had demonstrated the capability of our forces. Many technologies had become an integrated part of those forces. And the quiet internal revolution of the U.S. military in the decade of the 1980s had matured.

General Carns' instructions to us were quite broad. He wanted to know what constituted the minimum Air Force. That was both a simple question and a very complex one. For one cannot attempt to define the size of the Air Force without knowing what it is supposed to do and what the size and capability of the threats and the other services might be.

Because the question appeared so broad, we chose to take a very broad look at the issues and investigated well beyond the scope of the question. Fortunately, General Carns and RAND management gave us the time and latitude to analyze the problem. We started in earnest in September 1991 and completed the first round of our analysis by January 1992. The results were briefed to General Carns and the Air Staff in February 1992. He was quite interested in the results, but expressed two important reservations. First, he wanted to have the Air Force weigh the assessment quite carefully. Second, he was concerned about having a report sponsored by the Air Force which explicitly stated that the Nation's strategy could be supported by a force structure other than Gen. Colin Powell's Base Force. So we gathered up all of the nineteen copies of the original report and put them in a safe.

Then, the Air Staff sent a group of experts to work with us in evaluating the concepts and analyses contained in the work. At the time, this experience was not much fun, but it turned out to be a very valuable part of the overall effort. In retrospect, it is usually a good idea to have an outside observer provide an independent view of things. However, even the best advice from outsiders will always be a little bit off the mark. The result of this three-month exercise was that we sharpened our conclusions and aligned the basis of the analysis more closely with current Air Force programs.

By the fall of 1992, we published a classified draft of the analysis and briefed most of the Air Force decisionmakers. Nonetheless, not until June 1993 was the New Calculus approved for release. There were many interesting hurdles along the path, and by this time, the Bottom Up Review was well underway, but not complete. Of course, the numbers changed a little. But, the Air Force has accommodated itself and demonstrated great flexibility.

Now the "merry-go-round" has turned another revolution. We have gone from the Base Force to the BUR force. The planned, massive drawdown of the United States military is nearly complete. Also, the character of the post-Cold War world is pretty well understood. We see regional adversaries tottering, but challenges to stability remain. However, we are in a good position to adjust, especially because the accelerating pace of information technology offers great new possibilities.

The pace of change in technology will be an important factor for us in the future. Whereas technology in the past this was threat-driven, it can now be opportunity-driven. We have always thought we were in future shock, but now we are.

As an interesting sidelight, while I was thinking about this occasion I came across a December 1953 issue of *National Geographic* containing two articles: "Aviation Looks ahead on It's 50th Anniversary" and "Fact Finding for the Future." These articles predicted several things about developments in aviation. Some happened, others did not. It is interesting that the developments that came true did so at a much faster rate than the optimists predicted. Necessity drove some programs, while opportunity drove others. Some turned out not to be practical or possible or had good other alternatives.

The idea that strikes one as we look ahead at technology and the Air Force is that the Department of Defense and the Air Force have a lot of terrific plans underway that may not come true. But, that is all right, provided we understand the role and limits of these evaluations and provided that we are able to diverge smoothly from them as opportunities arise and use these plans as foundations from which to make coherent variations.

Now we need to look beyond the extrapolations to think a bit about what air power can be, for now we are reaching the point where the visions of the air power pioneers can come true. There is some hard evidence to support this assertion. The U.S. Army's "Left Hook" aside—Desert Storm showed that air power can be used to defeat an army in the field. Air power can be used to enforce the behavior of nations, for example, to establish a "no-fly" zone. It remains to be proven, but the decisive application of air power may be the instrumental factor in resolving the tragedy in Bosnia. And the rapid delivery of forces and supplies around the world has become a reality.

The New Calculus explored analytically concepts for major regional conflicts. Now we need to look beyond major regional conflicts to see how air power can be used as a primary force over the full spectrum of military operations that are likely to emerge in the twenty-first century.

What is new to make it all possible through air power? We need integrated planning, training, and rehearsals. Precision strikes, using both lethal and nonlethal tools, are available. We require the rapid massing of forces, air power's equivalent of maneuver warfare. And we need worldwide expeditionary forces, tailored to the particular situation.

Thus, air power can give us, as a nation, a host of new capabilities. When integrated with diplomatic and economic means, the real joint operations of the future will progress from deterrence to control and defeat of selected forces.

Much of this seems to be a blinding flash of the obvious. In the end we must know when to plan and act to make things happen. Of course, that means making tough choices among different courses of action.

Barton C. Hacker received his Ph.D. in history from the University of Chicago in 1968, the same year he published the first of over thirty articles to date. He has also published six books, the latest *Elements of Controversy: The Atomic Energy Commission and Radiation Safety in Nuclear Weapons Testing, 1947–1974* (California, 1994). Over a thirty-year career, he has held teaching posts at Chicago, Iowa State, and Oregon State and research positions with NASA, MIT, and the Department of Energy (DOE). In 1992 he became laboratory historian at the DOE Lawrence Livermore National Laboratory in California.

Nuclear-Powered Flight

Barton C. Hacker

Well, I now see what these people were talking about with the kleig lights staring you in the face. It is impossible to see the audience. It is a real pleasure to be here, and this paper is somewhat different, I think, than what we heard yesterday in that it focuses much more closely on the R&D process. In fact, it tends to skip over the relationship of the development of nuclear-powered flight to the Air Force.

There were three nuclear-powered flight programs in which the Air Force was joined with the Atomic Energy Commission and the National Aeronautics and Space Administration in various combinations over the years.

The idea of powering aircraft or missiles with nuclear reactors began immediately after actually World War II, well, at least the idea. There were, as I say, three major programs. One was nuclear-powered turbojets, the program known initially as NEPA, later as the Aircraft Nuclear Propulsion (ANP) program, which ran from 1951 to 1961. Another was the nuclear-powered rocket, which went under the name of Project Rover and from lasted 1955 to 1973. The one that is probably least well known was Project Pluto, the Nuclear Ramjet Program, which had a much shorter life, 1957 to 1964.

All three programs shared the key technical idea that a nuclear reactor could replace burning fuel as the source of heat for an internal combustion engine. Ultimately, all three programs shared the same fate: cancellation before ever being flight-tested.

Administratively linked though they were, each followed its own largely distinct course and failed for different reasons. I am talking about programs that did not succeed in the larger sense.

Unworkable technology may well account for the failure of ANP, the manned nuclear turbojet program, but technical shortcomings will not so easily explain what happened to Rover or Pluto.

That neither of those programs passed beyond R&D in fact seems puzzling to me, given their success in producing prototype flight-weight engines, to say nothing of high marks awarded both programs, not only by the people who worked on them, which may not be such a surprise, but also by Congress and the trade press.

Technology and the Air Force

They, in other words, had a considerable amount of support in the outside world. None of these programs have yet received adequate historical study, and obviously I am not going to remedy that situation today. But I do want to try to talk about how such a history might be framed. Why did R&D programs as successful as Rover and Pluto, in particular, appear to be meet the same fate as the obviously flawed ANP? Why did nuclear rockets and ramjets, if not nuclear turbojets, fail to achieve operational status or even to undergo flight testing?

It is these puzzles that I particularly hope to shed some light by exploring the interactions among technological innovation, institutional priorities, and bureaucratic politics. This essay really does center on the closely related Rover and Pluto projects, that is, the nuclear rocket and the nuclear ramjet.

Largely because those programs achieved their technical goals, the story really has to begin with aircraft nuclear propulsion, which did not. What made ANP look so promising was, of course, the potential endurance of several days, perhaps even weeks in the air without refueling. That was the appeal of nuclear-propelled bombers to the U.S. Air Force, and like earlier speakers, I will use Air Force generically. In 1946, the Air Force launched a new research program called Nuclear Energy Propulsion for Aircraft, NEPA.

The first step was a contract with the Fairchild Engine and Aircraft Company for a feasibility study. Fairchild's report highlighted three key technical problems. One, no available materials could endure the reactor's intense radiation. Two, shielding thick enough to protect crews could not easily remain light enough to fly. Three, potential operating hazards ranging from routine maintenance to crash landings might pose serious radiation risks, not only to those directly engaged but also to hapless bystanders.

Notwithstanding such unresolved problems, nuclear flight seemed feasible, and NEPA continued as an experimental program at Oak Ridge, the nuclear research laboratory in Tennessee. It was, in fact, the major program at the Oak Ridge National Laboratory well into the 1950s.

Military authorities were still doubtful, however, and sought additional scientific advice. With the help of the new Atomic Energy Commission, the AEC, they got it from Project Lexington, organized by the Massachusetts Institute of Technology for the summer of 1948.

Lexington, in fact, became the first in the series of influential summer studies. We heard about a few of the later ones yesterday from Mike Gorn. These studies brought together academic and scientific personnel with program officers in uninterrupted meetings over several months during the summer to brainstorm a specific specified military problem.

That was the pattern that obtained for many years. The Lexington panel, in its end of the summer report, much as Fairchild had done, judged some form of nuclear-powered flight feasible, despite the daunting technical problems.

Accepting the panel's heavily qualified recommendations to proceed with development, DOD in early 1951 ended NEPA and began ANP. The promise

192

of nuclear-powered flight, flight time measured in days instead of hours, outweighed all drawbacks.

All three nuclear engine programs, turbojet, rocket, and ramjet, overlapped administratively through a joint AEC-Air Force organization installed in AEC Headquarters, the Aircraft Nuclear Propulsion Office, ANPO. In mid-1952, the AEC and Air Force agreed on a single chief for both their programs. That was Gen. Donald J. Keirn, about whom we heard some discussion yesterday. The same Air Force officer heading the AEC headquarters office also held an Air Force staff position for nuclear systems development and directed work on aircraft reactors in the AEC Division of Reactor Development. Initially, the joint office managed only the nuclear turbojet R&D, that is, the ANP program. But when Rover and Pluto later became separate programs, nuclear rockets and ramjets also fell under the same administrative arrangements.

The Air Force and the AEC divided their responsibilities along nuclear lines. Non-nuclear development, that is, everything but reactor and shielding, belonged to the Air Force and its contractors. R&D contracts went to two industry teams, one that would put General Electric direct cycle engines, that is, engines that heated the air by passing it through the reactor in Convair airframes, the other for Pratt & Whitney indirect or close cycled powerplants, that is, that transferred heat to air by a sealed working fluid. Those were to go in Lockheed airframes.

Nuclear R&D was, by law, exclusively the AEC's charge. The Atomic Energy Act of 1946 had created the Atomic Energy Commission as the civilian agency to control both military and civil uses of atomic energy. And that mandate remained fully intact, despite some quite sweeping changes in the Atomic Energy Act of 1954.

Like the Air Force, the AEC relied on contractors to do the actual work—in this instance, its contract laboratory at Oak Ridge. The Congressional Joint Committee on Atomic Energy, another creation of the 1946 Act, strongly endorsed early flight testing, of what the Air Force defined in the 1955 weapon specification as a manned nuclear-powered bomber, subsonic in normal flight but capable of supersonic dashes.

By late 1956, however, budget cuts and still unresolved technical problems induced the Air Force to cancel the bomber program as such. Subsequent efforts over the next five years focused on a flyable powerplant. Developmental ups and downs lasted until 1961, when the Kennedy Administration finally canceled the program, leaving only a modest research effort directed at a workable small reactor.

After fifteen years and a billion dollars, as it was usually described, the nuclear-powered bomber dream foundered on just the problems that it had faced at the beginning, in what I have always regarded as a delicious piece of irony, the announcement of the cancellation of the program. It included a statement to the effect that all the problems had been solved, with the exception of three, and

The Convair NB–36H carrying an airborne nuclear reactor.

those three, of course, were the very three that the NEPA feasibility study in 1946 had pointed out: inadequate materials, excessive weight, and radiation hazards. Project Rover's rise coincided with ANP's loss of momentum in the mid-1950s. I do not know whether those were connected; I suspect not.

Speculation about nuclear rockets began in World War II, but enthusiasm for such vehicles tended to wane in a flurry of postwar paper studies that tended to show the potential advantages of a nuclear powerplant were largely offset by the difficulties of realizing one.

Interest revived in late 1954 when new calculations suggested that the performance of a nuclear rocket boosted to altitude by a chemical rocket might in fact have sharply improved performance. Suddenly, nuclear-powered intercontinental ballistic missiles able to carry very large warheads seemed feasible.

Although the name came somewhat later, this was the beginning of Project Rover. With AEC and Air Force backing, the AEC's two contract weapons laboratories, Los Alamos and Livermore, set to work. Both laboratories formed special units in 1955 to conduct systematic research into the basic physics and engineering of nuclear rocket engines.

Rocket thrust comes from hot gas rapidly expanding through a narrow nozzle. Chemical rockets burn propellant and oxidizer in a combustion chamber to produce the gas. Applying reactor heat directly to a propellant allows nuclear rockets not only to dispense with oxidizer and combustion chamber, but also theoretically to expel hotter gas of lower molecular weight than any product of chemical combustion. Of course, the higher the temperature and the lower the mass of exhaust gas, the larger the payload that can be thrown intercontinentally or lifted to orbit. On November 7, 1955, the AEC approved the R&D programs that Los Alamos and Livermore proposed. Four months later, in March of 1956, they became officially Project Rover.

194

The prospect of nuclear-propelled ICBMs had pretty much faded by 1956, as the issue was looked at more closely, but the idea of chemically boosting a nuclear stage still held appeal for space missions. After high-level review, DOD declined further support for nuclear-powered missile development, but did urge the AEC to focus on reactor research for rocket engines. A scaled-back Project Rover in January 1957 became purely a Los Alamos effort as Livermore switched to Project Pluto and nuclear ramjet research, to which I will return later. In March 1957, Los Alamos received the go-ahead for a series of Rover test reactors called Kiwi. Named after a flightless bird, for good reason, they were meant only to prove the basic concept in ground tests at the Nevada test site.

Area 400 on Jackass Flats housed the special test facilities that were constructed for Rover. Kiwi-A, the first reactor, was designed for 100 megawatts. It used gaseous hydrogen as a propellant, fed under pressure to and through a cylindrical reactor which was a model of compression, roughly four feet high and four feet across.

Though hardly flawless, the first full-powered test at the beginning of July 1959 succeeded spectacularly. Hydrogen heated in the reactor core shot hundreds of feet into the air, a flaming jet lit by methane torch as it left the upward-pointing nozzle.

Such setbacks as a hydrogen explosion during one test or a reactor core lost in another did not, in fact, reveal any basic problems, just normal test mishaps. A final full-power run in October 1960 concluded the Kiwi-A test series and had Project Rover off to a respectable start. Success, in fact, bred expansion.

Among the several former Air Force projects taken over by the new National Aeronautics and Space Administration in 1958 were non-nuclear engine and vehicle development for Rover. In 1960, NASA joined the AEC in forming the Space Nuclear Propulsion Office.

It was an uneasy union in some respects. The two agencies differed sharply about what adequate testing meant, for one example. Ultimately though, it was a fruitful relationship. The new office promptly launched two programs to exploit Kiwi's apparent success.

The Nuclear Engine for Rocket Vehicle Application (NERVA) would build on Kiwi technology to develop an actual space-going nuclear rocket engine. Contracts went to Aerojet General for engine development and to Westinghouse for reactors.

The second program, Reactor In-Flight Tests (RIFT), would prove the NERVA reactor under flight conditions. Nuclear rocket prospects brightened appreciably after May 1961 when President Kennedy committed the nation to Project Apollo, a lunar landing by the end of the decade.

Engineers could tailor NERVA specifically for an upper stage of Apollo's launch vehicle. All too soon, however, the foundations of this ambitious R&D structure showed cracks. Optimism bred of early success proved ill-founded.

Approximating a flyable reactor more closely, the higher powered Kiwi-B reactor series used liquid hydrogen in a redesigned core. The new design proved much more troublesome than the old. It was delayed until December 1961—the first test, that is—and it went very roughly, literally very roughly.

Vibration during firing and other problems that proved very difficult to solve prolonged the Kiwi-B testing program into 1964. Delays like this dimmed Rover's once-bright promise and imposed major program changes.

Reactor development after Kiwi proceeded along two paths: one, a series of research units and the other centered on NERVA. Trials of the first NERVA reactor, which was designated NRX, nuclear reactor experimental, began in 1964 and ended in December 1967 with the design goals achieved. In 1969, a prototype flight engine, Experimental Engine No. 1 (XE-Prime), ran at full power, but by then, it was really too late. NASA suspended production of the Saturn V launch vehicle in 1969, leaving NERVA without a booster.

Implicitly, at least, the United States abandoned the proposed manned mission to Mars as the next step after Project Apollo's lunar landings. That, in essence, eliminated the need for NERVA.

The nuclear rocket program formally ended in January 1973, a technical success according to one recent survey, obviated by changing national priorities. Star Wars, the controversial space-based system of defense against ballistic missiles—officially SDI — briefly revived prospects for nuclear-powered rockets in the late 1980s.

In concert with the Department of Energy, the AEC's present-day successor, and its contractors, DOD initiated a secret project to help solve, or at least ease, SDI's pressing logistics problem. It was code named Timber Wind. Its goal was developing a nuclear-powered booster for very large earth-orbit payloads.

Timber Wind, of course, was a far cry from the NERVA upper stage, although it did bring back thoughts about the immediate postwar thoughts of nuclear-powered ICBMs. Opening the way for NERVA technology much more directly was President Bush's Space Exploration Initiative of 1989, which of course brought Mars back into the picture. It also brought NASA into public partnership with DOD and the Department of Energy. When Mars once again became a potential goal for manned missions, a nuclear upper stage for the launch vehicle seemed just as necessary as it had two decades earlier.

Jack rabbits and tumbleweeds no longer had the former nuclear rocket development station at the Nevada test site to themselves, but, as it turned out, only for the time being. A Los Alamos team assessed the current value of the old test facilities that were still standing. At the same time, NASA dusted off its old data and made "proven NERVA technology" a program catch-phrase. But alas, all for nought: the Mars mission remains only a dream and SDI has vanished, or at least lowered its sights and changed its name. The chief legacy of this program probably resides more in promoting space-borne nuclear

196

reactors, not as engines, but as power sources for prolonged unmanned missions.

When Los Alamos took sole charge of Rover early in 1957, Livermore got Pluto, the nuclear ramjet. That story began late in 1955 when the Air Force first approached the AEC about nuclear ramjets. Chemical ramjets, of course, had excited military interests since the end of World War II with their promise of mechanical simplicity and Mach III speed.

Virtually without moving parts, ramjets—flying stovepipes, as some called them—needed only three major elements: a diffuser at the head of the pipe to slow the incoming air and produce ram, or stagnation pressure; a heat exchanger midway down the pipe where burning fuel raised air temperature still higher (like rockets, of course, the higher the better); and three, a nozzle at the tail end where rapidly expanding exhaust gas provided thrust.

Because a ramjet can achieve ram pressure only at supersonic speed, it also needs a booster of some kind to carry it from launch to self-sustained flight. Easier said than done. But unresolved problems did not prevent several ramjet-powered missiles from seeing limited service.

In a nuclear ramjet, reactor heat would simply replace combustion. Livermore's job in Project Pluto was showing whether or not a flyable reactor could be built. If it could, the Air Force would have within reach a powerplant for its proposed supersonic low altitude missile (SLAM).

A Southern California aerospace firm, Chance Vought, got the Air Force contract for the missile. A subcontract for the ramjet engine, which would incorporate the Livermore-designed reactor, went to Markhart, another Southern California company. Despite the name—and this again returns to some discussion we heard yesterday—SLAM was really less a missile than an unmanned bomber. Instead of a warhead, it was intended to carry several thermonuclear gravity bombs to drop on programmed targets.

Pluto, unlike Rover, always remained an Air Force-oriented program, though the AEC provided most of the reactor funding. The AEC Division of Reactor Development assigned Livermore as its contractor to design, build, and test the experimental reactors for Pluto. The reactors formed a single series called Tory 2. Tory 1, I will mention incidentally, was the canceled reactor that had previously been planned when Livermore was working on Rover in the earlier 1950s.

Whether installed in ramjet or rocket, the reactor's job was essentially the same: heating a propellant. The major difference, that rockets carried their own propellant while ramjets drew theirs from the air, required engineering solutions, not fundamental changes. A year and a half of engineering research for Rover thus gave Livermore a long head start on Pluto, but formidable challenges remained. Nuclear ramjets push the limits of current technology in several areas, notably high-temperature materials, nuclear physics measurement, and reactor engineering design.

Technology and the Air Force

Eventually, Livermore designed and built two Pluto test reactors, Tory 2–A and Tory 2–C. The 2–A version could not pretend to the role of ramjet power-plant with its undersized core, overweight reflectors, low power, unstabilized fuel, and external controls. It could, nonetheless, provide data and materials, physics and engineering for two vital purposes: first, to confirm the findings of detailed studies of each reactor component, and second, to ensure that the system as a whole performed as the sum of its parts and not in some surprising new ways. More broadly, it could also demonstrate the feasibility of a nuclear ramjet. Tory 2–C, in contrast, offered a realistic ramjet engine design, though lack of approved SLAM specifications meant a good deal of informed guess-work. Livermore shipped Tory 2–A to the Nevada test site in November 1960 after preliminary tests in Livermore. Pluto testing at the Nevada test site took place in Area 401, adjacent to the Area 400 Rover facilities on Jackass Flats.

Because ramjets function only at very high speeds, ground testing posed unique challenges. The static test unit required vast amounts of high-pressure air. Simulating Mach III flight conditions required an air system able to deliver roughly a ton of air per second at upwards of 1,000 degrees Fahrenheit and almost 600 pounds per square inch to the reactor test unit's intake. The air storage farm held 1.2 million pounds of air at 3,600 psi, enough for a five-minute test run at full power.

Important data also came from taking the highly radioactive reactor apart for study after testing, but disassembly required a special-purpose heavily shielded building fitted with remote controls. A battery-powered rail car, almost remotely controlled, carried the reactor over the two-mile track between test pad and disassembly building. Developing the static test facility, in fact, ranked with developing the reactor itself as one of Project Pluto's major accomplishments.

Tory 2–A's first run took place in mid May 1961. It lasted 45 seconds at 40 megawatts, roughly 25 percent of maximum power, equivalent to about 2,000 pounds of thrust. For once, the AEC's public utterances matched the internal assessment. A veteran of Nevada testing described it as, "the most efficient and smoothest operation I have ever been connected with." Upgrading test facilities for full-powered tests took several months, but in September and October 1961, three tests in rapid succession at 150 megawatts full power made Tory 2–A a resounding success. Not only did these tests amply confirm the earlier decision to skip an intermediate Tory 2–B stage, they also justified canceling the planned test of a second Tory 2–A reactor.

The technical doability was now beyond dispute, as one of the engineers said. Livermore could now proceed directly to 2–C, the prototype flight engine under study since early 1960. Whether or not the program would proceed though remained a question mark, despite what appeared to be an exemplary R&D program capped by nearly flawless ground tests.

Although the Air Force approved the next phase, development of a flight-rated engine, further support for full-scale ramjet development and flight testing remained uncertain. The final decision belonged not to the Air Force, but to Department of Defense, specifically, the director of defense research and engineering.

In late 1962, then DDR&E Harold Brown, ironically a former director of the Livermore Laboratory, decided against full development though he did endorse continuing work on Tory 2–C and some advanced research.

Livermore completed Tory 2–C in 1963, but unfinished work on the test facility in Nevada delayed shipment until February 1964. Checkout went smoothly; so did testing: an intermediate power run on May 12, followed by a full-powered test on May 20th.

Like Tory 2–A, three years before, Tory 2–C passed its first trials with flying colors. But unlike Tory 2–A, it had no successor. On July 1st, just six weeks after Tory 2–C's flawless full-powered test run, the Pentagon officially canceled both Pluto and what had now become LASV, low altitude supersonic vehicle, the appropriately renamed SLAM program.

Cancellation so sudden caught Livermore by surprise, though the laboratory had long known of Pluto's dimming prospects. Cutbacks had followed the 1962 defense decision against full-scale development and chances for even a modest flight test program continued to decline.

There was no real warning, however, of so abrupt a shutoff of all activity except mothballing. Livermore would not even be allowed to complete Tory 2–C tests, which left Livermore managers scrambling to cope with several hundred displaced workers, most of whom did find other jobs with the laboratory.

So why did DOD cancel Pluto so abruptly? By the end of fiscal year 1962, Pluto had cost approximately $133 million. Because the program to that point centered on reactor development, the AEC had footed over three-quarters of the bill. Another year or two at most of ground tests at similar expense, that is, about $20 million a year, again still chiefly AEC money, would almost certainly produce a flyable nuclear ramjet.

Flight testing, the logical next step, would raise the ante sharply. Estimates ranged up to $500 million through 1969, with the AEC moving to the sidelines and the Air Force forced to pick up a much larger share of the cost. And still to come, of course, would be an even higher tab for an operational system, several billion dollars at least. But just when such crucial funding decisions were coming due, the much delayed and sometimes troubled development of rocket-propelled strategic missiles at last began to produce results.

Having proved themselves in costly flight test programs, the intermediate-range Polaris became operational in 1960 and the long-range Atlas in 1962, soon to be joined by Titan II. Chemical rockets had ensured themselves a major share of the deterrent role.

Among the losers had been long-range jet-propelled missiles. The usual mental category for SLAM or LASV, despite its novel powerplant, again a subject about which we heard some yesterday. For the most part, SLAM seemed to have been perceived as the latest in a long line of abortive or ineffective air-breathing cruise missiles, from the First World War's short-range aerial torpedo, through the medium-range German V–1 of World War II, to the postwar Navaho missile designed for intercontinental range.

After ten years of R&D, Navaho flight tests began in 1956 and produced a string of failures that led to the program's abrupt cancellation in mid-1957. In retrospect, the timing could not have been worse for Project Pluto, which was just then beginning development of a nuclear reactor for the ramjet-powered SLAM. Nuclear-powered or not, SLAM looked like Navaho's successor in a failure-prone tradition. SLAM, LASV, and cruise missiles in general clearly lacked a constituency within the Department of Defense bureaucracies to match supporters of manned bombers or ballistic missiles by then.

The Pentagon, under Secretary of Defense Robert McNamara, required any new weapon system to meet an explicit mission requirement. No such mission requirement ever existed for SLAM or LASV, and that became the constant refrain during the early 1960s as defense officials resisted pressure from some Air Force and Navy elements and from Congress, especially the Joint Committee on Atomic Energy, which was much impressed by an R&D effort, as Pluto appeared to be. Speaking for the Department of Defense in 1962 hearings, the Secretary of the Air Force explained that DOD was willing to make the technological bet that nuclear ramjets were merely desirable, not critical.

Military funds accounted for a far larger share of the budget for nuclear turbojet development, that is, ANP, than it ever did for rocket or ramjet. Although Projects Rover and Pluto enjoyed significant military funding, most of the money nonetheless came from Pentagon sources.

Even if the AEC and NASA may have displayed more than a trace of military coloration, they were still ostensibly civilian agencies. In effect, they split nuclear rocket costs. While the AEC alone furnished the bulk of nuclear ramjet funds, by and large the lure of exciting new technologies and the internal dynamics of technological development drove the programs. Unlike ANP, which was largely developed and defended in response to some exposed external threat from Soviet Union, the nuclear rocket, of course, became largely a civil program after it was transferred to NASA.

Internal factors likewise account chiefly for the fate of the program. Since nuclear rocket and nuclear ramjet achieved their major design goals and were successfully tested in flight-weight prototypes, technological shortcomings clearly were not responsible for their failure. Delays in the programs probably account for some of it. For Pluto, in particular, timing was critical, given the very narrow window of opportunity afforded by problems in the ICBM development.

I might say that radiological safety problems, even with unmanned missiles, may well have blocked nuclear-powered flight of any kind without regard to other obstacles, although it probably would have taken longer to derail them. Flight testing also presented a problem because they could not have been tested over the continental United States. And testing at the Pacific Missile Range, that stretch of ocean between California and the Marshall Islands, would have meant the introduction of radioactive particles into the atmosphere, something that was studied and appeared not to be a serious problem.

But the Test Ban Treaty was even potentially a larger obstacle. Although testing these vehicles would not have caused violation of the treaty technically, in fact, there were potential problems in a test program that could have resulted in a treaty violation. A test vehicle exploding in flight, for example, would put the radioactive products of an explosion into the atmosphere outside the continental boundaries, which would have been a technical violation of the treaty and a source of discomfort, if not of anything worse.

As events transpired, institutional priorities and bureaucratic politics best explained the demise of Pluto and Rover. This point, I think, deserves real emphasis. Intrinsic technical flaws could not easily kill the nuclear turbojet, nor could the demonstration of a working technology save either rocket or ramjet. Technological characteristics and test results determine only in part the success or failure of R&D programs and not necessarily the largest part.

David C. Aronstein, currently an aerospace engineer at ANSER, has worked in weapons systems acquisition, aircraft design, and experimental fluid dynamics research. Previously, he worked at Boeing on the 777 and the High Speed Commercial Transport projects, including work on supersonic aerodynamics and low speed, high-lift system development. He has a Ph.D. in Aeronautics and Astronautics from the University of Washington, an M.S. from Stanford University, and a B.S. from Princeton University.

Albert C. Piccirillo is a principal aerospace engineer at ANSER. A retired USAF colonel, he has over thirty years experience in R&D, flight test and evaluation, and program management. He managed USAF's ATF program, heading the source selection for the YF–22 and YF–23. He headed a division in the Air Force Center for Studies and Analyses that analyzed advanced aircraft, weapons, and electronic warfare concepts. He has an M.S. in aerospace engineering from AFIT and a B.S. from Pennsylvania State University.

The F–16 Lightweight Fighter:
A Case Study in Technology Transition

David C. Aronstein and Albert C. Piccirillo

Introduction

The F–16 represented a major milestone in fighter aircraft development. It resulted from a new way of thinking about fighter aircraft design, combined with a set of technological advances in propulsion, flight control, crew systems, and aerodynamics that set the trend for fighter development for the next two decades. The F–16 emerged from the Air Force's Lightweight Fighter (LWF) Program, an innovative experimental prototyping effort that took place between 1972 and 1975. The LWF program was noteworthy for its rapid execution, innovative management strategies, and successful approach to technology transition. The purpose of this report is to document those aspects of the LWF program, focusing on the technological achievements embodied in the F–16.

The Environment

The ideas that led to the LWF program began to form in the mid-to-late 1960s, driven to a large extent by experiences in the Vietnam War. At that time, the Air Force lacked a dedicated air-superiority fighter with high agility for close-in combat. This was the result of an earlier perception that classical air combat would be replaced by the use of long-range guided missiles. Thus, the highly maneuverable air-to-air fighter had given way to more sophisticated (and heavier) multirole aircraft, some of which did not even have internal gun armament.

These heavier aircraft were seen to be at a disadvantage when faced with numbers of small, agile adversaries. The North Vietnamese MiG–17s and MiG–21s did not need long range and could choose when and how to fight because they were always close to home. Thus, they could hit and run or, at times, force the kind of close-in combat that the larger U.S. aircraft were not designed for. As a result, air-to-air kill ratios were significantly lower than expected, and at times even approached 1:1. The eventual overall average was approximately 3:1. This was disappointing in view of the perceived technological superiority — and much higher unit cost — of the U.S. aircraft.

In response, the Air Force began to formulate concepts for an uncompromised fighter, called the FX (Fighter Experimental), around 1965. There were even studies of a high-low force mix at that time. The appearance of the swing-wing MiG–23 and the Mach 3 MiG–25 at the Moscow Air Show in July 1967 focused attention on the high end of the fighter spectrum. The need to counter these threats required the FX (that became the F–15) to be capable of combat at a high Mach number at high altitude and to be equipped with long-range missiles and radar. Consequently, the F–15 was rather large and expensive, although still possessing excellent agility for dogfighting. As of 1970, no action had been taken to fill the low end of a high-low force mix. So there still appeared to be a need for a light, agile fighter that could be procured in large numbers, complementing a smaller number of the higher capability F–15s. There was beginning to be a desire to see just what form such a lightweight fighter might take.

Highly Maneuverable Lightweight Fighter

What kind of technology was needed to achieve the desired lightweight fighter capability and could it be implemented without excessive cost or complexity? A concept that surfaced in the late 1960s was the Energy Maneuverability Theory, developed by Maj. John Boyd, an experienced fighter pilot, and Thomas Christie, an Air Force mathematician. Their central concept was that the state of a maneuvering aircraft can be expressed as its total energy, which is the sum of its kinetic energy (due to speed) and potential energy (due to altitude). An aircraft that possessed higher energy would have more options available than one with low energy; so energy could be equated with combat agility. Any time the maximum thrust exceeded the drag, an aircraft could add to its energy; any time the drag exceeded the available thrust, the aircraft would lose energy. Thus, the ability to maneuver without losing energy required low drag at high-lift maneuvering conditions and high excess thrust (beyond what was needed for level flight). The most fundamental design characteristics that contributed to these qualities would be low wing loading (W/S) and a high thrust-to-weight ratio (T/W). The importance of these characteristics was not really a new discovery, but the Energy Maneuverability Theory provided a formal articulation for the idea and was an elegant methodology for quantitatively evaluating aircraft maneuvering performance.

Wing loading could be reduced without any new technology by sizing the wing larger than had been the practice on recent fighters that were actually optimized for interceptor or nuclear strike roles. Further benefits in drag at high lift could be achieved by tailoring the aerodynamic configuration to maneuvering conditions and not just to high-speed dash. This meant high maximum lift coefficients, low drag at high lift coefficients, and good handling qualities at high lift coefficients (no loss of control, and no deep stall or spin tendencies). Some existing fighters, most notably the Northrop F–5, especially the E version, embodied these aerodynamic characteristics in a limited way. All contemporary operational fighters, however, lacked the high thrust-to-weight ratio that was also needed for superior energy maneuverability.

Achieving a higher thrust-to-weight ratio required advances in jet engine technology. Experience showed that the takeoff weight of a fighter, when sized to carry a reasonable amount of fuel and armament, would be around 8 times the weight of its engine. Then current operational engines had thrust-to-weight ratios of under 6:1, which meant that the upper limit on aircraft thrust-to-weight was about 0.75:1. For example, the General Electric J79 used in the F–104 and F–4 had a thrust-to-weight ratio of 4.7:1, while the SNECMA Atar-9 that powered Mirage fighters at that time had a thrust-to-weight ratio of 5.6:1. To achieve an aircraft thrust-to-weight ratio approaching 1:1 at take off weight (or around 1.2:1 at combat weight) would require an engine with a thrust-to-weight ratio of 8:1. Such engines were becoming available. Pratt & Whitney and General Electric had both built demonstrator engines as part of the FX program with thrust-to-weight ratios in the neighborhood of 8:1. Thus the potential existed by 1970 to achieve energy maneuverability levels far superior to those of contemporary operational fighter aircraft.

The F–15 was the first fighter design to utilize the new high-technology engines to achieve a thrust-to-weight ratio greater than 1:1. This, in combination with a much lower wing loading than other contemporary fighters, gave the F–15 exceptional energy maneuverability. Features such as a conical-camber wing and a large, low-mounted horizontal tail provided superior handling qualities. There was no fundamental barrier to implementing these same ideas in a smaller and lighter design. Other technological advances were also envisioned for air combat fighters of the future, and there was interest in seeing which of these could be applied without exceeding reasonable levels of cost or complexity.

Composite construction using advanced materials was seen as a way to achieve further improvements in T/W and W/S by allowing the airframe structure to be lighter. Studies conducted around the time of the LWF program indicated that 15% to 25% reduction in gross weight and comparable savings in fly-away cost could be achieved by making extensive use of composites.[1]

Control configured vehicle (CCV) was becoming a popular phrase, and aerospace companies cited all kinds of benefits that would result from CCV technology. CCV concepts basically fall into three categories: relaxed static stability, maneuver load control, and novel control modes.

Relaxed static stability (aft center of gravity) reduces trim drag and increases usable lift by reducing the downward force needed at the tail to trim the aircraft. Cruise and maneuver performance benefits would result. An aft

205

The large, low-mounted horizontal tail of the McDonnell Douglas
F–15 Eagle is clearly visible in this view of the aircraft.

center of gravity makes an aircraft unstable in the pitch axis, so artificial
stability augmentation is required.

Maneuver load control means optimizing the design or providing the design
with the ability to be optimized via control surfaces for a maneuver condition
instead of a steady flight condition. An example applicable to bombers or transports is active wing bending load alleviation, by using automated control surfaces to modify the spanwise lift distribution. However, maneuver load control
is not always that exotic, and any design feature that enhances maneuverability
may be considered to fall into this category.

Novel control modes consist primarily of direct-force flight controls. Using
direct-lift or direct-side-force control surfaces, it would be possible to change
the flight path without changing the aircraft's orientation, or conversely, to
point the nose without changing the aircraft's flight path, resulting in offensive
and defensive advantages. Fly-by-wire control is not in itself a CCV concept,
but is the primary technology that enables the CCV concept of relaxed stability
to be implemented. It was necessary to define a high-g cockpit in which a pilot
could fully exploit the superior agility of an advanced fighter. Centrifuge experiments conducted by the Air Force indicated that increased seat reclining
angles and raised heel positions could increase pilot g tolerance by up to 1.5 g
and pilot tracking capabilities at high g levels by 30% to 55%. Simpler control
switch arrangements and improved field of vision for the pilot were also desired.

All of these technologies contributed to the vision of a small, agile fighting vehicle in which the machine and the pilot would function efficiently to their full combat potential. Basic research had already been conducted in most of these areas, and the next step was demonstration of an integrated set of technologies in a representative fighter configuration.

The political climate was amenable to exploring options for smaller and simpler fighter aircraft. The emerging F-14 and F-15 programs were being criticized for fostering solutions that some considered to be overly large and sophisticated. The general perception was that the entire process of requirements definition, program management, service testing, and technology development was not properly coordinated nor was it focused on achieving affordable solutions to realistic operational needs. Instead, DOD and the services were believed to be adding unnecessary gold plating to most new systems. The entire defense acquisition establishment was under severe attack in Congress and the press partly as a result of serious cost overruns, schedule delays, and technical issues in a number of programs including the F-14, F-15, F-111, and C-5.

The Total Package Procurement approach, under which full-scale development and production commitments were made on the basis of conceptual design competitions, without any hardware validation, was identified as a major culprit for two major reasons. First, the paperwork involved in this approach was enormous. Because source-selection decisions were based on paper proposals alone, the requirements had to be spelled out in great detail. Contractors had to produce vast quantities of documentation to prove that the proposed system would satisfy all of the requirements and relevant military specifications, then the government had to evaluate it all. Horror stories involving truckloads of proposals being delivered for source-selection evaluation boards abounded. Second, the accuracy of the analysis and documentation was questionable, because it applied to a system that had not even been built, but was expected to meet rigorous performance requirements and often used high-risk, leading-edge technologies. As new weapons systems fell short of their performance goals, fell behind schedule, and ran over budget, people began to realize that paper studies and computer analyses just were not enough. Total Package Procurement policies may have been based on the assumption that science and technology had given man a perfect understanding of the physical world and that prototyping was therefore unnecessary. That simply was not, and still is not, the case.

The criticism reached its peak with the release of the Blue Ribbon Panel report in July 1970. The panel had been chartered to examine the DOD weapons system acquisition process. Their report was highly critical of nearly all aspects of the process. The panel strongly recommended prototyping as a key element of its proposed package of acquisition reforms. Even prior to release of the report, Deputy Secretary of Defense David Packard issued a memo on weapons system acquisition in which he announced that DOD would adopt a fly before buy philosophy. Prototyping was incorporated into the ongoing AX program;

The Fairchild A–10, winner of the AX competition.

this would culminate in a flyoff between the Northrop YA–9 and the Fairchild YA–10 prototypes.

DOD formally implemented an Advanced Prototype Development Program in early 1971. In May, the Secretary of the Air Force proposed to DOD that the USAF develop a prototyping plan that would identify worthwhile new candidates. Deputy Defense Secretary Packard strongly agreed and urged that one or two of the resulting candidates be selected to start development in 1972. An Air Force prototype study group was immediately formed. By midsummer, they had recommended several candidates for prototyping, including an Advanced Medium STOL Transport (AMST), an RPV, a large tanker, an LWF, and a low RCS aircraft. The latter concept was pursued as a special project that became the HAVE BLUE program. Using small experimental prototype aircraft employing a faceted shape, this effort proved that stealth was feasible and led to the successful F–117A program. Meanwhile, the AMST and the LWF were officially approved to start in 1972 by a program decision memorandum released on August 25, 1971. Two days later, the Air Force Prototype Program Office (PPO) was created within the Aeronautical Systems Division at Wright-Patterson AFB, Ohio.[2]

Another aspect of the new DOD policy was streamlined management. This was articulated in a new directive, DODD 5000.1, "Acquisition of Major Defense Systems," issued in July 1971. The prototyping study group accordingly recommended the use of small program offices and greatly reduced demands for paperwork and documentation. This was to start with the initial request for proposals and include significant limitations on proposal length. Simplified performance goals would be used instead of detailed requirements to make the process less formal, focus effort on only the most important technical issues, and encourage innovative solutions that would meet the overall intent of the goals. Requirements would come later, when the technology had been proven; this would help avoid unrealistic specifications that had been a problem in programs such as the F-111.

The overall focus of the new DOD prototyping concept, which became known as experimental prototyping, was presented to industry in August 1971 by Deputy Secretary Packard and senior service acquisition officials. They outlined the intent and purpose of prototyping in the following terms:

Prototypes were to be experimental systems. They would precede, not be a part of, engineering development of any new weapon system.

They were intended to support anticipated future military needs with proven technology options, rather than specific weapon system designs.

In this context, experimental prototypes were not intended to form the sole basis for system procurement decisions.

They were to be focused on reducing cost and schedule, as well as technical risk, to future development programs.

They were intended to help achieve lower cost alternative solutions.

Experimental prototypes were expected to include some degree of technical uncertainty and risk. However, they should also have a reasonable chance of succeeding.

Experimental prototypes were to have a low relative cost, compared with potential follow-on development and procurement program costs.[3]

This new approach of Experimental Prototyping would emphasize early hardware demonstration instead of studies and analyses.

The Lightweight Fighter Program

The purpose of the prototype program, as reported in a July 1972 *Interavia* article, was "to determine the feasibility of developing a small, light-weight, low-cost fighter; to establish what such an aircraft can do; and to evaluate its possible operational utility."[4] The demonstration, if successful, would give the Air Force the option of complementing the F-15 with a light-weight, lower cost day fighter, although this would not necessarily be one of the specific designs being prototyped. "This development strategy provides us with proven alternatives, rather than paper analysis, for the prudent selection of future weapon system features."[5] It was intended that any feature or technology that was to come out of the program, once demonstrated, could be used with confidence in future weapon systems without necessarily buying one of the specific designs on which such a feature was demonstrated. "If it does [lead to an operational fighter], the resultant aircraft is likely to incorporate technologies from both designs."[6]

Effort was to be focused on demonstrating technologies that met the following criteria: make a direct contribution to performance; have moderate risk, but be sufficiently advanced to require prototyping to reduce risk; and meet cost, utility, and complexity restraints.[7]

The intent was not to try out every new thing, but to focus effort only on those technologies that were relevant to achieving the major objectives. The LWF prototypes would "use advanced technology to hold procurement and operating costs down, rather than as a lure for unneeded sophistication."[8] It was hoped that this would avoid the gold plating that was thought to have been encouraged by earlier development strategies and policies.

The prototype study group, when it initially recommended the LWF prototype effort, defined general objectives that included a gross weight of 20,000 pounds or less, mission essential avionics only, and basic air-to-air armament of a gun plus IR missiles. Although the maximum speed was to be Mach 2, emphasis was placed on superior maneuvering performance and handling qualities in the transonic, high-g regime.[9] Existing engines, or those in the final stages of development, were to be used. These objectives were included in the LWF RFP, supplemented by the specific performance goals of superior air-to-air performance, sufficient range, and design to cost.

The maneuver points selected for demonstrating superior air combat capability were maximum sustained turn (load factor) at 30,000 feet, at Mach 0.9 and at Mach 1.2; maximum instantaneous turn at 40,000 feet, Mach 0.8; acceleration time at 30,000 feet from Mach 0.9 to Mach 1.6.[10] Although additional performance data was naturally intended to be obtained, the above points were identified in the RFP as being of highest priority. Emphasis was also placed on improving the pilot's ability to function effectively while experiencing high instantaneous or sustained load factors, i.e., defining a high-g cockpit.

The LWF was to be able to fly at least 500 nautical miles, following a representative series of combat maneuvers that included seven 360 degree turns, three of them supersonic. For the purpose of meeting this goal, it was assumed that external fuel would be used on the outbound leg, to arrive at the combat area with full internal fuel. The complete RFP criteria for this aspect of the evaluation are presented in the table on the next page.

These criteria represented a departure from earlier Air Force practice, in that requirements prior to the F–15 had defined the combat segment simply as a fixed amount of time at full afterburning power, typically between 3 and 5 minutes. A fighter with a high thrust-to-weight ratio would be penalized because it would burn a larger portion of its internal fuel in that time than an aircraft with a lower thrust-to-weight ratio, other factors being equal. Defining a specific set of tasks, on the other hand, gave proper credit to an aircraft with high agility and/or a high thrust-to-weight ratio, which could accomplish the tasks in a shorter time than a less agile aircraft, thereby using less fuel.[11] To illustrate the magnitude of this effect, the turning and acceleration performance of the YF–16 would allow it to perform the RFP combat tasks in approximately

210

Lightweight Fighter Air Superiority Mission Profile

Condition	Criteria
Start	Full internal fuel, two AIM–9E missiles, and 500 rounds of 20-mm ammunition.
Combat	Four 360 degree turns at 0.9 Mach, 30,000 ft, MAX thrust. Three 360 degree turns at 1.2 Mach, 30,000 ft, MAX thrust. Accelerate from 0.9 to 1.6 Mach at 30,000 ft, MAX thrust. Expend two AIM–9E missiles and 50% of gun ammunition (97 lb).
Return	Climb from 20,000 ft to optimum cruise altitude, INT thrust. Cruise at optimum Mach and altitude for at least 500 nm.
Landing	Fuel reserve for 20 minutes loiter at sea level.

Source: Hicks, *Performance Evaluation of the YF–16 Prototype Air Combat Fighter*, p. 26.

---

two minutes less time in full afterburner than an F–4E would require to execute the same maneuvers.

The design-to-cost goal was new. The goal was a three-million-dollar unit cost, assuming a hypothetical production run of 300 aircraft over three years.[12] This goal was stated, not because production was intended, but because affordability was part of the demonstration. Any performance gains shown would have to be affordable in future production aircraft.

Sources differ, but it appears that the original RFP was either 21 pages long,[13] or just over 50, but with the technical meat contained in only 10 of those.[14] Either way, this represents a dramatic reduction from the 250-page RFPs that were typical of slightly earlier programs. Proposals were limited to 60 pages, 10 management and 50 technical, allowing all evaluators to get the big picture instead of each evaluator focusing on a small specialized aspect of the proposed effort. This allowed the source selection board to absorb the entire proposal instead of just an executive summary.

When the contracts were awarded, they were also kept as simple as possible. The program did not require full milspec compliance, only satisfaction of the overall intent of the relevant specifications. However, it was unlikely that this would compromise the quality of the aircraft that would be built, because they would be flying the most demanding air combat maneuvers with the contractors' reputations at stake. So it was left up to the contractors to apply good design practice.

Technology and the Air Force

To minimize risk both to the government and to the contractors, the contracts did not include a contract end item nor a tie to a subsequent production contract. Absence of a contract end item freed the contractor of any obligation to continue work after the contract money was spent. If, at that point, the system did not meet performance goals, the contractor would not have to spend his own money to fix it. This gave the contractor more freedom to take technical risk. Conversely, the government would be free to say, at the end of the program, "Okay, this wasn't really good enough, we're not going to buy a lightweight fighter." The risk for the government was also kept to a minimum.[15]

The PPO was kept as small as possible, with a minimum of full-time staff. However, it was not the intent of the program to eliminate government technical involvement in research and development. All contractor-generated data was made available to anyone at the Program Office who needed it. Any required technical assistance to the PPO was promptly provided from the Air Force Systems Command's Aeronautical Systems Division, Air Force Laboratories, and NASA. The idea was to replace massive documentation and viewgraph engineering (then called brochuremanship) with smaller quantities of higher quality hard data. Contacts were more face-to-face, with fewer formal reports. "This management concept created an environment in which the contractor and the Air Force LWF program personnel maintained a common goal of identifying problems or concerns and finding quick and suitable solutions that were most often made verbally and on the spot," according to H. J. Hillaker, the YF–16 Deputy Chief Engineer at General Dynamics.[16] This allowed maximum progress with minimum delay, while still keeping the customer in the loop.

The RFP was released to industry in early January 1972, and responses were due on February 18. Nine companies received RFPs: Boeing, Fairchild, General Dynamics, Grumman, Ling-Temco-Vought (LTV), Lockheed, McDonnell Douglas, North American Rockwell, and Northrop.[17] Five of these responded: Boeing, General Dynamics, LTV, Lockheed, and Northrop. There were actually six proposals, as Northrop submitted two: a twin-engine design (P–600) and a single-engine design (P–610). The Northrop P–600 was to use two General Electric YJ101s, while all other designs submitted would utilize a single Pratt & Whitney F100.[18]

Lockheed and LTV proposed design evolutions from their F–104 and F–8 fighters respectively, which may have offered an appealing marketing opportunity with a minimum of development effort from the contractors' point of view, but they simply were not what the Air Force was looking for.

The Boeing, General Dynamics, and Northrop proposals, while all based on earlier work by those companies, were not related to any aircraft already in production. Thus they had been designed from the ground up to meet the objectives of the LWF; and the selected advanced technologies were integral to the designs, rather than being tacked on. This was the approach that the PPO was looking for.

212

Another factor in the source selection was that bidders were required to submit wind tunnel data and wind tunnel models with their 60-page proposals, and the models would be tested to substantiate the data. This illustrates the Air Force's commitment to the idea of relying on hardware demonstration, even in the very early stages of a program. Two contractors' models failed to support their claims.[19]

The Air Force chose the General Dynamics 401 (to be designated YF–16) and the Northrop P–600 (YF–17) in part because of the differences between them. "The two aircraft . . . are as much complementary as competitive."[20] The Boeing proposal had actually rated very well in the evaluation, but it was too conceptually similar to the General Dynamics' proposal to offer much added value to the program in terms of validating a wide range of technology options.[21] The General Dynamics/Northrop combination covered one vs two engines, one vs two vertical tails, and fly-by-wire vs conventional flight control.[22] Contracts were awarded to the winners on April 14, 1972.[23]

The LWF flight-test approach and organization emphasized testing that would quickly and efficiently accomplish the overall objective, which was "to demonstrate the feasibility and operational usefulness of a highly maneuverable lightweight fighter."[24] Several new program concepts were implemented to accomplish this with a minimum of delay and unnecessary expenditure.

Each prototype design was to be evaluated by AFFTC and TAC pilots. Flight testing would include simulated air combat exercises against other U.S. and covertly obtained Soviet fighters, but the two designs would not be flown directly against each other; they would not necessarily even fly at the same time. Each prime contractor would pick the starting date for his respective one-year flight-test program. Col. William E. Thurman, PPO director, explained: "It is simply not in the interest of this program to exert competitive pressures in terms of schedule."[25] The test programs were to be two independent evaluations of the performance and combat potential offered by each of the alternative LWF designs.

Testing of each prototype would be accomplished under the direction of a Joint Test Team consisting of representatives of the airframe and engine contractors, the AFFTC, and TAC. This team would cooperatively prepare the test plan and share data during the test program. Teamwork was encouraged, rather than a strict division of responsibilities between the various members. The traditional phases of a test program — development testing, aircraft and systems evaluation, and operational evaluation — would not be executed in strict sequence. Instead, there was to be an "integrated program of development testing, aerodynamic and systems evaluation, and operational factors."[26] The evaluation tests and even the operational factors tests would begin as early as possible, before the entire flight envelope was cleared. Having the military user involved early provided an incentive to the contractors to make sure everything would work the first time, allowed the customer to get an earlier start on evaluation,

The Northrop YF–17.

and provided earlier user feedback into development testing activities and design improvements.

Thorough test plans were prepared well in advance by the members of the Joint Test Team and approved by the PPO, as well as by a LWF Safety Board. Nevertheless, it was realized that unforeseen problems could necessitate additional development testing and also that some planned test conditions might have to be modified. Provision was made to allow minor changes from the test plan as long as the intent of the test plan would still be accomplished. Colonel Thurman noted, "The test program will also be more flexible than in the past. We don't want to bog down in minor details."[27] To minimize delay, the Joint Test Team was vested with the authority to approve minor changes. Major changes still required the approval of the Test Management Council, which was chaired by the PPO director and included the AFFTC Director of Test Forces, the TAC Director of Fighter Requirements, and the Program Manager of the applicable airframe contractor.[28]

Testing was not intended to show that the prototypes would meet military specifications, but rather that the demonstrated technologies and design innovations could achieve substantial performance improvements at moderate cost. To provide this focus, the PPO issued seven Critical Test Objectives in October 1973 to be given priority during the test program:

Basic flight envelope clearance:

10,000 to 40,000 feet, and 0.6 to 1.6 Mach.

Energy maneuverability performance data:

Sustained turn at 30,000 feet, 0.9 Mach.

Sustained turn at 30,000 feet, 1.2 Mach.

Instantaneous turn at 40,000 feet, 0.8 Mach.

Ps envelopes

Acceleration time at 30,000 feet from 0.9 to 1.6 Mach.

High AOA flight characteristics:

Departure resistance, handling qualities, engine operation.

Tracking characteristics.

Weapons/engine/airframe compatibilities (AIM–9E and gun firings).

RFP mission performance (Air superiority mission).

Air Combat Maneuvering suitability.[29]

This illustrates that streamlined management was not misconstrued to mean lack of guidance. The test program was carefully laid out to insure that the desired advances in technology would be convincingly demonstrated.

The YF–16

Prior to the initiation of the program, General Dynamics began to develop the aerodynamic configuration of what became the YF–16 as they investigated a single-engine, low-cost concept to supplement the Air Force's FX. They had been testing wide forebody shapes for increased lift since 1966, and at that time rejected a sharp-edged forebody strake because it would lead to airflow separation. However, when they tested rounded forebody cross sections, they experienced loss of directional stability at high angles of attack. NASA aerodynamicists pointed out that at high angles of attack, forebody flow separation was inevitable; so rather than attempting to avoid separated flow, it was better to control and exploit it.[30] Sharp-edged forebody strakes produced a more stable flow pattern that generated significant body lift, improved directional stability, and shed vortices over the wings that delayed wing stall by continually mixing boundary layer air with freestream air. These qualities became known as controlled vortex lift. The Northrop YF–17 embodied similar aerodynamic concepts.

Adding a hinged leading edge flap to the wing further increased lift, reduced drag, and improved stability at high angle of attack. On the YF–16, the leading edge flaps deflected downward a fixed amount for takeoff and landing, and otherwise automatically deflected to an optimum position depending on the angle of attack, pitch rate, and Mach number. This feature would dramatically reduce the drag during high-lift maneuvers. The YF–16 was the first aircraft to use automatic leading-edge maneuvering flaps.

Technology and the Air Force

The wing of the YF–16 was blended into the body, producing a very thick wing root. This thickness provided extra fuel volume with a minimal drag penalty, and also reduced the structural weight by providing a large structural depth at the wing root, where bending loads are the greatest.

Construction was mostly conventional. The structural material usage in the YF–16 was approximately 80.6% aluminum, 7.6% steel, 2.8% advanced composites, 1.5% titanium and 7.5% other materials.[31] Composite usage consisted primarily of graphite/epoxy in the tail surfaces. General Dynamics claimed that the use of composites reduced the weight of these items by 30% relative to conventional construction.[32] However, this represents a very small fraction of total airframe weight, and even in the 1990s, the dramatic improvements predicted for composites have not been achieved.

The impact of high sustained load factors on the pilot was addressed by the use of a 30-degree seat back angle and a 6-inch heel elevation. To reduce pilot effort, the original design included a force-sensitive nonmoving sidestick controller. The YF–16 frameless canopy offered a full 360-degree view at eye level and above, and maximum downward view angles of 15 degrees over the nose and 40 degrees to either side.[33] The only canopy frame was aft of the pilot's head position, giving the pilot unrestricted forward view.

The YF–16 and YF–17 were the first aircraft to have a head-up display (HUD) designed into the initial versions. The HUD, combined with simplified control switches and instrumentation, would allow the pilot to perform all necessary tasks with a minimum of physical movement and without taking his attention away from the combat situation.

The most prominent flight control feature of the YF–16 was its negative, or unstable, static margin (the center of gravity was aft of the aerodynamic center, by a distance of 7% to 10% of the mean aerodynamic chord). For comparison, most contemporary fighters had a 1% to 3% stable static margin. This relaxed static stability was claimed to offer a 4% to 8% increase in maximum trimmed lift coefficient during subsonic flight, and an 8% to 15% increase during supersonic flight, due to the reduction in tail download needed to trim the aircraft.[34] With relaxed static stability and automatic maneuvering flaps, the YF–16 represented an implementation of some of the concepts that were popularly known as control configured vehicle technology. It should be noted that the YF–16 was only unstable in subsonic flight; in supersonic flight, the aerodynamic center moved aft sufficiently so that the YF–16 became statically stable.

Fly-by-wire control was necessary to implement the artificial stability augmentation that would make the YF–16 flyable with its negative static margin. The YF–16 used a quadruplex, analog fly-by-wire control system with no mechanical backup. The choice of four independent channels was based on the level of failure protection desired. Because active elevator control would be needed at all times on the YF–16, two-fail-operate protection was desired. This

The F–16's cockpit provides clear visibility in all directions.

dictated the use of four independent channels.* Mechanical backup would have been useless because, as noted above, the YF–16 would not be flyable without the artificial stability provided by the flight control computer via the fly-by-wire system. Quadruplex redundancy was also desirable for acceptance in the Air Force, because it represented a minimum change from the proven F–111, in that a triply redundant electrical stability and command augmentation system acted in support of the single mechanical control system.[35]

Other elements of the pitch control loop — the air data system that provided angle-of-attack feedback, and the actuators themselves — would be just as critical as the electronics. The hydraulic actuators (proven components from the F–111) only offered one-fail-operate protection against hydraulic or servovalve failure. This was accepted for all functions except pitch control, as discussed above. Therefore a crossover linkage was used between the right and left horizontal tails (each of which had its own actuator). Each actuator would keep working normally after one failure. If two failures affected the same actuator,

---

*Fail-operate means a control surface is functional after a failure; fail-safe means it could not operate but could be locked in a safe position. However, there is no such thing as a safe position when the aircraft is unstable and requires continuous corrective movement of a particular surface. Two channels are fail-safe but not fail-operate: if one of them fails, a discrepancy could be detected, but there would be no way of knowing which signal was still good. Three channels are fail-operate: if one signal disagrees with the other two it can be identified and disregarded. A second failure could be detected, but there would be no way to identify the one remaining good signal. Four independent channels are required to provide 2-fail-operate protection, in which a system remains fully functional after a second (electrical) failure.

the failed side would be depowered, and the linkage would then allow the one good actuator to power both horizontal tails. Thus the pitch axis control was two-fail-operate against hydro/mechanical failures as well as electrical failures.[36]

A quadruplex air data system was initially included in the YF–16 design, to match the four channels of the flight control computer. However, this would have required the development of a special nose air data probe; whereas the probes and equipment were readily available to provide three independent sets of air data signals. Triplex redundancy was therefore accepted in view of the known high reliability of the conventional air data instrumentation.[37]

The YF–16 used the Pratt & Whitney F100–PW–100 advanced afterburning turbofan developed for the F–15. Maximum sea-level static thrust was 23,830 pounds with afterburning, and 14,670 pounds dry. Weight was 3,068 pounds, giving a thrust-to-weight ratio of 7.8:1. Powder metallurgy techniques were used for manufacturing high-temperature core components; this allowed turbine temperatures of 2,500 degrees Fahrenheit, compared to 2,000 degrees in the TF30 that powered the F–111, and 1,800 degrees in the earlier J79.[38] Fan inlet guide vanes were of variable trailing edge design. Compressor stators on the first three high-pressure stages were also variable.[39]

The F100 was a more advanced engine than the relatively simple General Electric YJ101, the only other candidate engine. However, this went along with increased complexity and risk. The F100 encountered several development problems, including premature fatigue and a tendency to experience compressor stalls. These problems were aggravated by the high performance and maneuverability of the F–15 and F–16, which imposed more dynamic loads and a larger number of thermal cycles per flight hour than earlier fighters had done.

The YF–16 used a fixed-geometry inlet to reduce cost and weight relative to a variable inlet. This choice reflected increased emphasis on the middle of the flight envelope where most air combat was considered to take place, rather than the extreme high-speed regime where a variable inlet would be most beneficial. The inlet was located underneath the wide forebody to benefit from the high air pressure and relatively uniform flow in that region, particularly during high angle of attack maneuvering.

As described earlier, the YF–16 flight-test program was run by an on-site Joint Test Team at Edwards AFB, California. Responsibilities that traditionally would belong to one particular member of the team were shared. In particular, the test pilots from the different organizations flew missions interchangeably, without strict regard for which group would normally be responsible for a particular mission. For example, much of the control system development testing (modifying gains to achieve better handling qualities) was accomplished while sharing flights with other types of testing. The freedom to share flights allowed more efficient utilization of aircraft time than would be possible under the conventional type of phased flight-test program.

218

This view of the F–16 shows the fixed inlet under the nose,
the blended wing and thick wing root, and the leading edge flaps.

The first YF–16 was rolled out on December 13, 1973, and shipped to Edwards in a C–5A on January 9, 1974. There was an unscheduled first flight during a high-speed taxi test on January 20, when roll oscillations caused the left wingtip missile and the right horizontal tail to contact the ground. General Dynamics test pilot Philip Oestricher elected to take off to avoid further damage to the aircraft. This flight lasted six minutes. The first scheduled flight was made on February 2, less than 2 years after General Dynamics had submitted the proposal for the YF–16, or 22 months after contract award. The first supersonic flight was 3 days later, and the top speed of Mach 2 was reached on the 20th flight, on March 11. The second YF–16 was shipped to Edwards on February 27, but with no available F100 engine, did not fly until May 9. The test program was completed on January 31, 1975, with a total of 439 flight hours in 347 sorties.[40]

The most distinguishing feature of the YF–16, apart from its tremendous thrust-to-weight ratio, was the flight control system. The results of the YF–16 flight-test program would be critical to the acceptance of fly-by-wire control and relaxed static stability in the Air Force.

The maximum steady-state command roll rate was initially too low, which noticeably detracted from the YF–16's agility. This required the installation of a higher rate roll gyro. Nearly all other flight control functions, including the roll stick force gradient, were initially too sensitive. This contributed to the roll oscillations during high-speed taxi tests that led to the unplanned first flight. However, stick force gradients and control gains were adjusted during the test program to achieve generally excellent control characteristics about all axes. The fly-by-wire control system allowed these adjustments to be made quickly and easily. The fly-by-wire system also showed high reliability throughout the test program, which helped to gain acceptance for the use of fly-by-wire in operational aircraft.[41]

Early in the test program, the YF–16 experienced an aeroservoelastic instability, defined as an unstable mode produced by interaction of the active control system with the aeroelastic (flutter) properties of the airplane. Flutter analysis and testing had been performed during the design of the YF–16; however, this did not account for the effects of active control. Conversely, closed-loop control system tests had been performed on the ground, but the simulated aircraft dynamics used for these tests were based on a rigid airframe model. When the airplane was flying, the roll sensors picked up motions due to (otherwise stable) aeroelastic vibrations, and the active control system applied corrective aileron deflections that actually amplified the vibration. Two unstable modes were observed, both involving antisymmetric wing flexure. The problem was corrected by reducing the gain in the roll-control loop and by adding a filter in the feedback path that suppressed the high-frequency signals from structural vibrations. The rapid correction of this problem again illustrates the flexibility of the fly-by-wire system, and the quality of engineering talent that was available on-site during the joint test program.[42]

The design of the YF–16 resulted in outstanding performance throughout the flight envelope. The leading edge flap was so successful at reducing drag that it had to be detuned for the landing approach condition, to achieve an acceptable descent slope.[43] Handling qualities were good at angles of attack up to 25 degrees, the maximum permitted by an automatic limiter. This limiter was needed because directional stability would be lost at higher angles of attack.

As previously described, the control stick was originally nonmoving, the inputs being determined solely by the force of the pilot's hand on the stick. Although pilots were able to adapt to this feature, it did not give the pilot any indication of when he was applying the maximum command stick force. As a result, pilots would sometimes apply more force than necessary, causing tiring of the right arm. A stick with some motion was fitted to the YF–16 during the test program, although the inputs were still determined by the force exerted on the stick, not by its position.[44] The motion served only to provide the pilot with a more natural feel. Other high-g cockpit features were favorably received, and YF–16 pilots did not black out at 9 gs and experienced no decrease of vision capability at 7 gs.[45]

Several engine problems were encountered during the flight-test program. Afterburner ignition could only be accomplished within a limited envelope, which did not include low Mach numbers at high altitudes. Engine thrust levels were low during climbs and accelerations because the fuel control did not keep up with the changing demand of the engine during these maneuvers. Thrust levels at partial-power settings were nonrepeatable, and engine trim procedures were excessively complicated. The engine produced a visible smoke trail, which made the YF–16 easier to detect in visual combat.[46] In contrast, there were no engine problems specifically related to high angle of attack, even though the YF–16 flew at higher angles of attack than any of its predecessors. Thus the low-mounted engine inlet concept can be considered quite successful. The overall performance of the YF–16 was spectacular, although slightly below prediction in certain transient flight conditions because of the various engine troubles.

According to the Air Force's Performance Evaluation of the YF–16, "The YF–16 accomplished all of the objectives and successfully demonstrated the performance capabilities specified in the original Air Force Request for Proposal. The aircraft represents a significant milestone in fighter aircraft technology."[47] "The advanced technology features ... have all performed as designed and required a minimum of development tests."[48] These comments show that the YF–16 was outstandingly successful as an experimental prototype. Performance at the design conditions is summarized below:

Air superiority mission range was 648 nautical miles (goal was 500).

Acceleration time from Mach 0.9 to 1.6 at 30,000 feet was 66.5 seconds. (General Dynamics had predicted 61.2 seconds; discrepancy was blamed on engine lag. For comparison, the F–4E required 120 seconds.)

Maximum sustained turn at 30,000 feet, Mach 0.9, was 4.5 g.

Maximum sustained turn at 30,000 feet, Mach 1.2, was 4.6 g.

Maximum instantaneous turn at 40,000 feet, Mach 0.8, was 4.3 g.[49]

The performance of the YF–16 was within a few percent of predictions and far superior to contemporary U.S. and Soviet fighters.

Air Combat Maneuvering exercises were reportedly performed against F–106, F–4E, MiG–17 and MiG–21 opponents, and also between the two YF–16s. Not only did the YF–16 dominate its adversaries, but because of its higher energy maneuverability, it was able to do so while using dramatically less fuel in combat. On one sortie, a YF–16 flew repeated engagements against an F–4E, continuing each engagement until one aircraft or the other achieved a position that would be considered to result in a probable gun kill. After three engagements the F–4E reached minimum fuel and was replaced by a second F–4E. The YF–16 outlasted the second opponent as well, and won all of the engagements.[50]

The YF–16 test program was planned to take 22 aircraft months, over a period of 12 calendar months. Not counting the unscheduled first hop on January 20, 1974, the test program, including air-to-ground oriented tests added later for Air Combat Fighter (ACF) source selection and seven flights for Navy evaluation, actually took place between February 2, 1974 and January 31, 1975, or almost exactly the 12 calendar months originally planned. There were actually only 21 aircraft months, as the second aircraft did not begin flying until May 9, 1974.[51] All of the high-priority items in the test plan were accomplished, and the intent of all the lower priority items were satisfied. Thus, it was as an outstanding example of a successful and well executed test program.

Three main factors contributed to this achievement: the success of the Joint Test Team concept; careful planning of the test program in advance; and the quality of design, engineering, and construction of the YF–16 that exhibited high reliability in addition to its outstanding performance.

From YF–16 to F–16

Even prior to the start of the LWF program, NATO had been looking for a new fighter to replace F–104s and counter the growing quantity and quality of Soviet fighters in Europe. The U.S. Air Force in early 1972 began formulating the concept for an Air Combat Fighter that could utilize one of the LWF designs, while incorporating additional air-to-ground capability. A consortium of four NATO countries (Belgium, Denmark, the Netherlands and Norway) took an interest in this concept, and urged the Air Force to define the ACF as early as possible.[52] In April 1974, it was decided that one of the LWF designs would be selected as the basis for the ACF.[53] Partly in response to NATO pressure, the U.S. Air Force set the ACF decision date for January 1975.[54] This changed the LWF program from two independent best-effort technology demonstrations, into a direct competition between the YF–16 and the YF–17.

The transition officially took place in August 1974, with the addition of certain air-to-ground tasks to the flight-test programs and the revision of the test programs to be completed by January 1975. Both contractors were also requested to submit proposals for missionized operational versions of their respective designs. The air superiority day-fighter role was still prominent, and cost was still an important factor. The major departure from the original LWF design goals was added emphasis on air-to-ground capability. Increased avionics functions, enhanced radar, and a fly-by-wire flight control system would also be incorporated in the ACF.[55] The successful demonstration of fly-by-wire on the YF–16 earlier that year was crucial in establishing Air Force acceptance of fly-by-wire flight control.

The YF–16 was announced the winner of the ACF competition on January 13, 1975, and the four NATO countries that had originally expressed interest in the ACF placed an initial order for 348 F–16s on June 7.[56] Factors in the decision included the YF–16's superior performance in some areas, and lower expected cost that was largely due to its single engine. There were other benefits resulting from a common engine with the F–15, including F100 engine development costs already invested and the opportunity for common maintenance, support equipment and training. However, engine commonality also meant that any problems with the F100 would affect both the F–15 and the F–16, and this did happen at times. Development problems coupled with subcontractor labor strikes resulted in a shortage of engines, aggravated by both fighters using the same engine; and during 1979, F–15s were delivered to the Air Force without engines.[57]

The first of eight full-scale development (FSD) aircraft flew in December 1976. The FSD flight-test program was modeled on the LWF Joint Test Force concept. The first production F–16 flew on August 7, 1978, and the F–16 entered service in the United States and several other countries in 1979 and 1980.[58] These dates were exactly in accordance with projections published in 1974 immediately following the ACF decision.[59] This relatively rapid progression through FSD into operational capability illustrates how early prototyping can pay off by avoiding delays and problems during development. The F–16 had its first use in combat — ironically in the air-to-ground role, with F–15s providing fighter escort — on June 7, 1981, when Israel bombed the Iraqi nuclear facility at Osirak.[60] The production total as of 1994 was over 3,900 aircraft. Customers include the United States Air Force and 17 foreign countries. The United States Navy also operates some F–16s as aggressor aircraft for air combat training.[61]

The production F–16A is slightly larger than the YF–16, but very similar in its external shape as well as its structure. It has an improved production version of the F100 engine, along with extensive avionics. Because of the significance of the flight control technology used in the F–16, specific features of the F–16 flight control system are discussed below.

Although generally similar to that of the YF–16, the production F–16 flight control system incorporated several changes. The F–16 used a sidestick controller with some motion, as demonstrated during the later portion of the YF–16 test program. The central fly-by-wire system remained quadruply redundant, providing two-fail-operate protection against computer or electrical failures. New hydraulic Integrated Servoactuators (ISAs) were developed, and these replaced the F–111 actuators originally used on the YF–16. The ISAs were fail-operate/fail-safe against hydraulic or mechanical failures, meaning they would operate after one failure and automatically center after the second failure, independently of the flight control computer.[62] This was an improvement relative to the original actuators, which would operate after one failure but would not automatically center in the event of a second failure.

During development of the F–16 it was learned that, although unstable in pitch, the aircraft could actually be flown with only one horizontal tail operating and the other centered. This information, plus the additional centering feature of the new servoactuators, allowed the horizontal tail crossover linkage to be eliminated from the production F–16 design: after one local hydraulic or mechanical failure, both horizontal tails would function as usual; if a second failure occurred on the same side, the affected surface would automatically center, and the airplane would continue to fly using only the horizontal tail on the good side.[63] Thus, taken together, the two horizontal tail surfaces provided the desired two-fail-operate protection.

As previously noted, the YF–16 had an automatic 25-degree angle of attack (AOA) limiter because it would lose directional stability at higher angles of attack. No other limiters were used on the YF–16. On the production F–16, the AOA limiter was retained and a variety of other limiters were added. These included a roll rate limiter as a function of dynamic pressure, angle of attack, and horizontal tail position to prevent pitch departures due to rolling at low speeds and high angles of attack. Rudder authority was limited as a function of angle of attack and roll rate to prevent directional departures at high angles of attack. A yaw rate limiter became active at angles of attack over 29 degrees to prevent entry into a high-rate flat spin mode with unacceptable recovery characteristics. In addition, a manual pitch override could be selected when the angle of attack exceeded 29 degrees to permit the pilot to rock the airplane out of a deep stall that could occur at aft c.g. locations.[64] This illustrates that the aircraft could in fact be flown beyond the protection of the 25-degree AOA limiter under certain conditions. YF–16 flight tests were instrumental in identifying the needs for these additional limiting functions early in the F–16 development program.

The YF–17 and its descendant (F/A–18) provide an interesting contrast in the application of flight control technology. The YF–17, which used a basically conventional control system, was designed to be limit-free over the entire maneuver envelope. Unlike the YF–16, it was statically stable in pitch, and also had a tail surface arrangement that preserved lateral/directional stability up to and beyond the maximum trimmed angle of attack.[65] Partly as a result of the successful demonstration of fly-by-wire control on the YF–16, the production F/A–18 has a fly-by-wire flight control system. However, it relies on its natural aerodynamic qualities for stability and departure avoidance, rather than a battery of limiters. It has two different control system backups. The first is a direct electrical backup, in which the pilot inputs are transmitted directly to the control surfaces, bypassing the flight control computer but still using electrical rather than mechanical transmission of signals. The second backup is a mechanical linkage to the horizontal tails only.[66] The use of these backups emphasizes that the F/A–18 is statically stable and therefore not dependent on its flight control computer for full-time active stability augmentation.

Conclusions

The YF–16 demonstrated air combat performance far superior to anything in existence prior to the F–15. Information gained during the one-year test program was instrumental in supporting the rapid and painless transition of leading-edge technology into the F–16 FSD effort, with subsequent early achievement of initial operational capability. The F–16 went on to exceed all initial expectations for production and foreign sales and has shown high reliability and maintainability in the field. Combat performance in air-to-air as well as air-to-ground roles has been exceptional. The YF–16's counterpart, the YF–17, was developed into the F/A–18, which has had a similarly successful career.

Nevertheless, the development of one or more specific fighter aircraft designs, however outstanding, was not the original intent of the LWF program. The program was intended to validate selected technologies and make them available to future development programs at a low level of risk. Regarding this aspect of the program, Colonel Thurman noted that, "In addition to having developed two very promising aircraft . . . we have acquired a wealth of important design information for a fraction of the cost normally incurred."[67] The program was highly successful in transitioning a broad set of technologies into operational use, and nearly every fighter/attack aircraft that has emerged since the 1970s benefits from these accomplishments.

The YF–16 brought fly-by-wire control and relaxed static stability to a state of maturity and acceptance that made them viable options for production systems. The fly-by-wire system had demonstrated excellent reliability in the YF–16 flight tests. "Without the confidence gained with the YF–16, the Air Force probably would not have adopted a fly-by-wire control system in the production aircraft."[68] Two other new aircraft immediately adopted fly-by-wire control: the F/A–18 and the F–117A. Because of the successful demonstration of relaxed pitch-axis stability on the YF–16, the F–117A was allowed to be unstable about all axes. This gave the designers more freedom to concentrate on stealth, with minimal compromise for aerodynamics. Nearly all new combat aircraft utilize fly-by-wire control, along with negative pitch stability to achieve higher maneuverability and better supersonic trim.

Many other advances were brought about by the YF–16 and YF–17. In the area of aerodynamics, most modern fighters employ automatic leading-edge maneuvering flaps, along with some form of strakes or LEX (or canards, which have the similar effect of shedding a streamwise vortex that maintains attached flow over the wing). Not only do these features extend the high-angle-of-attack capabilities of an aircraft, but they dramatically reduce the drag during maneuvering at moderate angles of attack. Low-mounted inlets, which maintain good pressure recovery during maneuvering, have become nearly universal. High-g cockpit features, such as increased seat-back angles and HUDs, gained acceptance. The prototypes also continued a return (begun in the F–14 and F–15) to the use of high-visibility canopies on fighter aircraft.

There is no question but that all of these technological advances would have occurred far more slowly, one at a time, if there had not been a small, efficient, highly focused program with low legal and financial risk, specifically tailored to the purpose of advanced fighter technology demonstration.

"We have learned or relearned some crucial lessons. First, competition is vital. There simply is no other contractual or management incentive that is as effective as competition. Secondly, this program has demonstrated the importance of visibility, the need to bring problems out into the open the minute they occur. Both sides, the Air Force and the contractors, have been scrupulously honest, in a fiscal as well as a technical sense. And third, it is vital that prototypes lead requirements. The results are cost savings and better products," reported Colonel Thurman.[69] The last comment relates to the fact that an intelligent requirement cannot be formulated if the capabilities offered by the latest technology are not properly understood; this comment was made in the long shadow of programs such as the F–111 that illustrated the difficulty of trying to meet unrealistic requirements with unproven technology.

Many other flight demonstrator programs have been considered successful, but the technologies demonstrated simply sat on the shelf afterward. The unique technology transition accomplishments of the LWF program are due in part to the program structure and the way technologies were selected. The Air Force did not choose the technologies, but rather issued a set of performance goals and allowed the contractors to select technologies that could be integrated to achieve those goals. The prospect — but not the promise — of follow-on sales provided an incentive to take risks where necessary to achieve superior performance, but also to apply technologies in a practical, workable manner and to stay within realistic limits of cost and complexity. This kept the effort on technologies with real value and avoided the science fair project syndrome.

The LWF program was structured to facilitate open communication and cooperative relationships between contractor and government representatives. This contributed materially to the success of the program. In the design stage, H.J. Hillaker of General Dynamics noted that the confidence to use fly-by-wire with no mechanical backup on the YF–16 was established through "free exchange of experience from the Air Force laboratories and McDonnell Douglas 680J projects on the F–4 and from NASA's F–8 fly-by-wire research program. It is probable that such free exchange would have been severely constrained under a conventional contract."[70] In flight testing, the Joint Test Team concept facilitated a high degree of teamwork that allowed more tests to be accomplished in less time than otherwise would have been possible. It became the model for the F–16 full-scale development flight-test program, and many aspects of this approach have since become part of standard Air Force flight-test procedures.

The program reinforced the value of early prototyping to obtain actual flight-test data before entering into full-scale development of a particular weapons system. Issues associated with high angles of attack, dynamic engine performance, and automatic control systems were identified. For example, engine

226

design needed to address rapidly changing flight conditions and not just high steady-state flight speed, which had formerly been the area of emphasis. Early flight testing uncovered these issues and allowed them to be worked while subsequent development was ongoing. Otherwise, these problems would not have been discovered until a preproduction prototype had been flown, resulting in much higher subsequent risk to the program.

The YF–16 provided a very important technical lesson that is applicable to all aircraft that utilize active control. As aircraft and their control systems increase in complexity, there are exponentially more possibilities for unforeseen, potentially catastrophic dynamic modes and control interactions, such as the aeroservoelastic instability experienced by the YF–16. The YF–16 had an extremely simple control system consisting of only five primary flight control surfaces. Many newer designs have a larger number of control surfaces, plus thrust vectoring about one or more axes. Redundancy is necessary to protect against failure of flight-critical elements of the system, but redundancy also adds further to the complexity. "Redundant system design must be a careful balance between failure protection level and unnecessary complexity."[71] Good technical communication across disciplines and thorough analysis, ground testing, and flight testing are crucial to insuring safe expansion of the flight envelope.

The management approach used in the LWF program amounted to using a carrot instead of a stick to get the best possible effort from the contractors. The nature of the contracts minimized the financial or legal risk to the contractor in the event that performance goals were not met; on the other hand, the lack of tie to a production contract minimized risk to the government and made both contractors feel that they would have to do a superlative job to capture some of the large market that they both felt existed. Perhaps the best summary of the LWF program approach and the payoffs that it achieved was given by Northrop president Thomas V. Jones: "Turning loose the design experts is the best way to a better integrated aircraft. . . . It affords the opportunity for the United States to exercise creative people on real vehicles. The whole process is stimulating to all involved. . . . This is how we are going to beat the Soviet Union and other nations in remaining the leader on the cutting edge of advanced technology."[72]

Notes

1. Baumann, *et al, Advanced Design Composite Aircraft,* Final Technical Report, Rockwell International, Feb 1976, p 11; Forsch, *Advanced Design Composite Aircraft (ADCA) Study,* Final Technical Report, Grumman Aerospace Corporation, Nov 1976, abstract.

2. Gable, *Acquisition of the F–16 Fighting Falcon,* Report Number 87-0900, Air Command and Staff College, Maxwell AFB, April 1987, pp 7–8; Mott, *A Review of Events and Management Strategies in Fighter Programs (F–11, F–15, and F–16),* ANSER draft, Sep 1983, pp 3–4.

3. Hillaker, "The F–16: A Technology Demonstrator, a Prototype, and a Flight Demonstrator," Aircraft Prototype and Technology Demonstrator Symposium, AIAA Paper 83-1063, March 1983, pp 113–4.

4. "General Dynamics and Northrop to build Lightweight Fighter Prototypes," *Interavia,* Jul 1972, p 693.

5. Powell, "Prototyping: Competitors or Companions?" *Government Executive,* Apr 1974, p 37.

6. Ulsamer, "The Lightweight Fighter Halts the Cost Spiral," *Air Force Magazine,* Oct 1973, p 65.

7. Hillaker, p 116; Powell, p 40.

8. Ulsamer, "The Lightweight Fighter," p 65.

9. Gable, p 9.

10. Hicks, *Performance Evaluation of the YF–16 Prototype Air Combat Fighter,* Final Report, Air Force Flight Test Center, Edwards AFB, April 1975, p 1.

11. Letter, Wayne O'Connor to Albert C. Piccirillo, Jun 22, 1995.

12. Fitzsimmons, *et al,* eds, *The Great Book of Modern Warplanes,* Portland House, New York, 1987, p 208; Mott, p 5.

13. "Lightweight Fighter RFP's Out," *Armed Forces Journal,* Feb 1972, p 19.

14. Hillaker, p 114.

15. *Ibid,* p 115.

16. *Ibid,* p 115.

17. "Lightweight Fighter RFP's Out," p 19.

18. Mott, p 4.

19. Hillaker, p 114.

20. Ulsamer, "The Lightweight Fighter," p 65.

21. Miller, *The General Dynamics F–16 Fighting Falcon,* Aerofax Incorporated, Austin, 1982, p 19.

22. "General Dynamics and Northrop to build Lightweight Fighter Prototypes," p 693.

23. Gable, p 10.

24. Hicks, p 7.

25. Ulsamer, "YF–16 On Time, On Track, On Budget," *Air Force Magazine,* Jan 1974, p 54.

26. Ford, *YF–16 Lightweight Fighter Prototype Program Interim Flight Test Report,* vol 1, Introduction and Flight Test Program Summary, General Dynamics FZM-401-178, Nov 1974, p 10.

27. Ulsamer, "YF–16 On Time," p 54.

28. Ford, *Interim Flight Test Report,* p 19.

29. *Ibid,* pp 23–24.

30. "General Dynamics' YF–16: Design Considerations and Evolution," *Interavia,* Jan 1975, p 40.

31. Lyons, *The Search for an Advanced Fighter: A History from the XF–108 to the Advanced Tactical Fighter,* ACSC/EDCC Report Number 86-1575, Apr 1986, p 29.

32. "YF–16: General Dynamics' Lightweight Challenger Enters the Ring," *Interavia,* Feb 1974, p 156.

33. *Jane's All The World's Aircraft,* 1974–75, p 343.

34. Droste and Walker, *The General Dynamics Case Study on the F–16 Fly-by-Wire Flight Control System,* AIAA Professional Study Series, p 11.

35. *Ibid,* p 59.

36. *Ibid,* p 33–5.

37. *Ibid,* pp 68–70.

38. Fitzsimmons, *et al,* p 160.

39. *Jane's,* 1981–82, p 787.

40. Hicks, pp 216–33.

41. Ford, *Interim Flight Test Report*, p 6; Ford, *YF–16 Lightweight Fighter Prototype Program Final Flight Test Report*, supplement 1 to vol 1, Introduction and Flight Test Program Summary, General Dynamics FZM-401-178, Mar 1975, pp 19–21.

42. Peloubet and Haller, *Application of Three Servoelastic Stability Analysis Techniques*, General Dynamics, Fort Worth, Sep 1976, pp 1–2.

43. Ford, *Final Flight Test Report*, p 20.

44. Rider, "F–16 Pilot Report," *Aerospace International*, Jan/Feb 1977, p 14; Hicks, pp 229.

45. Ford, *Final Flight Test Report*, p 8.

46. *Ibid*, pp 16–18; Hicks, pp 1, 16, 18, 210.

47. Hicks, abstract.

48. Ford, *Final Flight Test Report*, p 9.

49. Hicks, pp 16, 20, 26.

50. Rider, p 13.

51. Ford, *Final Flight Test Report*, pp 9–11; Hicks, pp 216–33.

52. Mott, p 6; Fitzsimmons, *et al*, p 146.

53. "Everything Works: LWF to ACF to F–16," *Government Executive*, Mar 1975, p 33.

54. Gable, p 15.

55. Lyons, p 33; Ulsamer, "New Ways to Fly and Fight," *Air Force Magazine*, Sep 1974, pp 60–61.

56. Mott, p 7; Manning, "The Lightweight Classic," *Airman*, Sep 1975, p 47.

57. Fitzsimmons, *et al*, p XX.

58. *Jane's*, 1981–82, p 361.

59. Ulsamer, "New Ways to Fly and Fight," p 58.

60. Fitzsimmons, *et al*, p 188.

61. *Jane's*, 1949–50, p 569.

62. Droste and Walker, pp 48–50.

63. *Ibid*, pp 50–51.

64. *Ibid*, pp 96–110.

65. Fitzsimmons, *et al*, p 209.

66. *Ibid*, p 221.

67. Ulsamer, "YF–16 On Time," p 54.

68. Fitzsimmons, *et al*, p 144.

69. Ulsamer, "YF–16 On Time," p 54.

70. Hillaker, p 119.

71. Droste and Walker, p 113.

72. Powell, p 41.

R. Cargill Hall is the contract histories manager at the Air Force History
Support Office. Previously, he was chief of the research division and deputy
director of the USAF Historical Research Agency, Maxwell AFB, Alabama, and
a historian at Headquarters Military Airlift Command, Headquarters Strategic
Air Command, and at the Jet Propulsion Laboratory. He holds a B.A. degree
from Whitman College and an M.A. from San Jose State University. He
authored *Lunar Impact: A History of Project Ranger* (NASA), was editor and
contributor to *Lightning Over Bougainville* (Smithsonian), and was series editor
of the International Academy of Astronautics history symposia, *History of
Rocketry and Astronautics*, ten volumes (Univelt). He is a member of the
International Academy of Astronautics, and the International Institute of Space
Law, and serves on the board of advisors for the Smithsonian Institution Press
history of aviation book series.

230

The Air Force Agena:
A Case Study in Early Spacecraft Technology

R. Cargill Hall

An incredible amalgam of space technology affects our daily lives. But we yawn, rather than gape in wonder, at this technology that brings into our homes Olympic sporting events from the other side of the world and the close-up features of planets from the other side of the solar system. The same technology provides us with warnings of hurricanes, depicting their track on the evening television news, and guides our automobiles over unfamiliar streets in a strange city to a preselected destination. As a recent measure of its rapid change, consider this: NASA officials in 1972 secured approval of the Space Shuttle because it could perform a number of novel missions, key among them retrieving expired or malfunctioning satellites and returning them to Earth for refurbishment and reuse. By the time of the first Shuttle flight nine years later in 1981, satellite technology had progressed so rapidly and automated spacecraft had become so reliable that they operated unattended in orbit for ten years and more. It no longer made economic sense to retrieve and refurbish them many years after launch when their technology, especially the solid-state electronics, was fit only for display at the National Air and Space Museum. Needless to say, that mission disappeared from the Shuttle manifest and a corresponding increase appeared in the cost of Shuttle flight operations.

When the space age began in the 1950s, however, satellite technology was mostly undeveloped and certainly untried and untested in space. The two original American space flight entrants consisted of a civil and a military model. The first was a scientific satellite approved for the International Geophysical Year. Designed by a team at the Naval Research Laboratory to be launched atop a three-stage modified Viking sounding rocket, the satellite consisted of a sphere 20 inches in diameter pressurized with helium and housing a scientific experiment within a 12-inch diameter internal canister. To help maintain the temperature of the electronics within acceptable limits, the 22-pound satellite was spun to 180 rpm at insertion in orbit. A battery-powered transmitter operated for a few days, reporting internal package temperatures through a signal radiated from four antennas protruding from the satellite sphere in a turnstile arrangement. A solar-cell-powered transmitter operated over several months and radioed data from the scientific experiment through two antennas arranged as a dipole at the north and south poles of the sphere.

If the IGY scientific satellite weighed tens of pounds and was to be launched by a modified sounding rocket, the military satellite would weigh thousands of pounds and be launched atop an Atlas ICBM. This spacecraft, built for the Air Force by the Lockheed Missiles and Space Division of Lockheed Aircraft, would be stabilized on three axes, receive and execute commands sent from stations on Earth, and transmit information to these stations. In fact, between 1959 and 1987, this automated spacecraft, eventually known as "Agena," would be launched by various boosters and in various models would be employed in numerous American military and civil space applications. Because the original Agena featured the primary design characteristics of all attitude stabilized satellites and space probes to follow, we will consider the evolution of its technology in more detail.

I

The Agena was predicated on Air Force-funded studies conducted by the RAND Corporation after World War II. These studies began in 1946 with engineering analyses that established the technical feasibility of Earth satellites. They also identified military missions that satellites could accomplish in outer space and progressed through subsystem design and development contracts. The initial RAND work concluded in March 1954 with the release of a final report, *Project Feed Back*, that described a reconnaissance satellite and called on the service to proceed with its procurement. The Air Force did so. The Air Research and Development Command established a small project office at Wright-Patterson Air Force Base. In 1955 that office began a reconnaissance satellite design competition with the award of contracts to three firms: Lockheed Aircraft Co., Glenn L. Martin Co., and RCA. A year later, Lockheed won the Air Force design competition to build the satellites for what was now called the WS 117L Program.

When the Air Force awarded Lockheed a contract for its WS 117L military satellite in October 1956, the configuration for a "Pioneer" and "Advanced" version adhered closely to the prior RAND designs—a rocket-powered spacecraft with a cylindrical body and a conical nose. The nose cone served as a faring to provide environmental protection for the spacecraft and its payload. The Air Force established a diameter of 60 inches for the cylinder of the vehicle, which made necessary a compromise to accommodate spherical, nested propellant tanks, and a minimum weight for the spaceframe structure that was still able to sustain vehicle bending caused by air loads during ascent. The length of this vehicle, initially postulated as high as 21-1/2 feet, was reduced to 15-1/2 feet in a compromise between the height allowable for clearance between the Atlas booster and the gantry crane of the contemporary launcher, the min-

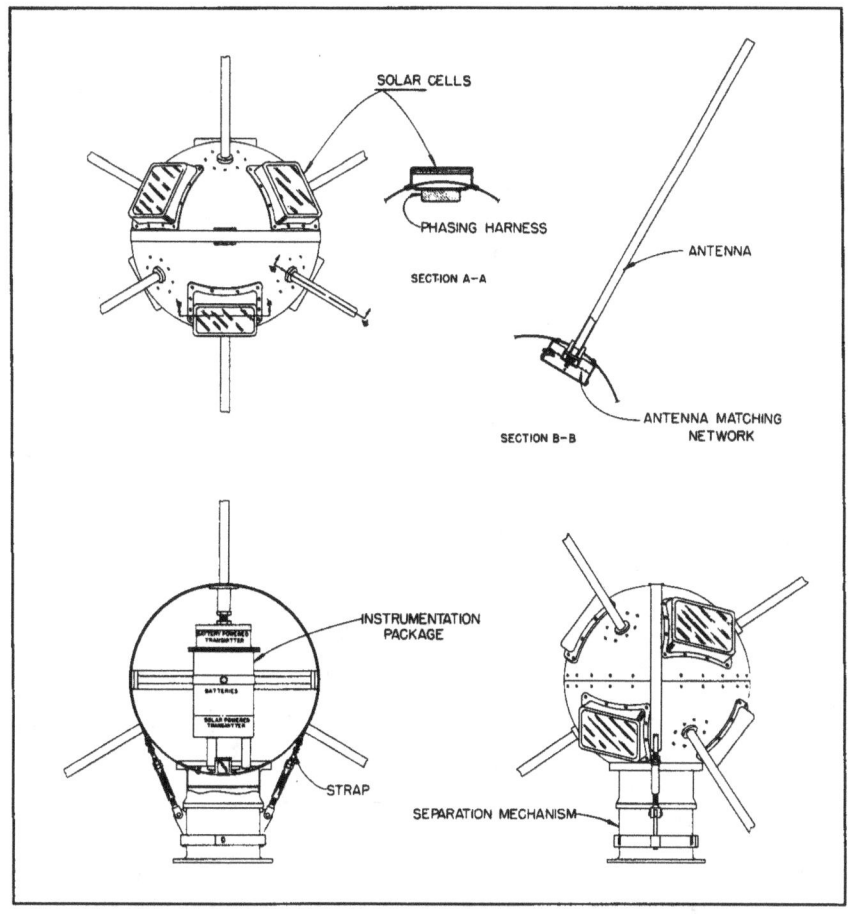

Drawing of U.S. Navy's Vanguard I.

imum structural weight, and a volume sufficient to enclose all the equipment and tankage. The Advanced vehicle, proposed to meet future requirements for increased payload weight and duration of operation in orbit, had the same diameter as the Pioneer (60-inches), but its overall length increased to 25 feet, then eventually to 37 feet, depending on the mission.

The fundamental challenges associated with the satellite spaceframe involved selecting structural materials with which to build it and developing a thermal control system to protect vehicle equipment that would operate in the extremes of temperature in space. Selection of material for structural application turned on weight considerations, coupled with producibility during manufacturing. Magnesium-thorium and beryllium were investigated, and a mag-thorium alloy was selected as the basic structural metal for use in all areas except the propellant tanks because of its light weight, strength, and performance under

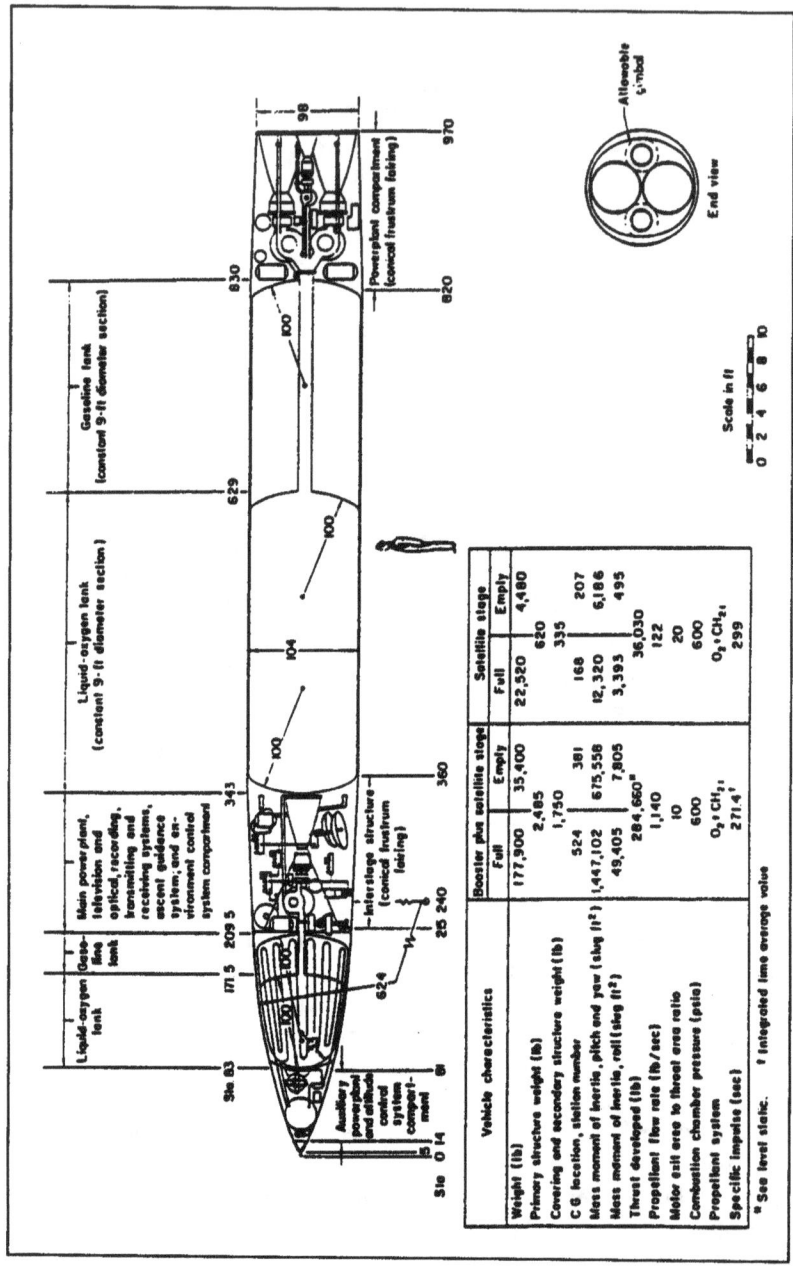

Schematic of satellite vehicle, RAND Project Feed Back, March 1954.

Vehicle characteristics	Booster plus satellite stage		Satellite stage	
	Full	Empty	Full	Empty
Weight (lb)				
Primary structure weight (lb)	177,900	35,400	22,520	4,480
Covering and secondary structure weight (lb)	2,485		620	
C G location, station number	1,750	381	335	207
Mass moment of inertia, pitch and yaw (slug-ft²)	524		168	
Mass moment of inertia, roll (slug-ft²)	1,447,102	675,558	12,320	6,186
Thrust developed (lb)	49,405	7,805	3,393	495
Propellant flow rate (lb/sec)	284,650ª		36,030	
Motor exit area to throat area ratio	1,140		122	
Combustion chamber pressure (psia)	10		20	
Propellant system	600		600	
Specific impulse (sec)	O₂+CH₃₁		O₂+CH₃₁	
	271.4†		299	

ª See level static.   † Integrated time average value

234

high temperature. Although initially considered for the skin only, new forging and sheet metal forming techniques for mag-thorium made possible handling of this material in large quantities, which permitted wider application.* Since thorium is radioactive, however, it also meant establishing methods for working it that did not expose anyone to radiation hazards. For the propellant tanks, Lockheed investigated stainless steel, titanium, and aluminum, ultimately selecting an alloy of the latter metal. Fabrication of the lighter weight aluminum propellant tanks through spin-forming allowed tank skin thickness to be increased without a serious weight penalty.

An effective environmental control system to protect vehicle electronic and mechanical equipment was essential for operations in orbit. Because the WS 117L satellite attitude was Earth center-stabilized on three axes in space, one side of the vehicle would be exposed continuously to the heating of direct sunlight during a portion of each orbit, while the entire vehicle would be plunged into the freezing shade of the Earth for another portion. To control these extremes in temperature, engineers conducted extensive studies of various skin coatings, compartmenting equipment racks, and heat transfer paths. The novel thermal control system that resulted from these investigations was completely passive. That is, it employed no artificial refrigeration or heat-producing sources. Environmental control was obtained entirely through conduction and radiation, without involving convection. This first passive thermal control system for an Earth satellite proved to be indispensable in the reliable operation of the Agena in orbit for prolonged periods. Variations of this passive thermal control system would be adopted and used on virtually all American satellites and deep space probes. Significantly, design bureaus in the Soviet Union chose to employ active thermal control systems in pressurized vessels, with generally unhappy consequences for long-term, reliable satellite orbital operation.

Lockheed's 1956 satellite proposal featured a pressure-fed Aerojet General Vanguard rocket engine in the second stage booster-satellite. By 1957, however, Cmdr. Robert C. Truax, USN (the WS 117L project deputy director), and Capt. James S. Coolbaugh, USAF, had convinced leaders of the firm to replace it with a turbo-pump-fed Model 8048 Bell rocket engine rated at 16,000-pounds thrust, almost twice that of its pressure-fed alternate. Designed to propel the detachable bomb pod of the B–58 Hustler bomber, the Bell "Hustler" engine, as it was known, burned a noxious hypergolic combination of unsymmetrical dimethylhydrazine (UDMH) fuel and inhibited red fuming nitric acid (IRFNA) oxidizer. Later models of the Bell engine employed in the advanced Agena B were capable of being re-ignited during ascent for a second burn. In contrast to single burn, where a satellite separates from the booster and coasts to apogee before

---

*The development of techniques for working mag-thorium later carried over on the Polaris submarine ballistic missile and proved of some assistance in the rapid achievement of program objectives.

its engine is fired, in dual burn, the satellite stage ignites right after separation and burns just long enough to provide a begin-coast speed sufficient for the long, shallow climb required for high efficiency. At apogee, half-way around the Earth, the satellite stage rocket is restarted to provide orbit injection. The greater begin-coast speed afforded by dual burn reduced the total amount of propellants required in a satellite stage of a given gross weight, and this weight savings could be exchanged for increased payload.

The performance gains offered by the dual-burn Bell Model 8096 engine took advantage of the larger Agena B, lengthened amidships with new integral propellant tanks with a capacity twice that of the Agena A (13,255 pounds). Altogether, the Pioneer Agena A at separation from its booster weighed some 10,000 pounds and the Advanced Agena B about 17,000 pounds. The first Agena A, delivered to the Air Force in October 1958, was launched successfully from Vandenberg AFB, California, on February 28, 1959. Work on the advanced Agena B commenced in June 1959 and culminated in a successful first launch on November 12, 1960. The name "Agena," incidentally, identified the star nearest our solar system and was adopted for the spacecraft in the fall of 1958,[*] about the same time that the Advanced Research Project Agency separated the Air Force military satellite program into separate flight projects: SAMOS (near real time visual reconnaissance), MIDAS (infrared missile early warning), and a new Thor-boosted portion identified as Discoverer.

II

We know today that the Discoverer Project, ostensibly aimed at recovering biomedical research capsules from Earth orbit, served as a cover for the CIA-Air Force Project CORONA which returned exposed film taken by a 24-inch focal length panoramic camera manufactured by the Itek Corporation. Planned as a hurry-up interim system to be employed while the SAMOS film-readout system was developed, CORONA soon replaced SAMOS altogether. Originally designed for a long-life mission at an atmospheric drag-free altitude of 300 miles, plans called for the SAMOS Pioneer Agena to employ gravity gradient stabilization. That is, the vertically oriented Agena moved in orbit with its fixed, nose-mounted Eastman Kodak strip camera pointed toward the Earth, thus aligning the long axis of the satellite's mass distribution radial to the Earth. The gravity gradient stabilization scheme eliminated the expendable weight required for gas jets and has been used in other satellites. With the gravity approach to

---

[*]An ARPA special committee apparently selected this name in mid-1958, in keeping with Lockheed's tradition of naming aircraft and missiles after stellar phenomenon, e.g., Vega, Constellation, Polaris, etc. Lockheed officials, who at first objected that the star Agena (otherwise known as Alpha Centauri) was not readily visible in the night sky, soon agreed to the name.

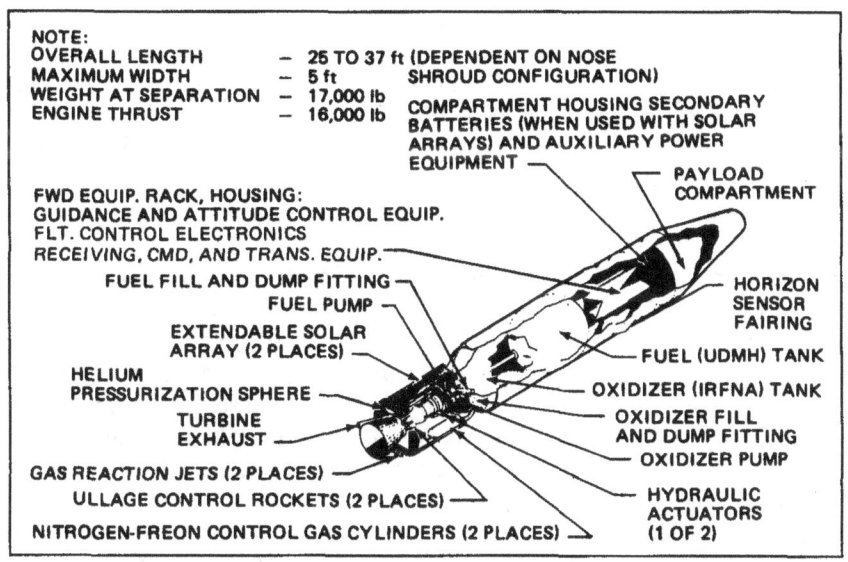

NOTE:
OVERALL LENGTH — 25 TO 37 ft (DEPENDENT ON NOSE
MAXIMUM WIDTH — 5 ft   SHROUD CONFIGURATION)
WEIGHT AT SEPARATION — 17,000 lb
ENGINE THRUST — 16,000 lb
COMPARTMENT HOUSING SECONDARY
BATTERIES (WHEN USED WITH SOLAR
ARRAYS) AND AUXILIARY POWER
EQUIPMENT

PAYLOAD
COMPARTMENT

FWD EQUIP. RACK, HOUSING:
GUIDANCE AND ATTITUDE CONTROL EQUIP.
FLT. CONTROL ELECTRONICS
RECEIVING, CMD, AND TRANS. EQUIP.

FUEL FILL AND DUMP FITTING
FUEL PUMP

HORIZON
SENSOR
FAIRING

EXTENDABLE SOLAR
ARRAY (2 PLACES)

FUEL (UDMH) TANK

HELIUM
PRESSURIZATION SPHERE

OXIDIZER (IRFNA) TANK

TURBINE
EXHAUST

OXIDIZER FILL
AND DUMP FITTING

OXIDIZER PUMP

GAS REACTION JETS (2 PLACES)

ULLAGE CONTROL ROCKETS (2 PLACES)

HYDRAULIC
ACTUATORS
(1 OF 2)

NITROGEN-FREON CONTROL GAS CYLINDERS (2 PLACES)

Schematic of Agena B.

stabilization, only electricity was required for momentum wheel damping with rate-sensing gyroscopes, which could be supplied with solar cells.

But the three-axis attitude control system fashioned to meet the horizon sensor-referenced vertical pointing and stability requirements for a long-life SAMOS reconnaissance mission at an altitude of 300 miles proved readily adaptable to a short-life higher resolution CORONA reconnaissance mission. Operating in a horizontal position with respect to the Earth at about 100 miles altitude, an inertial reference package (IRP) consisting of three gyroscopes, two horizon sensors, and proportional (later pulse) micro-jets using cold gas (a nitrogen-freon mixture) provided attitude control for the Agena and its oscillating panoramic Itek camera.[*] Two hermetic integrating gyro units sensed pitch and roll, and one miniature rate gyro unit determined yaw error by sensing orbital rate. The pitch and roll gyro errors were corrected from the horizon sensors, with the gas jets actuated by signals from the inertial reference package. This early attitude control system sensed vertical attitude to approximately one-tenth (0.1) degree accuracy and provided a yaw pointing accuracy on the order of one degree, adequate for SAMOS and CORONA cameras.

Later Earth-viewing payloads with narrow view angles and longer focal lengths required better stabilization and pointing to achieve fractional foot

---

[*] The number of pulses (at a specific thrust level) fired by a gas jet or set of gas jets was proportional to the magnitude of the disturbing torque, sufficient to cancel the momentum without over-correcting or under-correcting. For the early proportional control gas jets, the thrust level varied in proportion to the magnitude of the disturbing torques.

resolution. Subsequent Agena attitude control systems were improved with sun and star trackers to meet these demands. Today, the future appears to lie in an onboard all-radio frequency-directed attitude control and self-tracking positioning system that eliminates Earth sensors, sun sensors, and gyros. Designed by Space Systems/Loral, the unit is driven by signals from the Global Positioning System (GPS) satellite constellation. Called GPS-Tensor, it measures pitch, roll, and yaw output ten times per second and determines spacecraft attitude accuracy to about 0.1 degree Root Mean Square. By using GPS for an attitude reference and simultaneously establishing its own position and ephemeris onboard, it also eliminates the need for elaborate and expensive tracking nets around the world—which represents a real breakthrough.

The Discoverer/CORONA Project called for ejecting a film capsule from orbit for recovery on Earth, and that introduced a new set of challenges. The capsule had to separate from the Agena at the proper attitude and spin-up to a specified rpm, survive the slow steady heat pulse generated by atmospheric friction during a flat reentry trajectory, and be recovered on Earth. The recovery capsule finally designed, constructed, and flown on Agena A and B satellites was a conical-shaped reentry body that somewhat resembled a thimble with an ablative shell at the forward end providing reentry thermal insulation. This reentry body, built by the General Electric Corporation, was mounted in the nose of the Agena and had solid-propellant spin rockets at its periphery and a center-mounted retro rocket which afforded controlled reentry, with parachutes added to ensure a slow final descent for aerial intercept. Recovery of the film capsule was programmed to occur above or on the high seas, in this case the Pacific Ocean, thus avoiding unwanted complications with commercial air traffic and national boundaries.

The Air Force established an air and sea recovery task force in Hawaii that initially flew C–119 and later C–130 aircraft modified with special equipment to permit snagging the capsule parachute during its decent, whereupon the capsule was reeled into the aircraft. In the event air recovery failed, the capsule impacted the water and would float long enough for backup sea recovery to be effected by ships, helicopters, and para-rescue teams zeroing in on the position by tracking the radio beacon. This recovery sequence began with the Agena in near polar orbit, high above the aircraft. Moving north-to-south in a horizontal position with respect to the Earth, at a given command the Agena performed a yaw maneuver, turning 180 degrees so that it was positioned backwards to the line of flight. Upon receipt of a programmer signal to eject the capsule, the Agena pitched down 60 degrees and the reentry body was released by a set of pin pullers and springs. Shortly after separation, two solid-propellant spin rockets fired, and the capsule was spun to approximately 60 rpm, at which time the retro-rocket ignited for about 10 seconds. Following retro-burn, two remaining solid-propellant spin rockets ignited and de-spun the capsule to 7.5 rpm for aerodynamic reentry. At this point, to increase aerodynamic stability during reentry, the thrust cone and retro-rocket ejected and, after the capsule had descended to about 60,000 feet altitude, the aft thermal cover was hurled out of the capsule by a drogue gun, drawing a parachute out after it. Upon parachute deployment, the payload capsule separated from its ablative shell, the recovery system's radio beacon and flashing light activated, and the beacon antenna erected.

A neatly designed recovery system, everyone thought. But at first it didn't work. As you know, the Discoverer/CORONA Project experienced twelve consecutive failures between February 1959 and July 1960. After a variety of technical difficulties with the Thor booster and Agena spacecraft had been solved, attempts to recover the film capsule proved unsuccessful. The limited instrumentation that provided telemetered diagnostic data did not pinpoint the cause of the reentry capsule malfunction during reentry. Engineers knew with certainty only that a malfunction was occurring during the spin-up and retro-fire sequence. An investigation of the characteristics exhibited by a reentry capsule that ejected from the Agena but went into another orbit determined what was actually happening: some of the solid-propellant spin rockets were exploding on ignition (caused by aging of the solid-propellant), causing the reentry capsule to spiral into another orbit. The four solid-propellant spin rockets were removed and replaced by two opposing micro gas jets employing compressed nitrogen-freon to provide the capsule spin and de-spin torques. This technical change confirmed the engineers' diagnosis when a capsule of exposed film ejected by Discoverer/CORONA 14 went through its spin cycles and successfully reentered the Earth's atmosphere and, for the first time, was retrieved by an aircraft flying over the Pacific Ocean on August 18, 1960.

III

A survey of early spacecraft technology developed for the Agena A and B models would be incomplete without considering the auxiliary power subsystem and its electronics. Perhaps more than any other of the numerous technical elements that must function together, the measure of success of any satellite is determined by the performance in space of its electrical and power system. To provide electrical power for the Agena, engineers sought lightweight power sources and reliable power inverters and regulator equipment. The WS 117L program office at Wright-Patterson Air Force Base, led by Lt. Col. William G. King, had investigated various power sources before and during the Agena design study competition in the mid-1950s, including batteries, open- and closed-cycle chemical powerplants, solar cells, nuclear reactors, and radioisotope powerplants. For the early, short-duration CORONA flights, Lockheed selected silver peroxide-zinc batteries because of the simplicity of a battery-operated power supply and the state-of-the-art permitted their rapid adaptation. Secondary batteries also could be employed to store electrical power derived from solar cells. Engineering improvements consisted of a pressure-sealed silver per-oxide-zinc battery, increased battery capacity, and a battery voltage-regulator system that allowed conversion of battery power to well-regulated alternating current at a high efficiency. Modified silver peroxide-zinc batteries were flown on Agenas for many years.

Because of the poor quality of transistors found in contemporary solid-state inverters, Lockheed engineers considered 400-cycle rotary inverters for use on the Agena. But the rotary inverters were heavy, inefficient, and would significantly limit the operating life in orbit. Instead, efforts were directed toward improving transistor quality in solid-state inverters; the Lockheed bridge-type 2,000-cycle inverter was one example of a design driven by the lack of high-quality inverters. Incidentally, when I refer to "solid-state" electronics at the close of the 1950s, I am talking about 3/8 x 3/4-inch-long transistors with their three protruding wires, inserted and soldered on plastic circuit boards that contained etched-in copper wiring. Engineers quickly found these solid-state electronics, mounted in chassis and joined in equipment racks, to be highly susceptible to short circuiting caused by debris floating in the vacuum of space on Agena spacecraft, stabilized on three axes. At that time, no one imagined to-day's solid-state integrated circuits etched through optical lithography on tiny wafers of silicon, creating micro-circuits that integrate all the components with the wiring.

The primary satellite power system planned for the future consisted of the Satellite Nuclear Auxiliary Power (SNAP) 1 and 2 nuclear-drive turbine electric powerplants that employed Plutonium 210 with a 50-day half-life. These plans for auxiliary power accounted for the selection of 2,000 cycles for the Agena because future SNAP systems were to provide 2,000-cycle alternating current. In the meantime, much effort was devoted to improving solar cells because their low efficiency could not meet Agena's high-power demands. Between 1957 and 1959, working through subcontractors, the efficiency and producibility of solar cells improved markedly, and the first Agena using solar cells on extendible solar arrays, coupled with nickel-cadmium secondary batteries, was launched in 1960. Solar cells continued to be improved and array designs permitted their employment in fixed positions or rotated on two axes for sun-tracking. Indeed, the efficiency of solar arrays increased so greatly that they would be used to power virtually all spacecraft operating in the inner solar system. SNAP nuclear power systems would be developed and used only on those spacecraft sent to the outer planets in the solar system, far beyond the range of the Sun's energy available for solar power. (Soviet officials, I should add, elected to employ SNAP systems on some satellites in Earth orbit, only to suffer the consequences when a few of them went out of control, reentered the atmosphere, and crashed with resultant radioactive contamination.)

Thus equipped, the Agena became the first American spacecraft capable of being commanded from Earth to perform a variety of functions in orbit. Track-

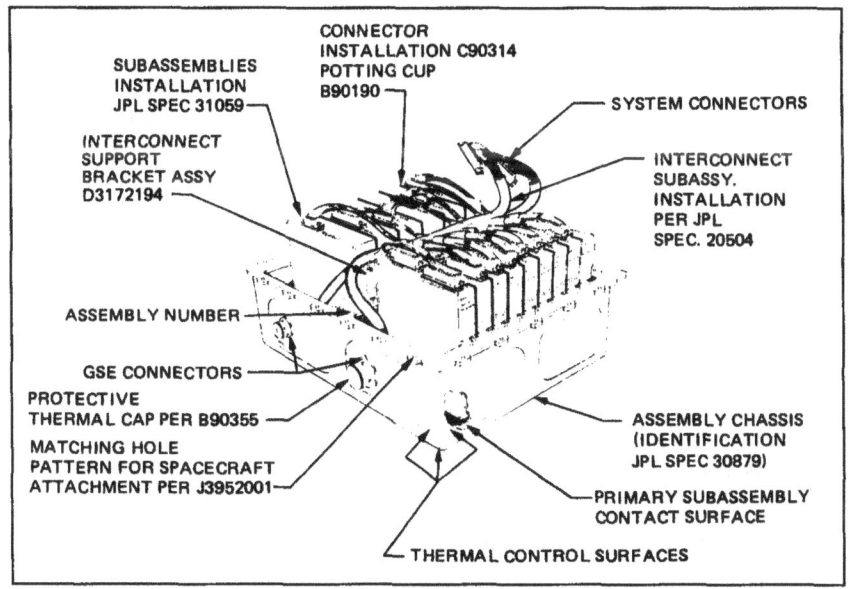

Typical spacecraft electronics module from late 1950s to early 1960s.

ing and commanding of the spacecraft was accomplished through the S-band beacon and a cavity back-spiral antenna that greatly resembled a bed-spring wrapped around a rod. To execute commands, the Agena A and early B models mounted an electro-mechanical timer-sequencer that operated with punched Mylar tape and, via ground command of onboard stepping switches, allowed choices of image motion compensation, an altered attitude of the Agena, or a reset of the basic orbital timer and adjusted initiation of the capsule recovery sequence, etc. (It was, to be sure, a far cry from the powerful computer-sequencers in use today that can program, reprogram, and override one another.) Vehicle housekeeping multiplexed telemetry channels were transmitted to Earth over two monopole whip antennas mounted on opposite sides of the Agena, thus keeping one in view of the Earth at all times. Later Agenas also mounted a "lifeboat" package that included an additional antenna and other backup electronic systems that could be activated in the event of an emergency failure of a primary system in orbit. Altogether, the Agena proved to be a versatile automated spacecraft that in the years between 1959 and 1987 would be purchased and used extensively by the Air Force and the National Aeronautics and Space Administration.

IV

The Air Force and industry team that built and flew the Agena truly "pioneered" spacecraft technology and its management for subsequent auto-

241

Agena D Characteristics

Length (payload interface to booster interface)	23.25 ft
Diameter	5.0 ft
Section Weight	
Forward Section	245 lb
Center Section (based on dual burn Agena)[1]	310 lb
Aft Section (based on dual burn Agena)	493 lb
Adapter Section	305 lb
Guidance module	114 lb
Telemetry module	17 lb
Total vehicle empty	1484 lb
Propellant weight IRFNA, VDMH (dual burn)	13,553 lb
Mixture ratio: oxidizer/fuel	2.55
Burn time	243 sec
On orbit burnout weight less payload (dual burn)	1,277 lb

Booster Characteristics

	Thor	TAT[2]	Atlas
Length (including Agena to payload interface)	76.4	76.4	89.8
Diameter	8.0	8.0	10.0
Typical liftoff weight	107,229	135,396	260,918
Booster thrust at sea level	155,500	317,050	388,300
Booster model	LR 7911	LR 7913	LR 89/ LR 105 (MA-5)
Propellants	LO_2/RJI	LO_2/RJI	LO_2/RJI

[1] Standard Agena is equipped with dual burn engine; single burn or multi-burn are optional.
[2] Improved thrust augmented Thor; conventional Thor booster augmented by three solid rocket motors that increase its thrust by approximately 153,000 pounds.

mated satellites and space probes. Often failing on the first try, they learned about reliable methods of testing space hardware, about sensible project management, about proper ascent sequencing, about space physics in orbital operations, and about reentry dynamics and sequencing. By the early 1960s, at the prodding of Clarence "Kelly" Johnson and Fred O'Green, the Agena itself had evolved from a "job-shop" Agena B to a "standard" Agena D* that featured

---

*An "Agena C," which featured an increased diameter, was designed, but never built.

a common bus onto which project-peculiar equipment and payloads could be readily integrated. Agena became this country's "DC–3 of space vehicles" for a variety of reasons. Lockheed engineers in the 1950s prepared a solid design with substantial margins, and they selected manufacturing processes that for the most part were well understood. After some early difficulty with solid propellant pyrotechnics in the reentry system, the spacecraft proved itself dependable and reliable. Because Lockheed Missile and Space Company over the years specialized in integrating satellite hardware, these Agenas were assembled, tested, and used by knowing hands. Indeed, 362 of them were manufactured, sold, and launched—a record that also made Agenas relatively inexpensive as "orbital trucks," which was, after all, an original goal of the Air Force satellite program and its Lockheed designers.

The Agena experience in developing and applying early spacecraft technology offers practical lessons for Air Force leaders charged with our future space architecture. Specifically, in the design and building of automated spacecraft, important "Agena axioms" still apply. First, employ a conservative design with margins for growth. Second, use materials and manufacturing processes that are well developed. Third, avoid pyrotechnics—large or small. Finally, delegate to a project manager authority commensurate with his responsibility and maintain continuity in the engineering-management team. Recalling these lessons when procuring new satellites, whatever their size and shape, should contribute significantly to extending, sustaining, and ensuring reliable space operations in the years to come.

Acknowledgments

I am indebted to David Bradburn, John Copley, James Coolbaugh, and Jack Herther for sharing their recollections and for their comments, corrections, and recommendations in the preparation of this study.

George W. Bradley, III, is the Air Force Space Command Chief Historian. He has a B.A. in history, Canisius College, M.A., Ohio State University, completed ABD for the Ph.D., Ohio State University, 1979. He is the author of *From Missile Base to Gold Watch: An Illustrated History of the Aerospace Guidance* and *Metrology Center and Newark Air Force Station* (GPO) and has authored or co-authored numerous other books and articles regarding Air Force history. He is a regular presenter at Air Force symposia, many times emphasizing Air Force Space History subjects. He is a member of the Society for Military History, Air Force Historical Foundation, and the Air Force Association.

Origins of the Global Positioning System

George W. Bradley III

Location, location, location. No, this is not the beginning of a commercial for Century 21. Rather, it is the introduction to a paper on an innovation that will have a significant impact on the 21st Century. I am referring, of course, to the Global Positioning System (GPS). In this paper, I would like to take a look at the historical origins of GPS. The primary focus of my paper will be to determine exactly how and why decisions were made that led to the development of this satellite system. While there will be some discussion of the evolving navigation technology, I will try to concentrate on the decisions leading to system approval. While most people are familiar with GPS, many from its use in the Gulf War, most do not know the legacy of this technology. And, this is important technology, technology that is a fundamental part of a revolution that the Chief of Air Force History, Dr. Richard Hallion, referred to in a presentation he made recently on the future of the Air Force: the precision revolution. While much still remains to be done in that revolution, its impact is already upon us. Understanding the origins of this aspect of the precision revolution is a useful prelude to understanding its future.

It's hard to precisely determine when man first began navigating by using objects in a fixed position. Pilotage, the oldest and simplest form of navigation, consisted of fixing one's position by using familiar landmarks as reference points. This method was used by early coastal mariners such as the Phoenicians, and it has remained an important navigational tool over the years. Early aviators, for example, navigated from one town to another using landmarks. Antiquity, however, offered celestial navigation, which did not have to depend on terrestrial objects but used the observed motions of the sun and stars. Its effectiveness increased over the centuries with advances in instrumentation such as the astrolabe, sextant, and accurate portable timepieces. This system, combined with dead reckoning (which determined location, based on speed, time, and direction through utilization of the magnetic compass), was the basis for seafaring out of sight of land. It too, remained an important tool over the years. Charles Lindbergh, for example, used dead reckoning and pilotage to fly the *Spirit of St. Louis* from Long Island to Paris in 1927.[1]

More sophisticated navigational instrumentation was developed in the 20th century, with radio navigation coming in the 1920s. The earliest systems were based on the ability of a radio receiver with a loop antenna to determine the direction of a radio signal and its relative bearing to the transmitter. Over the years, use of radio signals has become much more sophisticated. In World War II, for example, MIT Radiation Laboratory scientists introduced the Long Range Navigation system, or LORAN, which was based on measuring the difference in time of arrival of signals from synchronized pairs of transmitters at different locations. Today, a modernized version of LORAN has become a worldwide standard for aircraft and coastal marine navigation.[2] In the 1950s and 1960s, scientists developed inertial navigation for aircraft and missile systems. This system is not constantly dependent on other references such as radio signals or heavenly objects. The methodology for inertial navigation involves extremely accurate instruments to measure the acceleration of a vehicle in all directions and computers to calculate acceleration information to obtain velocity and position. It is standard on military systems such as submarines and missiles and is used on both commercial and military aircraft. What all these systems have in common is that they provide the user with location information, and military theorists from Sun Tzu to Clausewitz have emphasized the importance of knowing the exact location of friendly and enemy forces.[3]

While ground-based radio waves proved an important innovation in navigation, there were some problems. Low-frequency radio waves are not easy to modulate and are subject to errors because of factors like the ionosphere and weather turbulence. High-frequency radio waves are limited to line of sight, necessitating many fixed-site transmitters. Moreover, it was impractical to place a fixed-site transmitter at sea. Like ground-based radio navigation, celestial navigation also had problems. Mariners had traditionally been faced with problems in using celestial navigation during periods of intense cloud cover or dense fog. What was needed was a way to receive radio waves from a fixed point in the heavens. A solution began to present itself as early as 1955.

Even before the launch of Sputnik in 1957, scientists were attempting to develop a system to track proposed U.S. satellites. In the mid-1950s, the Naval Research Laboratory had proposed a system called Minitrack, which was built to track the movement of the Navy's proposed Vanguard satellite and other early man-made orbiting objects by using the signals they transmitted. The successful launch of Sputnik allowed a practical test since Sputnik emitted an active signal. Scientists at the Navy's Applied Physics Laboratory at Johns Hopkins University demonstrated that they could establish the ephemeris of Sputnik by measuring the Doppler shift of its continuous wave transmitter. Conversely, a year later, they reasoned that if you knew the ephemeris of a satellite, from the observed Doppler shift of its transmitter, you could then determine your position on earth.[4]

Passive satellites, those that emitted no signal, however, were beyond Minitrack's capabilities, and a different methodology was needed to deal with them. Roger I. Easton, Don Lynch, Al Bartholomew, and others at the Naval

Research Laboratory began working on a system to track such passive objects. According to Easton,

> We bought an FM transmitter and moved it . . . to Fort Monmouth, New Jersey. . . . Later, we used it to illuminate the Sputnik satellite when it passed over. This experience led us to conclude that if we could radiate a fan-shaped continuous wave beam, we could detect anything that passed through it."[5]

This was the early beginnings of the Naval Space Surveillance System or NAVSPASUR. Easton's group then turned its attention to devising a navigation system based on information obtained from orbiting satellites, a problem which, according to Easton, he had been thinking about for some time. "At NRL," he later explained, "we realized that a satellite navigation system could be established using satellites having synchronized clocks transmitting signals to users on the ground who could synchronize their own clocks to those in the satellites. The range measured to each satellite produced a line position, just as if one had obtained a sextant sighting on a star."[6] Essentially, Easton's system was still based on celestial navigation.* He substituted a satellite in a predicated and measured orbit for a star and synchronized clocks for the chronometers used by seafarers since the 18th century.[7]

Others reached the idea for a navigation satellite system through a different series of events. Following World War II, a number of variants based on the LORAN system discussed earlier were proposed. To provide better world coverage with fewer transmitters, a low-frequency system called Omega was developed. While Omega used continuous wave radiation rather than pulses, as did the original LORAN, in effect, it gauged the difference in time interval from ground stations by measuring the relative phase angles of transmitters from pairs of stations. Unfortunately, the accuracy of Omega was limited. To increase accuracy, LORAN–C, using ground-wave propagation, was developed for tactical aircraft. Widely used during Vietnam, LORAN–C had an accuracy of about

---

*Actually, the idea of a navigational satellite had been around for some time. Although it is unknown as to whether Easton and his colleagues knew of it, the concept may have dated as far back as 1838. It was the subject of discussion by two young college men, Nathan and Edward Hale. The two actually drew up a number of plans and sketches. In 1870, Hale published a fanciful story, in the Jules Verne genre, "The Brick Moon," in the *Atlantic Monthly."* In Hale's story, the protagonist proposed putting up a series of brick structures about 200 feet in diameter that would orbit the earth at fixed positions. Like the moon, they would give off light that could be seen on Earth (hence the title "Brick Moon"). Since the location of these moons would be fixed and known, they could be used for navigation. While it is unclear if this story had any impact on any of the GPS originators, it is interesting to note how early this concept was developed. The story is included in *The Brick Moon and Other Stories*, by Edward E. Hale, reprinted in 1970 by Books for Libraries Press. For a fuller discussion, see Keith Smith, "GPS Divined," *The Downlink*, March 1966, p 3–4.

100–200 hundred meters compared to Omega, which only had an accuracy within 2,200 meters. Unfortunately, all the early LORAN systems, as well as Omega, were essentially two-dimensional. In effect, they located by latitude and longitude, but not by altitude — the third dimension.[8]

This leads us to the ICBM and the Aerospace Corporation. The Aerospace Corporation was established in 1960 at the request of the Secretary of the Air Force to apply "the full resources of modern technology to the problem of achieving those continued advances in ballistic missile and space systems which are basic to national security."[9] Among the many projects worked by the Aerospace Corporation was a concept to use satellites for aircraft navigation. Obviously, the higher an aircraft went, the more important it was to obtain its location in all three dimensions. In addition, engineers at the Aerospace Corporation were working on the issue of the vulnerability of the land-based sites of the intercontinental ballistic missile system whose locations were well known by the Soviet Union, essentially making them vulnerable to a first strike. One solution to this problem was an Air Force proposal to develop a mobile version of the Minuteman ICBM mounted on railroad cars. On rails, Minuteman missiles could be moved at will to reduce their vulnerability to a preemptive attack. However, while mobility was a desirable trait, it complicated the guidance of the missiles. The vertical guidance systems in place on the Minuteman were dependent on knowledge of the precise location of the launch point, as well as on such other factors as the orientation of the zenith and the azimuth of the launch. This was not a problem as long as those factors could be calculated beforehand at a fixed site. Moving the missiles made it more difficult to achieve those location accuracies. Thus, a three-dimensional system was needed for both aircraft and missiles.[10]

In response to this problem, in February 1960, the Raytheon Corporation proposed to the Air Force a concept for a three-dimensional type of LORAN system called the Mobile System for Accurate ICBM Control or MOSAIC. Basically, MOSAIC used four continuous-wave transmitters at somewhat different frequencies, with their modulation locked to atomic clocks and synchronized via communication links. During its flight, the missile would continuously compute its position by using signals from MOSAIC. Through this application, the guidance of the missile would be less dependent on precise knowledge of the launch point.[11]

A central figure in this proposal was Dr. Ivan Getting, who was the first president of the Aerospace Corporation. Before he took that position, however, he had been vice president of engineering and research at the Raytheon Corporation at the time of their proposal. The concepts developed were crucial steps in the evolution of GPS. Whether MOSAIC would have ever worked is unknown, however, because before it could be developed, Secretary of Defense Robert McNamara canceled the mobile Minuteman program early in 1961. Nonetheless, many of the personnel involved in its conceptual beginnings, including Ivan Getting, were now working for the Aerospace Corporation. This was an important factor, since DOD had selected the Air Force as its executive agent for space launch, and the Air Force had contracted with the Aerospace Corporation to be its system engineer. Over the years, these two figures, Ivan Getting from the Air Force and Roger Easton from the Navy, would play crucial roles in the evolution of the U.S. satellite navigation system.[12]

It is now necessary to step back a few years. On October 4, 1957, with the launch of Sputnik, the Soviet Union had proved that the technology was available to orbit a satellite. The United States finally launched its own first successful satellite, Explorer I, on January 31, 1958. The interesting fact to note was that the issue of navigation was so important that only months after the successful launch of the first U.S. satellite, the U.S. Navy contracted with the Johns Hopkins Applied Physics Laboratory to design the first navigational satellite. That system, dubbed TRANSIT, actually represented the oldest U.S. military space system. The first TRANSIT satellite was launched on April 13, 1960, and the system became operational by 1965. The system was designed to meet the Navy's need for accurately locating ballistic missile submarines and surface vessels. It was the first satellite system to introduce corrections in the velocity of propagation of radio waves through the ionosphere by transmitting a signal at two frequencies. However, TRANSIT had some limitations. It was slow, intermittent, and two-dimensional, and it was subject to errors with even the slightest motion of the observer. In short, TRANSIT, while a big step forward in radio position location, was impractical for use on aircraft or missiles.[13]

Therefore, the Air Force continued to work on its own system. By 1963, the Aerospace Corporation, using knowledge gained from MOSAIC, convinced the Air Force that its navigation needs for missiles and aircraft could be met by a system that involved measuring distances to satellites with known positions. It was in this navigation study, directed by Aerospace's Phil Diamond, that the concept known as the Global Positioning System was born. In October 1963, the Air Force directed the Aerospace Corporation to pursue its navigation satellite study and named it Project 621B, "Satellite System for Precise Navigation." However, Diamond's GPS acronym outlasted the numerical designation. From the beginning, the system included the capability of supplying accurate all-weather position data anywhere on or near the earth to an unlimited number of users. Early on, developers planned to achieve accuracies to within 50 feet in three dimensions. By mid-1966, successful studies of the concept led the Air Force to decide to award a system hardware design contract to Hughes Aircraft Company and TRW Systems. From 1967 to 1969, further design studies proposed a global network of 20 satellites in synchronous inclined orbits using atomic clocks synchronized with a master system clock. Since the satellites would be placed in orbit one after another during the development stage, the system could achieve a limited operational capability even before the entire constellation was deployed.[14]

Technology and the Air Force

The Air Force interest in satellite navigation spurred the Navy to continue its own advanced research, and our focus turns again to Roger Easton and the Naval Research Laboratory. In the spring of 1964, Easton and others at the NRL initiated a dialogue on their ideas for a navigation satellite (as described earlier) with the Bureau of Aeronautics. Supported by the bureau, a project evolved under the name TIMATION, derived from the words "time" and "navigation." On October 16, 1964, the Bureau of Aeronautics issued a work order for the first TIMATION satellite, which was designed and built by the Techniques Branch at the NRL. The branch was headed at the time by E. L. Dix and Peter G. Wilhelm, while Roger Easton remained at the head of the NRL's Space Surveillance Branch, which was responsible for the satellite's internal electronics. The first satellite, TIMATION I, was launched at Cape Canaveral on May 31, 1967. Two years later, in 1969, TIMATION II was launched. To test the system, ground receivers were located in trucks, small boats, and some aircraft. One of NRL's engineers, Jim Buisson, later recalled "piloting" a truck around Washington's I–495 beltway and down the Dulles access road while its position was determined via signals from the TIMATION satellites.[15] By 1971, the Navy and RCA, its prime contractor, were proposing a system of 21 to 27 satellites in inclined eight-hour orbits. The satellites would carry sophisticated crystal oscillators and rubidium atomic clocks transmitting UHF signals for ranging and time transfer. At about the same time, the Army entered the picture with a system dubbed SECOR for Sequential Correlation of Range. Clearly, some mechanism needed to be established to correlate the three systems. As a result, in 1968, a triservice committee, later called NAVSEC for Navigation Satellite Executive Committee, was tasked to coordinate the efforts of the services' navigation satellite systems. However, while the NAVSEC was able to foster discussion and the sharing of information, it had no authority to enforce recommendations or choose a system, and the services continued independent development efforts.[16]

In 1969, the Air Force awarded contracts to four companies — TRW Systems, Magnavox Research Laboratories, the Grumman Aerospace Corporation, and the Boeing Company — to refine the design and determine a cost for the proposed 621B (or GPS) navigation system. During 1971 and 1972, tests of operator equipment at White Sands Proving Ground using ground- and balloon-carried transmitters achieved accuracies to within 50 feet. However, the DOD, because of service concerns, would still not commit to the Air Force program. A new figure emerged who would provide a solution to the deadlock. In late 1972, Gen. Kenneth Schultz, then commander of the Air Force Space Division, appointed Col. Brad Parkinson as the Air Force 621B program manager. Col. Parkinson immediately opened talks with the Navy with the express purpose of combining 621B's GPS with the Navy's TIMATION. On April 17, 1973, Deputy Secretary of Defense William Clements sent a memo to all three service secretaries directing a joint development program to test satellite navigation

Launch of a Navstar GPS satellite.

systems for future acquisition. In a key move, Clements named the Air Force as executive agent of the joint undertaking, and the project was called the Defense Navigation Satellite Development Program. The Air Force Chief of Staff, Gen. John D. Ryan, directed Systems Command to set up a joint program office at Space Division to manage the program. All three services, as well as the Marine Corps and the Defense Mapping Agency, participated. By September 1973, the Air Force and Navy had reached a compromise program that used elements from both the Navy and Air Force systems: the Air Force's signal structure and frequencies, and the Navy's satellite orbits. The system would also use atomic clocks, which the Navy had already successfully tested in its TIMATION program. On December 13, 1973, the joint program office pre-

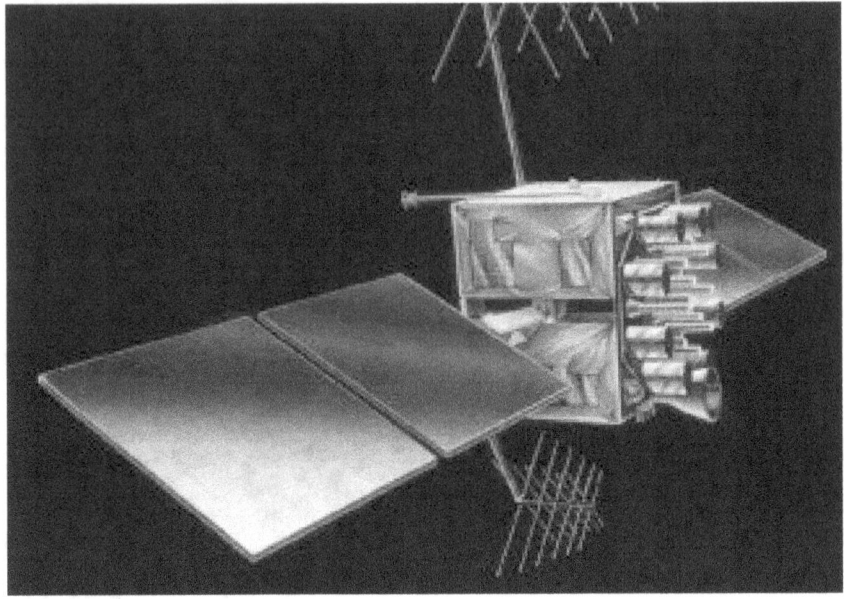

Artist's depiction of a GPS satellite in orbit.

sented a development concept paper to the Defense Acquisition Review Council. Deputy Secretary of Defense Clements approved the development proposal and authorized the first phase of a three-phase development effort. The initial four-year validation portion had a four-satellite configuration. On May 2, 1974, HQ USAF issued the program management directive and renamed the system the Navstar Global Positioning System.[17] It took twenty years, but in March 1994, the launch of the twenty-fourth Block II satellite completed the GPS constellation. Without a doubt, it was the combination of near-concurrent efforts of the Navy and Air Force, the efforts of visionaries like Getting and Easton, and an eventual spirit of compromise that made this revolution in precision a reality.[*]

---

[*]Although this paper does not attempt to discuss the development or deployment of GPS, there are a number of useful sources. One of the most detailed and useful secondary sources is Maj. Dennis Alford's ACSC Report, "History of the Navstar Global Positioning System, 1963–1985." Chapter II of that paper covers the Navstar Space Segment, Chapter III deals with the Control Segment, and Chapter IV concludes with the User Segment. There are also a number of useful journal articles covering development and deployment. See Lt. Col. John F. Scheer, "Navstar GPS: Past Present and Future," *The Navigator*, Winter 1983, pp 16–19; David A. Boutacoff, "Navstar Forecast: Cloudy Now, Clearing Later," *Defense Electronics*, May 1986, p 90–100; Christopher H. Clarke, " . . . And A Star to Steer By," *Defense Electronics*, June 1989, pp 57–64; "Navstar GPS Constellation Complete — 24th Satellite in Orbit," *Orbiter*, 16 March 1994.

Notes

1. Maj Dennis L. Alford, "History of the Navstar Global Positioning System (1963–1985)," Air Command and Staff College Student Rpt 85–0050, p 1; Aerospace Corp, "The Global Positioning System: A Record of Achievement," (1994), p 3.

2. Ivan Getting and John Darrah, "Global Positioning, Time Transfer and Mapping," unpublished paper, 1996, p 1.

3. Aerospace Corp, p 3.

4. *Ibid*, pp 5–6.

5. Dean Bundy, "Time, Navigation, and Global Positioning, *Space Tracks Bulletin,* Nov–Dec 91, p 8.

6. *Ibid*, p 10.

7. Bundy, "Time, Navigation, and Global Positioning," p 8.

8. Ivan A. Getting, "The Global Positioning System," *IEEE Spectrum* (Dec 1993), p 37.

9. Aerospace Corp, p 6.

10. Getting, p 37.

11. *Ibid;* Aerospace Corp, p 7.

12. Getting, pp 36–37; Aerospace Corp, pp 6–7.

13. Alford, pp 2–3.

14. *Ibid,* pp 4–5; Getting and Darrah, p 74.

15. Bundy, p 10.

16. Alford, p 4; Getting, p 44.

17. Aerospace Corp, pp 8–9; Bundy, p 10; Alford, pp 5–8; Getting, pp 43–44.

Gen. Thomas S. Moorman, Jr., is Vice Chief of Staff of the Air Force. Commissioned through AFROTC as a distinguished military graduate, he has served in various intelligence and reconnaissance related positions in the U.S. and worldwide. While stationed at Peterson AFB, Colorado, in 1982, he became deeply involved in the planning for and organizing of the Air Force Space Command. During his Pentagon tour in 1987, he also provided program management direction for Air Force satellites, space launch vehicles, anti-satellite weapons and ground-based and airborne strategic radars, communications, and command centers. He additionally represented the Air Force in the SDI program. As commander and vice commander of Air Force Space Command, General Moorman was responsible for operating military space systems, ground-based radars and missile warning satellites worldwide, as well as maintaining the ICBM force.

The Space Revolution

Thomas S. Moorman, Jr.

I am delighted to have the opportunity to attend this luncheon and deliver another stirring speech on space. Today, my assignment is to discuss "The Space Revolution" in light of your theme, "Technology and the Air Force: A Retrospective Assessment."

Before I begin, I would like to take the opportunity to continue to encourage the Air Force Historical Foundation and you historians to capture the Air Force's colorful space history. The Air Force is not only blessed with superb historians like Jack Neufeld, Dick Hallion, Cargill Hall, and those at the commands and elsewhere, but also with a space program that is rich with achievements, extraordinary leaders, and first class technology. Space has earned a rightful place on the history shelf, we just need some books to fill up the space.

Last month, I addressed the Historian's Dinner Banquet, and my wife Barbara said I spoke too long. I promise to keep my prepared remarks to a reasonable length so that I can field some questions afterwards. In that regard I am reminded of Mark Twain's assessment of a successful preacher. If he spoke for five minutes, Twain said he would give two dollars to the offering. If he spoke for twenty minutes, he would give $5 dollars to the offering. If he spoke for thirty minutes, he would put nothing in the offering. You can take a good thing only so far. I hope to leave here with a few bucks credit.

As I said, I am going to spend a few minutes discussing "The Space Revolution." Not an easy task following two days of the experts discussing technological advancements in pivotal systems like jet engines, ICBMs, and munitions. This morning Cargill Hall gave a good summation of early space operations and George Bradley spoke on GPS. So let me offer the "revolutionary view" of space from a different perspective.

Folks often try to compare the birth of space with aviation's labor pains. I think they are fundamentally different, even though there are some similarities, to be sure, like nurturing fledgling industries, honing brave pioneers, and debating how to use assets. But, in my opinion, the space revolution continues to differ from aviation history because space has simply evolved differently. Our struggle has not been one of more satellites, faster satellites, or even bigger satellites. In fact, there are those who criticize the space program for being too

255

conservative and for having too few satellites that are not launched fast enough and are not big enough.

The key difference has been the politics of space and the legacy of classified reconnaissance, intelligence collection, weapons in space, and a diffuse customer base — policy makers, war fighters, cartographers, meteorologists, and many others.

With that as a backdrop, let me concentrate on what I believe are four discrete struggles that have punctuated space's development. These struggles, or perhaps creative tensions, have existed over the focus of the Nation's and Air Force's space program: technology or requirements, research and development or operations, Washington or the war fighter, and DOD or the commercial sector. Now, before the computer internet lights up with some creative summation like, "Tom Moorman says, 'DOD and commercial space are locked in death struggle for supremacy,'" what I am trying to convey is that the national security space program's evolution was greatly influenced by the dialectic between these competing forces. So, in the next twenty to thirty minutes I want to share my view of the space "revolution" through the eyes of these four struggles.

Technology Versus Requirements

The Royal British Astronomer who, in 1956, called space travel, "so much bilge" could not have been more wrong. Not long after this statement, the USSR's *Sputnik* launched the U.S. and the Soviet Union into a race for the "High Ground."

In my view, beginning in the 1960s, and into the mid 1980s, national security space programs were dominated by "technology push." To address the urgency attendant to a superpower Cold War, fears of nuclear annihilation, and shifting geopolitical alliances, space pushed a technology explosion in semiconductors, transistors, integrated circuits, onboard computers, focal plane technology, propulsion systems, and launch systems. This was, indeed, the stuff of rocket scientists.

Technology pushed us from low-resolution, to medium-resolution, to high-resolution film-based optical systems, and eventually into a totally new phenomenon called charge coupled devices — CCDs — which record pictures electronically. This saved spacecraft weight and power, increased mission capability, and increased a spacecraft's life span from weeks to years. Most notably, it permitted the first "near-real-time" spaceborne reconnaissance capability. By the way, CCDs are the mainstay of the Pentax pocket camera.

It is hard to imagine, but in 1953, the B–29, in its many roles as a bomber, reconnaissance, and observation aircraft, relied on four Pratt & Whitney reciprocating piston engines and *one thousand* vacuum tubes to power its equipment. Ten years later, Hughes flew the first geosynchronous communications satellite (SYNCOM II) with integrated circuits, thermal and radiation protection systems, and solar cell power. The "Global Village," connected by instant communications, was about to become a reality. Technology push led us to the pinnacle of space power from the moon race to intelligence collection.

However, beneath these and many other spectacular accomplishments, lurked a changing resource picture and a maturing view of how to more fully exploit space systems to support the war fighter. While the youngsters in the audience may think that the 1960s and 1970s were high cash-flow days, the truth is that Defense budgets were constrained and disconnected from requirements and needs. Remember Secretary Robert S. McNamara's efforts to link planning, programming, and budgeting into a system called the PPBS? Remember, too, that the war in Southeast Asia gobbled up Defense dollars, which forced the cancellation of the Air Force's Manned Orbital Laboratory, SAINT, and Blue Gemini space programs. Then, in the aftermath of the Vietnam War, the U.S. suffered from a debilitating inflationary spiral in the late 1970s. Despite tighter dollars, the budget for space programs was increasing in the mid- to late 1970s, driven by the development of Global Positioning System and the fielding of advanced communications systems, such as the Defense Satellite Communications System (DSCS).

In addition to the rising space budget, the Air Force was beginning to foresee a time when we would become more dependent on space systems. The Defense Meteorological Support Program (DMSP), once a highly classified system, had been employed in Vietnam and was becoming the source for tactical weather forecasts to the battlefield. The Defense Support Program (DSP) was providing increasingly valuable data on an aggressive Soviet ICBM buildup. On top of that, the Soviets had an operational antisatellite system which threatened our low-altitude satellites. Finally, with slight increases in the space budget and heightened space dependency, the military leadership began to realize that the Air Force space program needed a new paradigm — in a mission area that had owed its advances to the civilian side of the bureaucracy.

Given all this, the era of technology push was coming to an end as the decade of the 1980s began. The ground had now been prepared for the era of "requirements pull," which would more equitably balance space costs and performance with customer needs. "Requirements pull" in the Air Force was born with the creation of Air Force Space Command in 1982.

This shift in focus reflected a growing belief within the Air Force leadership that it was time to "normalize" space. A revolution took place as space was organized under standard Air Force processes and practices. A requirements regime was established and contact with the war fighter initiated through the TENCAP and space application initiatives. Now that was revolutionary. To ask the war fighter, "How can space support you," rather than, "here is my system, here is how it will help you, and you *will* like it!" Space crew selection and space operations were normalized. Eventually space research, development, and acquisition mirrored aviation with space laboratories, product

Two military communication satellites — DSCS II and DSCS III — are launched on a TITAN 34D/IUS from Kennedy Space Center, Florida.

centers, and acquisition specialties which provided engineering and support, not operations and requirements.

Over the years, this struggle has settled into a healthy relationship that taps the creative energy of a full spectrum of Air Force people, from technologists, to developers, to air crew members. While the government was once the stimulus for new technologies, always pushing the envelope, today we have more technologies than we can exploit. We are prioritizing and refining our structured requirements process to make the right choices. I would say that identifying space as a "core" Air Force mission solidified the transition to a requirements pull space program. Today there are over 30,000 people working in Space Command, and they help execute an Air Force space budget of about $5 billion.

Research and Development Versus Operations

In this Homeric struggle, the space program listened to the siren's R&D song long enough to lay the ground work. Then we had to put wax in our ears and sail on to operations. What were the dynamics that made an R&D based space program so attractive for many years? I will cover the reasons that explain the R&D focus, which I believe was necessary at the time. It served us well and, with the proper balance, will continue to serve us.

For centuries, man has gazed at the "man in the moon," and Jules Verne spurred the imagination of millions of would-be space travelers. But credit for real space exploration rests with the extraordinary engineering skill and vision of people like Wernher von Braun; Gen. Bernard Schriever; Gen. Curtis LeMay; Maj. Gen. David Bradburn; Bill King; Simon Ramo, Dean Wooldridge, and Charles Thompson (founders of TRW); Arthur Clark from Hughes; and many great scientists and engineers from NASA. Even the early test pilots were engineers, as were the early Atlas, Titan, and Minuteman launch crews. Why was R&D so critical?

Systems were complicated and ponderous by today's standards. Successful operations and checkout hinged on the "Tyranny of Tubes." Then as now, systems were continually improved and refined to incorporate the latest technological advancements which would allow us to perform more missions, for longer periods, affordably. Every space event was a happening, an experiment; nothing was routine. Consequently, we used engineers in white coats, not checklists. Each satellite and launch vehicle was different. Again, engineers in white coats — not checklists. I know, you are thinking, "So what's changed?" Well, notwithstanding space launches, quite a bit, and I will get to that in a minute.

An outgrowth of the R&D mindset were large satellites, beginning with Telstar's 117-pound low-earth-orbit satellite. Systems are now referenced against 30,000 pounds to low-earth orbit and 10,000 pounds to geosynchronous orbit. It is Parkinson's Law of Spacecraft Design. While not terribly launch responsive, large satellites have proven to be cost and mission effective. One of the reasons we built them so large was to offset the increasing cost of launch. Technological improvements resulted in longer lived satellites, which pushed replenishment rates to the right, lowered the production rate of launch vehicles, ultimately pushing up the price per launch. The sublime paradox was established: the better we got, the more it cost.

Now before I am accused of being an apologist for the past, it is instructive to note that we are making progress in downsizing spacecraft. DMSP and GPS satellites, at about 2,000 pounds, are well within the medium class envelope and even DSCS and DSP are considered small "heavies." By the way, SBIR is being designed to fly on the medium-class EELV, I hope.

During this period, the military-industrial complex thrived. During the 1960s, new technology companies were founded at a rate of 400–500 per year.

Artist's depiction of a DMSP satellite in orbit.

Competitors strove for revolutionary and incremental advantages. Major advancements in communications, optics, digitization, lasers, radars, and propulsion systems helped spur and maintain a decided space R&D flavor. It did not hurt that Air Force Research and Development Command (later Systems Command) controlled the funding, programs, and program offices.

A real key issue was certainly classification. If anyone has reviewed the early copies of *Aviation Week* — the late 1950s and early 1960s versions — you will have noted that the Air Force and the other services were proud of their many space programs and technologies and openly discussed their war fighting and reconnaissance programs. Scaring the Soviets almost seemed to be the approach.

But that all changed under President John F. Kennedy, when he, Defense Secretary McNamara, and Secretary of State Dean Rusk agreed to classify virtually all space programs, launch dates, missions, and capabilities. I cannot quarrel with the foreign policy, diplomatic, and national security reasons at the time; however, the decision helped insulate the space community from potential operational users; and while probably a necessary sign of the times, I cannot overemphasize how the "Green Door Syndrome" retarded the development of space applications for the war fighter. So it is refreshing to see some of those walls finally coming down, as was done with the Corona program recently.

The Space Warfare Center, Colorado Springs, Colorado.

As predicted, an operations emphasis certainly brought a different focus to space and its applications. We codified the transition from R&D to operations back in 1982 by standing up Air Force Space Command, followed by the Army and Navy commands and the United States Space Command. Over the last thirteen years, the commands have overhauled space requirements and operations efforts, standardized space operations, and drawn together the operator and the war fighter; and we have stood up the Space Warfare Center in Colorado Springs. Space Command and the unified commands have space teams that go to the field and work the space issues and assist the theater commanders in chief to integrate space into their operations plans.

But the defining event for space operations was clearly Desert Shield and Desert Storm in 1990. That six-month and 100-day exercise not only proved the operational value of space support and space products, it solidified space as a core Air Force mission. It was shortly after Desert Shield/Desert Storm that our Chief of Staff, Gen. Merrill A. "Tony" McPeak published the Air Force mission statement: *Defending the United States through the control and exploitation of air and space.*

Now that operational requirements for space systems are supreme, we are beginning to look at a host of projects that can satisfy war fighter needs, while ensuring headquarters policymakers are not shortchanged. For example, with the EELV (Evolved Expendable Launch Vehicle) program we are trying to improve launch responsiveness, while drawing down the costs to get to space. Ultimately, RLVs offer an extraordinary opportunity to leapfrog into a true launch-on-demand regime. Satellite size and weight continue to reduce, driven by two engines — SDIO/BMDO and the telecommunications industry.

Technology and the Air Force

I expect the Space and Missile Tracking System to be the first widely deployed Defense "smallsat." Ground weather terminals are no longer Mark IV vans the size of C–141s, but Small Tactical Terminals the size of HUMMM–Vs. The entertainment industry has driven an explosion in bandwidth. An example is the Global Broadcast System, similar to Hughes' six hundred dollar commercially available systems. The military version will use eighteen-inch terminals, instead of eleven-meter DSCS terminals, and will deliver twenty-four gigabytes of data ranging from intelligence, weather, medical, logistics, air tasking orders, and a host of other products.

I could go on for a long time, but the point is that with the operational flavor and the proper role of the military user in the requirements process, space systems are more responsive, user friendly, and better integrated into our force structure. That is the leading edge leverage that no other country can match.

Washington Versus the War Fighter

As with many large technology-driven efforts, national security space programs have carried a Washington flavor which has befuddled the services for years. Let me briefly share my view. This struggle had its roots in the very survival of the nation in the 1950s and 1960s. The U–2 shoot down, other intrusive and dangerous airborne reconnaissance missions, the missile and bomber gaps, "Uncle Joe's" nuclear tests, and insurgency and guerrilla wars across the globe all contributed to a sense of urgency to field space-based warning, detection, and reconnaissance systems and to make the products of those satellites available to policymakers and intelligence agencies daily.

These systems not only kept the United States in the space race with the Soviets, but provided an increasing strategic and tactical advantage during the war in Southeast Asia and the SS–20 missile buildup in Eastern Europe. Perhaps space's greatest contribution was its critical underpinning of the strategic arms control initiatives with the Soviets.

In addition, Washington support ensured that our space programs — civil, military, and intelligence — contributed to our national strength. The United States developed a corps of astronauts, walked on the Moon, and sent satellites to the farthest reaches of the solar system. The irony may be that a Defense Department developed navigation system, like GPS, has infinitely more civilian or commercial uses than military ones.

Nevertheless, like a rite of passage, without Washington-oriented systems providing strategic intelligence collection, analysis, and arms control monitoring for twenty-five years, it is possible that these systems would not have garnered the Congressional support and focus needed to make the transition to more directly support military users years later. As I said earlier, Desert Storm clinched the operational niche, and the race has been on for the last five years to strengthen the war fighters' interface with space systems.

Military Versus Commercial

The last trend I would like to comment on is the relationship between the DOD space program and the commercial space program. The 1958 Space Act created a civil NASA and the national security space program. With creation of the National Reconnaissance Office in 1961, there were, effectively, three government sectors. But commercial space entrepreneurs were not recognized until President Jimmy Carter's national space policy did so by identifying a fourth sector — the commercial sector. U.S. commercial communications satellites quickly dominated the market and opened the world to the "Thrilla in Manila" boxing match, Moon walks, and paperless commerce and trade — all were pipe dreams just a few years earlier. Today, America produces 80 percent of all the space-based communications satellites and is expected to increase this share as we migrate into personal communications, digital communications, and direct broadcast.

Commercial endeavors were greatly enhanced by Reagan and Bush space policies. Today, there should be little doubt that the commercial sector is here to stay. Estimates put U.S. domestic commercial satellite sales at $2.5 billion per year, and a total commercial space market at about $4.5 billion. However, this growth came with a cost — the Challenger accident.

A legacy of that flight was a new policy preventing commercial payloads from flying on the Shuttle and the resuscitation of the comatose Expendable Launch Vehicle (ELV) line. Marco Polo was the first commercial launch, in 1989, on a Delta II from Cape Canaveral, and the Air Force has been cooperating with the U.S. commercial launch industry in making government factories, launch teams, launch base processing, launch facilities, and range support available to commercial users on a shared basis with the active government missions ever since.

These cooperative arrangements, based in Presidential policy and federal law, have enabled the U.S. commercial space launch industry to do about $500 million in business each year. Declining government launch rates in the next few years will allow them to do even more business, since more capacity of our infrastructure is available to support commercial missions. In fact, 1995 is the first year when commercial ELV missions outnumbered government launches, and we expect that to continue for several more years to come.

In response to this increasing commercial activity on our bases, we are nearing completion of a comprehensive policy that addresses how we assign use of the capacity of our launch pads for both government and commercial use. Our objective in this policy is to optimize use of active Air Force space launch complexes, associated infrastructure, and ranges to accommodate national security, civil, and commercial users.

Another aspect of successful Air Force cooperation with the private sector and state governments resulted from the awarding of $20 million in dual-use

Launch of an Atlas IIA carrying a commercial broadcasting satellite.

space launch infrastructure grants in Fiscal Years 1993 and 1994. As a result of this program, in which state governments and private industry provided literally millions of dollars in matching investments, we now have three state spaceports under construction, with two more in the planning stages, a new commercial payload processing facility in operation on Vandenberg AFB, California, and some substantial improvements to our current processes that support both government and commercial missions. To facilitate these development projects and their commercial operations, we have instituted real property instruments to arrange for use of land and facilities on Air Force launch bases and simpler agreements for Air Force launch base and range support.

The development of state government-sponsored spaceports on our launch bases, side-by-side with a variety of commercial space launch industry activities, some of which were built using private investment, has opened a new vista on policy questions surrounding use of Air Force resources and the commercial industry. In response, we have developed detailed guidelines regarding how to make real property available for commercial use. As defined by the President's National Space Transportation Policy, our role is to encourage private sector, state, and local government investment and participation in development and operation of space launch systems and infrastructure. Our objective is to ensure that when we make Air Force property available to support state spaceports or commercial space activities, we do so without providing an unfair competitive advantage to any one organization over another.

New policy issues also exist as U.S. commercial launch providers team with foreign space launch and satellite houses. Lockheed's partnership with Proton and Boeing's possible alliance with Ukraine's Zenit reflect an emerging cooperative international flavor in a heretofore protected business base. What are the space policy issues and will there be unacceptable impacts on DOD's space processes and infrastructure?

I do not have time to go into all the potential capabilities and policy implications of relying on commercial remote sensing, imagery, environmental monitoring, communications, commercial launch, and perhaps satellite command and control for military and intelligence space operations. But one can probably infer that the Air Force faces a critical debate in the near future over how we can integrate commercial products and services, while at the same time maintaining a core military space capability to fight wars and provide uninterrupted support during national security emergencies. This is not a trivial Air Force exercise. If commercial satellite builders can provide capable and affordable systems and products, what does the Air Force acquire and operate organically, and how does the Air Force configure its commercial assets to support military space operations? The struggle to find the right balance continues.

Closing

Well, let me end on a more upbeat note. The "space revolution" has occurred. It has been a thirty-year effort to align the programs, policies, organizations, and resources. Whenever I pause to think that the major tasks are done, new challenges come along, like accommodating commercial providers, integrating space products into weapons system designs and ultimately into the cockpit, and planning for Reusable Launch Vehicles. What this means is that space is an exciting place to work and will be for some time to come.

The words of Gen. "Bennie" Schriever, when he spoke of the many "naysayers" who can always come up with reasons why a new idea will not work, come to mind. But, he said, "The people who produce progress are a breed apart. They have the imagination, the courage and the persistence to find solutions." Our space community typifies this spirit.

Col. George K. Williams, at the time of this symposium, was Commander, Air Force History Support Office. He is now a member of the faculty at the National War College. After graduating from West Point, he earned an M.A. from Cornell and a Ph.D. in history from Oxford University. His Army service includes a Vietnam War tour as a company commander with the 1st Cavalry, Americal Division. He was an instructor at West Point and the Air Force Academy. In 1977 he transferred to the Air Force, serving as an air weapons controller, director, and AWACS battle staff member. He was Chief of Staff and Deputy Director, Operations and Requirements, 28th Air Division. In 1987 he became Vice Commander USCENTAF FWD, Saudi Arabia, and later ELF ONE Commander. In 1989 he was Chief, NATO AEW at SHAPE in Belgium. In 1992 he became Deputy Air Force Historian.

AWACS and JSTARS

George K. Williams

I am reminded at the outset of what Gen. Curtis LeMay said. It may be apocryphal, but he said, "Little airplanes are more fun. Big airplanes are more important." And if you think of AWACS, you soon realize it is not necessarily just an Air Force asset, it is also a national asset. Wherever the AWACS goes in the world, it is generally welcome because it is nonthreatening. The E–3 has no offensive weapons aboard, and no spy systems. It has a glossy paint job, and it is widely acknowledged as a major commitment of the United States whenever it is deployed. During my talk I will refer fleetingly and in no great detail to William of Occam and Jean Paul Sartre, with a little bit of Woody Allen thrown in.

One advantage of being the commander of the History Support Office is that the Air Force declassification team works for you. And I was able, with some urging and modest requests, to get them to declassify some documents on AWACS operations. In fact, some of this material was classified until about a week ago. I also must thank Col. Frank Welty, formerly with the NORAD detachment performing airborne battle staff duties aboard AWACS, for reviewing this draft.

I will look very quickly at the history of aerial surveillance and then at AWACS, the Airborne Warning and Control System. There were dual requirements written originally because AWACS was contemplated to fall under two commands, Air Defense Command and also Tactical Air Command. At the time, both commands thought they knew what they wanted. However, with some experience in development, the requirements were changed as the concept of operation was formulated. We will look at the major requirements and then, over time, the major scenarios and the operational history of the airplane and how that impacted enhancements, the upgrade programs. Then we will take a blind stab at what might be available in the future.

The evolution of U.S. military reconnaissance, highlighted by the introduction of observation balloons in the Civil War, the airplane just prior to World War I, and airborne surveillance radar during the final phases of World War II, in a sense paved the way for the development of an airborne warning and control system by the United States Air Force in 1972.

Technology and the Air Force

Holding the high ground, whether a hill, a tower, or a castle, has a number of advantages both for offensive and defensive roles. With unimpeded surveillance, one's antagonist can be observed to determine the strength of his forces, dispositions, and movements, along with some potential insight into his intentions.

In 1783 Etienne and Joseph Montgolfier, sons of a French paper manufacturer, flew a balloon over Paris for twenty-five minutes, traversing a distance of five miles in November 1783. During the American Civil War, balloons using hydrogen or coal gas, rather than hot air, could remain aloft almost indefinitely. Once helium was discovered, its fireproof qualities were appealing, but its near-astronomical cost — $1,100 (1918 dollars) — per cubic foot and limited production made hydrogen the only feasible medium for inflation. The Great War added airships and airplanes to the assortment of platforms that could be exploited for visual or photographic observation, albeit without benefit of radar.[1]

The Korean War did not see a fielded radar capable of distinguishing moving targets from the ground clutter of surface returns. The 1950s saw the advent of improved radar-mounted aircraft, most notably the Royal Air Force "Avro Shackleton," the USN WV–2 and the USAF EC–121 "Warning Star," a modification of the Super Constellation civil airframe.[2] Vietnam was little improved; additionally, strong enemy air defenses magnified the tactical problem. In large measure, this hostile environment provided the impetus to obtain a much better airborne surveillance, command, and control.[3]

All of these airborne early warning (AEW) systems in operation prior to the 1960s were overshadowed by the Navy's E–2A "Hawkeye" early in the decade. With five tons of specialized avionics, including the AN/APS–96 radar system with its antenna in a rotating radome — "rotodome" — some eighteen feet in diameter, the Hawkeye employed a computerized surveillance capability linked to the Naval Tactical Data System. However, as with its predecessors, the Hawkeye suffered from a severe limitation: it could provide successful surveillance coverage over water, but lacked the ability to detect and track targets amid ground clutter over land.[4]

Fortunately, a radar technology emerged that could in fact detect and track airborne targets through surface clutter. Called pulse Doppler radar because it processes pulses and detects the Doppler frequency shift — up or down — of the moving target, it was first used operationally by the USAF in the Boeing BOMARC IM 99–B Interceptor Missile in the late 1950s. It has been used in the radar of the USN McDonnell F–4J aircraft, in the Hughes APG–63 Radar aboard the McDonnell F–15 Eagle, and in the Westinghouse APG–66 Radar of the USAF F–16 Fighting Falcon, as well as in the AWACS. The principle is obviously here to stay.[5]

Neatly paralleling this effort during the mid- to late 1950s was the realization of the need to detect and track low-flying targets for surveillance, com-

EC–121 aircraft at Tan Son Nhut Air Base, Vietnam, in 1965.

mand, and control. Earlier work on airborne navigation systems had focused on using continuous wave Doppler techniques to detect and track an object's velocity. Some of these techniques appeared to offer promise as an approach that would be useful in solving both the low-altitude-target air interceptor and surveillance problems. Efforts in this direction proved successful when pulse techniques were combined with the Doppler approach.[6]

At about the same time, the USN was planning a new fighter system for fleet defense. Their concept required a long-time-on-station combat air patrol (CAP) aircraft equipped with a track-while-scan radar and long-range air-to-air missiles. The system was to have a multishot simultaneous attack capability to minimize the number of aircraft needed at the extended ranges of the CAP. Because potential enemies could fly low, it was necessary for the proposed interceptor to be able to look down. This system, later cancelled in the budget planning process, became known as the Long Range Missile Fighter, and its missile and fire control was known as the Eagle Missile System.[7]

Proof of the technical feasibility of this and other related technologies generated considerable interest in the USAF for developing a new airborne surveillance platform. In separate documents of qualitative operational requirements, both the Tactical Air Command and the Air Defense Command defined systems similar in basic concept, but considerably different in their specific target requirements. Because of their continental U.S. defense needs for long-range detection and tracking of large numbers of inbound, maneuvering targets, the ADC version was the more technically demanding of the two approaches. ADC called their scheme the Air Defense Command Post and TAC called theirs the Airborne Tactical Command and Control System. After sessions with Systems Command, TAC and ADC personnel agreed on a compromise joint Specific Operational Requirement (SOR). In 1963, this joint TAC/ADC SOR 206 entitled "Airborne Warning and Control System" was issued.[8]

Released on January 12, 1963, the joint SOR for TAC and ADC addressed six system requisites that would form the foundation of the future AWACS: provide quick response for airborne warning and control in conjunction with overseas operations; search for, detect, identify, track, and direct weapons against enemy threat aircraft; supply vector information for close air support, tactical reconnaissance, troop and cargo drop, and air interdiction missions; extend tactical ground warning and control coverage to areas where tracking by ground sites is impossible; furnish ultra-high frequency radio relay; and replace or augment ground-based Control and Reporting Centers and Control and Reporting Posts. Thus, the USAF delineated the general objectives of an AWACS but not the technology which could put it into effect; as yet, no such system existed.[9]

On July 12, 1963, Secretary of Defense Robert S. McNamara directed an investigation of AWACS technology. The Electronic Systems Division (ESD) of the USAF Systems Command accordingly set up a system program office and dutifully began to identify candidates for developing the airborne platform: the Lockheed EC–121, the Douglas DC–8, and the eventual winner, the Boeing 707. Similarly, four candidates for overland radar techniques were selected: Westinghouse (the winner), Hughes, Raytheon, and General Electric.[10]

Three years later, on September 1, 1966, TAC and ADC issued their joint Required Operational Capability Report with these specific requirements for a usable AWACS and its airborne platform: quick response capability to developing threats; ability to relay early warning and air surveillance information; onboard facilities to direct and control defensive and offensive weapons; augmentation or replacement of TAC and ADC control elements; ten-hour continuous operation, 1,000 miles from home base; a minimum cruise altitude of 35,000 feet; an electronic counter-countermeasures capability; a range resolution of one nautical mile; a specified height accuracy; ability to detect and track targets at speeds of at least Mach 4.5; and a high order of ground mapping, crew comfort, beacon mapping, data processing and display, navigation, communications, system reliability and supportability, and nuclear/EMP survivability.[11] ESD was given, for the first time, primary responsibility for an aerial vehicle, the AWACS system.

What emerged to satisfy all these operational requirements was the E–3A. Based on the commercial 707–320 airliner, some thought was tentatively given to using eight engines, but this did not survive initial consideration. In 1975, the first Westinghouse AN/APY–1 radar was built into the first E–3A. The most striking external feature, the rotodome — "Frisbee" — housing the main radar antenna and the identification-friend-or-foe (IFF) antenna, is mounted on two titanium struts above the fuselage. Appearances are deceptive: at the hub the rotodome is six feet thick, with a diameter of thirty feet. The dome weighs

An E–3 AWACS out of Kadena AB, Okinawa.

nearly 3,500 pounds and is angled downward 2.5 degrees to minimize aerodynamic moments. This assembly rotates at two speeds, 1/4 revolution per minute (the "idle," used to keep the bearings lubricated) and 6 revolutions per minute (once every ten seconds) while on station. The radar beam is electronically scanned in elevation; azimuth scan is, of course, achieved by the physical rotation of the rotodome.[12]

The phenomenon of Doppler shift allows the AWACS radar to distinguish between radar energy reflected from moving targets and the energy reflected from the earth's surface (ground clutter). When energy from a radar pulse is reflected from a target, a change in radar frequency occurs if there is motion relative to the AWACS radar. If the target in the radar beam has a component of its velocity vector moving away from the E–3, the radar pulse returns at a frequency lower than that transmitted; the converse also applies: approaching targets, higher reflected radar frequency. The AWACS onboard computer uses the Doppler shifts, coupled with the E–3's own velocity, heading, attitude and position data from the navigation system to generate target reports for those targets moving relative to the ground, rejecting the rest as clutter. For design purposes, some lower limit of velocity has to be selected as a cut-off speed for the moving target indicator (MTI). This figure balances the trade-offs between operational requirements to detect slow movers such as helicopters and technological feasibility — the computer state of the art in the 1970s.[13]

At long ranges, the Doppler-shift technique becomes unnecessary, since clutter diminishes or becomes nonexistent beyond the horizon. In such surveillance modes the AWACS radar can function as a conventional pulse radar for greater efficiency. For the E–3, five basic operating modes for the radar are possible, and the radar can change its mode from scan to scan or from one sector to another within a scan. For maximum long-range performance, a beyond-the-horizon mode may be used, but if good range resolution is required, the shorter range, pulse-Doppler, non-elevation scan can be used to obtain elevation data on the target of interest. Also, a passive nonradiating mode can be employed, generally in an ECM environment, for passive tracking of emitting targets. In the radar maritime mode, the velocity threshold of the moving-target indicator circuitry can be reduced approximately to zero, so that slow-moving surface targets or even stationary objects can be displayed. Among other techniques, a very short pulse is used in this mode, reducing the amount of sea or ground clutter in the return signal.[14] The region of greatest clutter, the land-sea interface (the shoreline) can be electronically blanked within the computer by using stored maps of land areas. Upgrades to the onboard computer memory were a prerequisite to enable the maritime mode of detection to be used.

No raw sensor data of any kind is provided directly to the onboard mission crew; everything is first processed by the IBM System 4Pi computer. Computational power thus represents a critical parameter for success. On the E–3, the 4Pi computer occupies the center of the information web, correlating sensor and data link inputs with its own geographical position data from the onboard navigational system, whether INS (inertial) or satellite (Omega), to present a coherent, accurate, near real-time situation display to the mission crewmember. This computer in its current version — the CC–2 — has a core memory capacity of approximately 640,000 words, a five-fold increase over the original version. In the late 1970s, the Computer Data Management Technician on the E–3 took great pride in pointing out that his wall-locker-sized computer had the same capability as the old NORAD computer that had occupied a entire floor in the region blockhouses. (An operator at a single console on the E–8C JSTARS has at his position more computer capacity than an *entire* E–3 airplane and mission system. Time marches on.)

As technical development proceeded, the Air Staff named TAC as the single manager of the future U.S. AWACS fleet, programmed to include thirty-four aircraft operating from a single main operating base — later chosen as Tinker AFB, Oklahoma. "The mission of AWACS," the new concept of operations stated, "is to provide worldwide responsiveness in the employment of its unique capabilities for all-altitude surveillance, warning, and aircraft control in a variety of tactical, strategic, and special mission applications."[15]

The CONOPS also estimated peacetime AWACS requirements for various locations, particularly its contemplated role in Europe: "The AWACS in Europe will complement, supplement, and provide additional capabilities that do not currently exist within NATO. This includes deep-look surveillance, extended low-level coverage, and interface with external systems." For Europe, the Air Staff computed that five aircraft could provide a ground alert force to fly daily training and surveillance sorties and still be capable of supporting operations twenty-four hours a day when required. A planned detachment of three AWACS

272

at Keflavik Air Station, Iceland, would extend CINCLANT coverage in the Greenland-Iceland-United Kingdom gap. These E–3s would replace EC–121s, which had covered the strategic gap between Greenland and Norway continually since 1968.[16] For a major conventional war in Europe, this in-theater AWACS force would be augmented from the CONUS main operating base to provide four or more additional surveillance orbits. Through interfaces with the NATO ground and naval systems, AWACS would act primarily to extend the range of the surface networks, mainly long-range surveillance and low-level detection.[17]

USAFE was also interested in AWACS to respond to potential contingencies in the Mediterranean and Middle East areas, conceivably "out of area" for the NATO Alliance. To respond to a medium-intensity conflict in the Middle East, TAC's studies indicated that an AWACS force could be deployed within twelve hours, compared to a minimum of eighty-three hours' arrival time for deployment of ground tactical air control elements. On this basis, TAC found the AWACS a more cost-effective investment than spending funds on additional ground control capability, such as the 407L ground-based system.[18]

By 1973, TAC and USAFE had identified these potential uses of AWACS in the European Theater: extend high- and low-altitude radar coverage; monitor Warsaw Pact airpower for early warning; improve peacetime and wartime intelligence gathering; fill gaps in low-level radar coverage; transmit radar picture to control posts with disabled antennas; give advanced tracking data to NATO SAM radars; indicate location and status of friendly ground forces through beacons; monitor location and status of friendly surface vessels; detect and track enemy ships over large area; vector friendly naval and air forces to reconnoiter or to attack enemy ships; track aircraft despite chaff or electronic jamming; provide a backup to air traffic control facilities; control emergency airlift to remote disaster areas; monitor sensitive reconnaissance and special interest flights; assist in deployment of rapid reaction forces; serve as initial command and control system during contingencies; assist tactical airlift forces; help with air refueling operations and rendezvous; assist ground control agencies by assuming control of air battle sectors; enhance survival of friendly strike and attack forces; provide airborne control of remotely piloted vehicles; direct combat air patrols and provide threat warnings; identify location of downed aircraft and direct rescue efforts; assist in reconstitution of forces after a nuclear exchange; help resolve problems of radar control center interoperability; and give senior commanders an overall picture of the ground/air battle.[19] One should keep in mind that this list of functions — now over twenty-two years old — is for the USAF AWACS in Europe and not for the NATO AEW fleet.

Reflecting the growing awareness of the importance of an airborne warning and control system within the North Atlantic Alliance, USAFE's concept of employment enhanced the concept that NATO procure its own AWACS force with common funding from the member nations. In principle, the AWACS would merely serve as an airborne extension of NATO's existing air defense system, which had been welded into an integrated command structure in 1960. The idea that NATO operate its own multilateral force of AWACS aircraft offered obvious military advantages, but also posed political and economic problems. For one thing, the USAF concept of operations did not cover the issue of transferring operational control of the USAF AWACS deployed to Europe to NATO, even though most of USCINCEUR's other national forces transferred to the Supreme Allied Commander Europe and his subordinate wartime headquarters.[20]

However, by mid-1976, NATO's defense ministers concluded without much debate that their respective governments could not afford the force of AWACS their military advisors said they needed to shore up Alliance defenses. Cost had plagued AWACS since its inception. In terms of the Alliance, two implications followed from the $75 million per airplane cost. First, the NATO acquisition effort would have to be collective, rather than relying on nationally owned weapon systems. The high price also implied a high political cost, since few democratically elected leaders could afford the risk of seeking funds without convincing arguments. Value, after all, is a matter of opinion. Even though NATO's Military Committee eventually identified AWACS as a priority-one requirement, their judgment rested primarily on the basis of military imperative.[21]

Over the next few years, the shape of the NATO AEW program began to emerge. The British decision in 1977 to make the Nimrod Mark 3 its contribution to the NATO AEW force meant that the Federal Republic of Germany's role in the E–3A program became even larger. The issue of industrial benefits — the Alliance insistence that Allied defense purchases should follow a "two-way street" and not solely benefit the United States — made the AWACS issue even more complex. Before the dust settled, German sales of their 120-mm main tank gun, similar contracts for E–3 subsystems among Alliance nations, and the selections of the NATO AEW main operating base and deployed/dispersal airfield locations in Norway, Italy, Greece, and Turkey would shape the final agreement. Altogether, the expense of acquiring the NATO E–3A with its related equipment and facilities totalled slightly more than $1.8 billion, not including the costs for continuing operations and support or any contemplated upgrades.[22] Unquestionably, the interrelationships between operational requirements and industrial benefits to be negotiated among the member nations shaped the overall dimensions of NATO AEW. Considerable encouragement was given to NATO to procure their own fleet.

When the NATO AEW program began to emerge, the British, not unexpectedly, decided to go it alone, with an improved system postulated as the Nimrod Mark 3. This was a system which the Royal Air Force had yet to invent, but they assured NATO that it would be cheaper and more powerful than buying an AWACS. There was considerable industrial benefit as well as national pride

A NATO E–3 AWACS aircraft.

associated with it. At that, the Nimrod Mark 3 seemed to hold considerable promise in the design. It had a big radar in the nose and a big one in the tail and a computer in the middle. But every time the boffins sent radar impulses to the computer, they would get clouds of white smoke. (I think frankly it was just too ugly to fly.) Eventually the UK bought their own AWACS, the Boeing E–3D model, coming into the program at a time to take advantage of some of the operational improvements.

In monetary terms, the largest element of the overall program was the acquisition of the eighteen NATO E–3As. These aircraft are based on the Boeing 707 airframe, with mission equipment built to a standardized US/NATO design. Additionally, the second major element of the NATO AEW program was the extensive upgrade and automation of 40 existing NATO Air Defense Ground Environment sites ranging from northern Norway to eastern Turkey, with four locations in the United Kingdom. This upgrade, the AEW Ground Integration Segment (AEGIS) was accomplished over the years 1979 to 1988 at its own cost of over $350 million.[23] Aside from the USAF AWACS fleet, NATO has the largest stake in the E–3 community and is a major partner. Other national AWACS forces include the British RAF; the French element; the Royal Saudi Arabian Air Force (RSAF), with matching KE–3 aerial tankers; and, lately, the Japanese, using the Boeing 767 aircraft.

To make sense of all the political, operational and technological influences on the AWACS concept, it is best to simplify, or oversimplify somewhat, what it has been required to do to justify and establish itself. From its inception, and indeed throughout its twenty-year career, the AWACS system has had to satisfy three broad requirements (surveillance, communication, and arrive/survive) and three major mission scenarios (continental air defense, conventional conflicts, and contingency responses). The requirements first — and this is where Columbus, William of Occam, and Woody Allen figure.

Surveillance

AWACS has to survey a worthwhile volume of airspace to make sense of the airborne objects it detects, tracks and identifies therein. For radar surveillance of a flat earth, the main design problem would be the transmitted power of the system. Any surface system could see as far as any other postulated radar, since the horizon would lie, conventionally, at infinity. Christopher Columbus established — to nearly everyone's satisfaction — that the earth is indeed spherical. Therefore, the higher one goes, the farther one can see, in accordance with the approximate formula, "Distance equals 12.3 times the square root of the Flight Level, expressed in hundreds of feet." Specifically, at 29,000 feet (the elevation of Mt. Everest), the horizon shadow line falls at approximately 210 nautical miles. Increases in altitude do not significantly extend this horizon; at 31,000 feet the horizon lies at approximately 216 miles from the observer. Most of the world's current AWACS aircraft operate within this broad band. With the 707-based AWACS in the USAF and NATO fleets, design and physiological factors begin to offset any considerations of very high altitude orbits.

To make sense of the airborne objects it detects and tracks, the AWACS has to reduce operational uncertainty and ambiguity, generally by identifying or otherwise categorizing them. One must keep in mind that targets in the AWACS surveillance volume do not fall neatly into two categories according to, "If we cannot positively establish that air track as FRIENDLY, it must be HOSTILE." Negating this seductively attractive approach is the fact that a very large classification, a third category, exists — that of simple UNKNOWN. Sorting out the tracks of interest from these unknowns, in such a fashion that an appropriate response can be marshalled, remains a continuing challenge. The AEGIS cruiser *Vincennes* incident in 1988, when the Iranian airliner was mistakenly shot down, and the more recent shootdown of two U.S. Army Blackhawk helicopters in Iraq merely underscore the point. In a promotional pamphlet, Boeing touted the E–3 as "the alternative to uncertainty." This in some ways merely updates the fourteenth-century pronouncement of William of Occam, who intoned that "Entities should not be multiplied unnecessarily" — an observation that incidentally calls to mind Sherlock Holmes and the U.S. Army Infantry "KISS" exhortation — "Keep it simple, stupid." AWACS employs a variety of identification means, including IFF passive systems and information from other sources to reduce operational ambiguity.

Communication

Second, the AWACS has to communicate with other aerial and surface command and control elements within some coherent system. The E–3 must receive as well as transmit data to be effective; it is almost inconceivable that it could undertake a strictly autonomous mission. In fact, it is doubtful that an AWACS has ever flown either a training or an operational sortie without engaging in an exchange of information with other agencies. The multiplicity of links — computer, voice, teletype — also has a multiplicity of advantages and limitations. Not the least of these is the technical challenge of mounting so many HF, VHF, and UHF antennas for JTIDS and Have Quick frequency-hopping algorithms in close mutual proximity on a single airframe, under a high-power rotating radar and IFF radome. The preferred term for all this electromagnetic aggravation seems to be "co-site interference," which increases almost logarithmically as the number of onboard emitters proliferates.

For the world of acquisition, the number of candidates qualified and likely to bid on a proposal for any system to be fitted to the E–3 fleet approaches unity, that one being the Boeing Airplane Company. Similar considerations, plus costs, drive all the respective AWACS fleets on this planet more or less willingly toward compatibility, if not outright interoperability.

If the AWACS is embedded in a command-and-control system, one can assume that somewhere in the system somebody is making decisions. It is almost a military cliché of the twentieth century that — though it has never been fully revealed — the Big Picture exists, with an informed commander and his technologically omniscient staffs conscientiously examining its electronic entrails for omens and portents. Much of that picture comes from the AWACS. The Big Picture may, in fact, be best viewed while airborne. From the parochial perspective of AWACS, placing that commander, replete with supporting battle staff, aboard the E–3 as an Airborne Command Element is an attractive option. In an onboard battle staff arrangement, many times the probability of reliable communications with other net elements is enhanced, for the same reason that the sensor surveillance volume increases with altitude, subject only to the inverse square law of physics. The same consideration of varying ranges of sensors and communications devices dictates the location and employment of the AWACS relative to other participants in the operation, including other airborne E–3s on station.

One of the current challenges with AWACS is keeping it downwardly compatible with older surface systems, both afloat and on land. This requirement — as in the NATO AEGIS ground C^2 system — exerts a retarding force on AWACS even as it upgrades its onboard systems to cope with emerging challenges and roles. The flood of data being transmitted from the E–3 generally overwhelms these older systems, which then must filter or otherwise discriminate among all the categories of information available. New net architectures, as in JTIDS or

MIDS, must also be accommodated. A subset of this issue is the cryptologic security of the data communications networks — a problem whose solutions are by no means obvious when dealing with Alliance allies or sovereign foreign nations. In the apotheosis of C^2, every compatible system could contribute and share, in real time, all relevant bits of significant tactical information. Information exchange thus constitutes the second major requirement for an effective AWACS.

Arrive/Survive

Finally, the AWACS vehicle has to arrive and survive. As Woody Allen observed, "Eighty percent of the secret of success lies in just showing up." In a potentially hostile environment, the French philosopher Jean Paul Sartre, somberly remarking on the universe in general, postulated that, "Existence precedes essence," that is, one must first be able to survive before one can consider abstract discussions of philosophy or tactical effectiveness. For AWACS, timely availability and physical survivability are somewhat simplified by the venerable Boeing 707 airplane, an aging, but remarkably reliable platform. New engines have been a perennially proposed upgrade for the USAF and NATO airframes. A feasibility study by the latter estimated that, with all the obvious advantages in performance (especially thrust reversers) and maintenance, the payback point to amortize the cost of fleet retrofit would not occur for approximately twenty-four years. Against that funding commitment, the prudent course has been to live with the current engining. Some national E–3 forces, namely the French, British, and the Saudis, purchased their airplanes with the CFM 56 high-bypass fan jet engines installed.

To counter hostile airborne threats, possibly the safest seat in the house is aboard an orbiting E–3, an aerial high-powered surveillance system dedicated to tracking and identifying such objects. While in orbit at altitude, if an E–3 detects an inbound high-speed threat and immediately flies directly away from the inbound threat, it will run the enemy interceptor out of fuel (as well as its own GCI coverage), leaving it at a severe disadvantage with respect to fighters on protective CAP for the AWACS orbit. If the interceptor manages to close to missile-firing parameters, it is moot whether the large multiengine E–3 has chaff, flares, or a guaranteed fail-safe "last chance maneuver" to amuse and divert the enemy fighter pilot during the end game. Until an effective all-mode, all-hemisphere defensive system is developed, there seems to be little practical point to hanging active defensive countermeasures on the AWACS. Obviously, the most vulnerable phases of an E–3 sortie occur during the take-off/climb-out and in final approach and landing, when it comes within the range of man-portable surface-to-air missiles (SAMs), rather than while on orbit.

In summary, the three essential requirements for the system are that it be able to conduct surveillance in a specified volume of airspace, that it be able to communicate effectively with other C^2 elements, and, finally, that it be available and survivable on the day.

Let us now examine the three major scenarios in which AWACS has operated since its operational inception. Unsurprisingly, these major scenarios reflect the dual genesis of the E–3, grounded on one hand in ADC's focus on the strategic air defense of the North American continent, and on the other in Tactical Air Command's responsibilities for tactical airpower in all its manifestations from close air support through air interdiction to air superiority, including timely response to overseas contingencies.

Continental Air Defense

The first scenario places a premium on sensor systems, particularly the Doppler radar, since it is extremely unlikely that inbound bombers would be squawking IFF modes and codes or, indeed, using any active emitters. The detection problem centers on detecting and tracking high-speed targets, manned aircraft, or cruise missiles of ever-decreasing radar cross sections at ranges that permit response by one's own defensive assets. In the pre-AWACS era, an elaborate system of interceptor basing in the United States and Canada controlled by hierarchies of ground radars linked by an elaborate system of communications to centralized staffs assessing the situation inside huge hardened blockhouses in each air defense region did exactly that. The stakes involved, survival of the Free World, and the means employed, nuclear weapons, left no room for error and encouraged interlocking fail-safe standardized procedures to control the whole lashup from Cheyenne Mountain in Colorado Springs. Exercises to test the efficacy of the system depended on elaborate exercises to practice Armageddon, complete with such safeguards as "faker monitors" and "trusted agents" to monitor and assure the safety of those aircraft simulating hostile bombers. When AWACS came on the scene, its influence progressed through several distinct phases: from employment as just another radar set, albeit atop a 30,000-foot mountain, to limited participation in controlling friendly interceptors out of range of the ground C^2 nets (employing faker monitors aboard the E–3), to flying with certified NORAD battle staffs with authority to launch as well as control interceptors. A tactical analogue of this sort of strategic scenario is the mission in Iceland of tracking airborne objects over the North Atlantic. As the continental air-breathing threat diminished and the threat from the Soviet Union declined, this scenario has lost much of its immediacy. However, it remains as one of the main factors helping to drive radar and sensor upgrades.[24]

Conventional Conflict

The second major scenario has been one near and dear to the hearts of all who believe in air power: employment of the E–3 as a C^2 asset in conjunction with other elements of an in-place tactical air command and control system to fight a fluid, dynamic air war, whether in the first days of World War III in Europe or to thwart aggression in South Korea, South Vietnam. In this scenario, the E–3 could be tasked with a multiplicity of responsibilities and roles, from airborne early warning to close control of friendly air assets, all the while operating in concert with an established in-theater system. This scenario corresponds roughly to an operational level of AWACS employment, similar to the NATO system. The demands focus primarily on communications links and timely exchange of data, rather than detection at extreme ranges.

Contingency Response

Finally, the AWACS has proven itself in what might be termed the autonomous role. In the early days of the late 1970s, every one of its on-scene arrivals heralded the appearance of an exotic and unknown capability, whether at air defense exercises in CONUS, at the Red Flags at Nellis, or in the NATO environment. Nobody, including those aboard the E–3, was quite sure what this new system could do to reduce uncertainty and clarify the situation. Over time, a generally harmonious mutual accommodation transpired, so that scenario number two, working in conjunction with other local assets, really governed. However, many AWACS deployments to contingencies of indefinite duration in the Middle East or the third world during the 1980s took it to locations where it was the only friendly air defense asset in the theater. Again, as ground air defense elements arrived, the system could evolve accordingly.

The USAF AWACS deployment to Riyadh, Kingdom of Saudi Arabia, from 1981 to 1989 is perhaps the paradigm of this scenario. After the border war broke out between the Yemens in 1979, Saudi Arabia responses concentrated on two main courses of action, each centered on the E–3. First of all, apparently impressed by an initial AWACS deployment that spring, a few months later (in February 1980) the Saudis requested a purchase of E–3 AWACS and KE–3 tanker aircraft, as well as an F–15 enhancement package of conformal fuel tanks and multiple ejector bomb racks (MER–200s). This program, Peace Sentinel, provoked a bitter national debate which involved two administrations — Carter's and Reagan's — before the U.S. Senate consented in October 1981. At a total projected cost of $3.5 billion, the deal included five E–3 AWACS and eight KE–3 tanker aircraft, with associated facilities construction, training and support services. It also included 1,177 AIM–9L air-to-air missiles and 101 sets of conformal fuel tanks for the RSAF F–15 fleet.[25] The Senate Armed Services Committee observed that "The presence of an AWACS-compatible air defense network in Saudi Arabia would greatly facilitate deployment of U.S. forces and is a critical element of U.S. strategy."[26]

Second, the Saudis approved the long-term deployment of USAF AWACS into their Kingdom to strengthen the area's air defenses. On September 30,

1980, an initial package, code-named ELF ONE, of four E–3As from Tinker and 365 personnel, including six aircrews, deployed to Riyadh, flying their first surveillance mission the following day. As the deployment developed into an around-the-clock flying operation, AWACS crews were pulling 180 to 220 days of TDY annually. By the time the last of the E–3 AWACS aircraft departed the Kingdom on April 15, 1989, ELF ONE had flown more than 86,500 flying hours and 34 million miles. Its tankers had completed over 6,800 aerial refueling sorties. The E–3 aircraft fleet had been flown at its projected wartime utilization rates for over eight years, at the end of a 6,597-statute-mile logistic lifeline stretching over twenty hours flying time on the weekly C–141 rotator aircraft. Over the years, over 47,000 personnel, both crews and support teams, were sent to ELF; personnel turnover averaged some 440 individuals per month.[27] Despite those challenges, the USAF considered this ELF ONE deployment extremely successful. The bitter Iran-Iraq war did not spread to Saudi Arabia. No serious air or surface attacks were ever mounted against the Kingdom during the period. In fact, the most noteworthy incident was the shootdown of two Iranian Air Force F–4s by RSAF F–15s on June 5, 1984.

These three major mission scenarios — continental air defense, conventional, and contingency — have typified nearly all the AWACS operations since it entered the USAF inventory. In conjunction with these three mission scenarios, one can use the essential requirements — to conduct surveillance, to communicate, and to arrive and survive — to develop a 3 x 3 matrix. This matrix is useful because it allows one to map in general terms where the original operational requirements, those from January 1963, as well as those from the joint ADC/TAC Required Operational Capability of 1966 are distributed in this scheme. More specifically, the potential uses of AWACS in Europe that were highlighted in 1973 can also be located with some degree of precision. Finally, this mission scenario-requirement matrix also helps to assess the major upgrades, actual and contemplated, to the E–3 weapon system over the years.

Predictably, most attention centered on improvements to the sensor and communications systems, since the airplane itself proved itself to be a reliable, maintainable platform. Certain modifications to incorporate new equipment for air traffic control or safety of flight will be necessary, in conjunction with other upgrades. The GPS (Global Positioning System) represents a likely candidate. The extent to which such an accurate navigation system is integrated into the E–3 can proceed in at least two stages. It can simply be a nav aid for the flight crew in the front end, with some implications for the perennial debate whether a human navigator is now necessary or desirable, and it can also be integrated into the mission system to reduce or eliminate positional and parallax errors when the AWACS is providing data within a large net, and accurate positioning of all net participants is a prime requirement. Whatever its eventual extent of integration into the AWACS system, GPS will provide a higher degree of positional precision than is possible with other systems.

Aside from sporadic proposals to upgrade the E–3 engines, the most consistent demand for upgrade to the air vehicle centered on the addition of a second latrine to improve the crew's quality of life on long-duration sorties with a full onboard complement of thirty-five crewmembers, battle staff, and observers. Boeing's initial configuration, which provided a urinal — a.k.a. "the navigator's sink" — on the starboard side of the bulkhead just aft of the flight crew compartment, proved unworkable. During a mission out of Keflavik in late 1978, outflow from the system plugged and iced over the static ports for the air speed indicators, causing some consternation. The urinals are now inoperable, and the area is used as a crew stowage area for pubs and kit bags. The second full-up latrine has yet to be approved.

Of course, nothing is ever free. A new release of mission computer software is probably the most benign improvement, aside from the initial trial periods and patches it sometimes dictates. In the case of the E–3, nearly every other enhancement adds weight, and occasionally drag, as well as power and cooling demands to the vehicle. For unrefueled sorties, the added burden decreases the AWACS' time on station. Enhancements also add demands on the training unit and simulators as the crews develop a level of proficiency with new equipment. For minor upgrades, control of the fleet configuration has to be closely monitored from airplane to airplane, with some influence on the tail numbers selected for complex or sensitive mission deployments. For major upgrades, in which a number of aircraft are taken off-line and upgraded with a block of improvements, not only fleet configuration, but also the size of the operational force can be significantly affected.

When the USAF purchased the AWACS force of thirty-four E–3s, one airplane remained in Systems Command as Test System 3 (TS–3) so that future improvements could be planned and tested empirically. At least one hardpressed commander of the AWACS unit at Tinker AFB has schemed to restore TS–3 to operational status to help solve his scheduling and deployment problems. However, any such decision would effectively freeze the AWACS operational capabilities in place, simply by forestalling any further R&D.

The first major block upgrade to the E–3, the so-called Block 20/25 Upgrade during the mid-1980s, dealt with the obvious. The IBM computer was upgraded in memory and speed, with additional provisions of timing and sizing for future use. More UHF radios were installed (from fourteen to twenty). The total number of computer consoles increased from nine to fourteen, with the installation of color monitors at the surveillance, weapons, and battle staff sections of the mission crew. Within the radar system, improvements enhanced the AWACS ability to track slow-moving aerial targets and to monitor the health of the system relative to its detection abilities and extant ECM.

Because of economies of scale, the upgrade to the NATO AEW fleet closely paralleled the USAF Block 20/25 program. NATO also opted for much the same upgrades to sensors and onboard communications, and additionally requested location-dependent equipment and facilities so that the NATO AEW force could support NATO's Mobility Deployment Concept. Because of the decreasing radar cross-sections of targets of interest, NATO was particularly interested in a radar with increased sensitivity to maintain the desired detection range. This also would permit detection of smaller targets at the same range. NATO also specified improvements to height accuracy, helicopter detection, and maritime detection. Within the Alliance, additional issues of relative shares of funding and percentages of industrial benefits to be provided to each member nation also had to be resolved.

Further block upgrades to the AWACS fleets have been affected by the collapse of the Soviet Union and the consequent uncertainty in the international military environment. Funding has become less firm, and the urgency to act has largely dissipated. To an extent, the performances of the USAF AWACS during Desert Shield and Desert Storm has validated the earlier concepts of operation and upgrades to the E–3 fleet in the crucible of actual hostilities.

In the current climate, particularly in NATO, it has been suggested that the goal of any further enhancements to the AEW Mixed Force ("mixed," now that the RAF E–3D is operational) should exploit the capabilities that have been upgraded by the hardware improvements already added to the airplane. One line of approach immediately suggests itself, using software-expanded man-machine interfaces in sensor integration and decision aids to increase mission crew effectiveness in the airplane, in the mission and flight simulators for training, and at deployed locations.

Current platform effectiveness is highly skill dependent, especially for NATO, with multi-national crews, several languages and previous background training and experience levels. As the system matured, crew quality appeared to decline as more junior service members entered the crew force, and excessive TDY rates affected training and individual/crew proficiency. The airplane has become even more complex, owing to the enhancements already incorporated into its flight and mission systems. (One must note here that the Air Force Chief of Staff recently announced that, for the first time, part of the AWACS mission will be given to the AF Reserve to help reduce deployments for the overtaxed crews at Tinker AFB, Oklahoma. Within two years, the 507th Air Refueling Wing at that location will get six crews, but no planes.)[28]

Software engineering and decision-making algorithms provide a number of means to multiply the capabilities inherent in the hardware upgrades. Optimum man-machine interfaces have the same practical impact as adding more computer capacity, consoles or communications links. Now that passive sensor systems (e.g., ESM) have been added to the radar and IFF systems, a real need exists for sensor integration to relieve the operator workload and to accommodate future sensor systems. These software potentialities must also be extended to AWACS simulators to increase training effectiveness, as well as to deployed locations, especially for timely, reliable data reduction.

Technology and the Air Force

A second, emerging trend for future upgrades runs counter to the long-range, integrated, and painstakingly coordinated approach. This trend simply identifies timely candidates to satisfy operational requirements that can be quickly augmented onto the AWACS airframe or incorporated to solve pressing needs. A recent example is the "Eagle" infrared missile launch sensor, a $50 million package now scheduled to be added to the fleet starting in mid-1997. An *Aviation Week* article noted that the prototype is to be tested on TS–3, the perennial E–3 test bird. In the event a ballistic missile launch is detected, sensor data collected by the AWACS would be correlated with GPS positioning information and transmitted into the theater warning net via JTIDS.[29] The relationship to the Gulf War and the continuing issue of timely response to theater ballistic missile threats seems obvious.

Operational Parallels, JSTARS and AWACS

Up to now, very little has been said about the Joint Surveillance Target Attack Radar System, Joint STARS, or JSTARS for short. Part of that oversight reflects my relatively limited experience with the system. Part of it also lies in the remarkable similarities between JSTARS and its older sister, the AWACS, as airborne sensor and surveillance systems and also in the sort of information they each provide to the theater commander.

Apocryphally, a U.S. Army four-star general, Max Thurman, gets most of the credit for persevering with the concept of a ground surveillance system. At one point, he allegedly clarified the concept by telling his listener to think of the JSTARS as "an upside-down AWACS airplane," an excellent image and a concept that an infantry officer can readily grasp. In terms of functioning as elements in a C^2, or C^3, or C^3I, or even a C^4I, system, the two platforms are roughly complementary in their sensor surveillance volumes, and nearly identical in the demands on their onboard communications systems. The AWACS looks after airborne targets; the JSTARS at ground targets. Both look at stationary objects with a finely resolving synthetic-aperture radar and at moving objects via the MTI (Moving Target Indicator). The latter really extends the low-velocity detection mode of the AWACS radar into the speed range of ground tactical wheeled and tracked vehicles. To say that is not to slight the tremendous technical and design problems and the integration and computational issues that had to be overcome to bring the E–8 into being. The performance of the two test aircraft deployed to the Gulf War virtually guaranteed the survival of JSTARS into the post-Cold War era.

A recent article in *Defense News* further underscores this similarity between JSTARS and AWACS. NATO is now investigating with its member nations the different ways "to create an airborne ground surveillance (AGS) system for the alliance." the article notes that "NATO needs an AGS system capable of gathering intelligence information over territory over 200 miles in diameter."

The Boeing E–8 JSTARS aircraft.

Further, the "system should be capable of synthesizing data drawn from stationary and moving objects on the ground, much in the same way that Airborne Warning and Control System aircraft (i. e., the AWACS E–3) process data from airborne targets."[30]

Right now, the Alliance is considering a prototype AGS system of about 32 airplanes and helicopters, with an estimated cost of $23.2 billion (U.S.) to procure and operate. JSTARS is, of course, a major contender, and the French and Italians have a rotary-wing (helicopter) candidate. In a burst of historical *deja vu,* the British Astor concept "exists on paper only," reminiscent of the ill-fated Nimrod Mark 3 proposed to NATO in the 1970s as the RAF equivalent of the E–3 AWACS. As with the NATO AEW E–3 fleet, joint procurement appears to be likely, with the nations "deciding which collaborative approach would give the most operational flexibility and operational benefits."

By now, AWACS is considered within the USAF as a mature system, perhaps a bit sexier than a C–130 Hercules transport, but not nearly as exotic as the emerging JSTARS. Both are essential elements in the scheme of airborne command and control and will remain so in the coming decades of military expeditionary forces and regional contingencies of indeterminate duration.

The challenge is to create an architecture that can make best use of all the information now available. A combination JSTARS/AWACS, complete with an onboard Airborne Command Element to keep track of the Big Picture, would be one solution. Such an approach, while technically feasible, runs the risk of creating a single asset so valuable and so sensitive that it could never be realistically deployed. Perhaps a better concept would be to keep the JSTARS and AWACS platforms as discrete operational modules that can be combined in a package tailored with other reconnaissance or satellite assets, all under the operational control of an Airborne Command Element aboard its own dedicated command and control platform, in essence an upgraded Airborne Battlefield Command Control Center, which is now on a C–130 airframe.

Indeed, we may be moving in that direction. Again, the same issue of *Defense News*, under the heading, "USAF Prototype Aircraft is Data Command Post," notes that a KC–135 aerial tanker aircraft had been remodeled by simply changing the electronic equipment in the aircraft. The aircraft known as "Casey 01" is the first tangible manifestation of the concept called "Air Force C⁴I Architecture." It can now "transport a wide array of information and communication systems, enabling it to fill a variety of different missions." The article notes that "the objective of the plan is to provide aircraft with a communications framework that can share information with other Air Force assets and easily accommodate new technologies." As the U.S. Air Force moves toward its first half-century of institutional independence, the roles of AWACS and JSTARS seem assured. Thank you.

Bibliography

Benson, Lawrence R. "Sentries Over Europe: First Decade of the E–3 Airborne Warning and Control System in NATO Europe." HQ USAF Europe, Office of History, February 10, 1983.

Breslin, Vincent C. "Development of the Airborne Warning and Control System and the E–3A Brassboard, 1961–1972." Hanscom AFB, Mass: AFSC ESD History Office, June 1983. In AF Historical Research Agency Archives, Maxwell AFB, Ala, as Call No. K243.016.

Compart, Andrew. "Reserve to get a piece of the AWACS mission," *Air Force Times*, October 2, 1995.

Cooper, Pat. "USAF Prototype Aircraft is Data Command Post," *Defense News*, October 16–22, 1995.

Fulghum, David A. "AWACS to Carry Missile Launch Sensor," *Aviation Week and Space Technology*, September 4, 1995, 42–43.

Gross, Charles J. "Silent Partners: The U.S. Air Force and Saudi Arabia, World War II to Operation Desert Shield." Washington: Center for Air Force History, February 1994.

Gunston, Bill. *An Illustrated Guide to Spy Planes and Electronic Warfare Aircraft*. New York: Prentice Hall, 1983.

von Kospoth, Edward. "NAPMO—The NATO AWACS Managers," in *NATO's Sixteen Nations*. Utrecht: Boekhoven-Bosch, Special Edition, 1990.

Sun, Jack K. "AWACS Radar Program: 'The Eyes of the Eagle.'" Westinghouse Corporation, June 1, 1985.

Tessmer, Arnold Lee. "The Politics of Compromise: A Study of NATO AWACS." Washington: NDU Research Directorate, March 19, 1982. Document 1272A, Archive No 0171A.

Notes

1. Jack K. Sun, "AWACS Radar Program: 'The Eyes of the Eagle,'" Westinghouse Corp pub, Jun 1, 1985.
2. Vincent C. Breslin, "Development of the Airborne Warning and Control System (AWACS) and the E–3A Brassboard, 1961–1972" (Hanscom AFB, Mass: AFSC ESD History Office, Jun 1983, in AF Hist Res Agency Archives, Maxwell AFB, K243.016).
3. Sun, p 2.
4. Breslin, p 1.
5. Sun, pp 2–3.
6. *Ibid*, p 3.
7. *Ibid*, p 4.
8. *Ibid*.
9. Breslin, p 2.
10. *Ibid*, p 4.
11. *Ibid*, p 5.
12. Gunston, Bill, *An Illustrated Guide to Spy Planes and Electronic Warfare Aircraft* (New York: Prentice Hall, 1983).
13. *Ibid*, p 105.
14. *Ibid*.
15. Lawrence R. Benson, "Sentries Over Europe: First Decade of the E–3 Airborne Warning and Control System (AWACS) in NATO Europe" (HQ USAF Europe, Office of History, February 10, 1983).
16. *Ibid*, p 25.
17. *Ibid*.
18. *Ibid*, p 24.
19. *Ibid*, p 21.
20. *Ibid*, p 24–26.
21. Arnold Lee Tessmer, "The Politics of Compromise: A Study of NATO AWACS" (Washington: NDU Research Directorate, March 19, 1982 doc 1272A, archive 0171A).
22. Benson, pp 52–56.
23. Edward Von Kospoth, "NAPMO — The NATO AWACS Managers," in *NATO's Sixteen Nations* (Utrecht: Boekhoven-Bosch, Special Edition, 1990).
24. The author would like to acknowledge the expert contributions and keen insights of Lt Col Frank Welty, USAF, to this summary of AWACS and the continental air defense of North America.
25. Charles J. Gross, "Silent Partners: The U.S. Air Force and Saudi Arabia, World War II to Operation Desert Shield" (Washington: Center for Air Force History, February 1994).
26. *Ibid*, p 58, fn 42.
27. *Ibid*, pp 72–73.
28. Andrew Compart, "Reserve to get a piece of the AWACS mission," *Air Force Times*, October 2, 1995.
29. David A. Fulghum, "AWACS to Carry Missile Launch Sensor," *Aviation Week and Space Technology*, September 4, 1995, pp 42–43.
30. Pat Cooper, "USAF Prototype Aircraft is Data Command Post," *Defense News*, October 16–22, 1995.

John D. Anderson, Jr., is a professor at the University of Maryland. He has a B.S. from Florida and a Ph.D. in aeronautical and astronautical engineering from the Ohio State University. He served as a scientist in the USAF at Wright-Patterson AFB's Aerospace Research Lab and headed the Hypersonic Group at the Naval Ordnance Lab. Dr. Anderson occupied the Lindbergh Chair at the National Air and Space Museum in 1986–87. He has published over 100 scientific papers and six books, the latest being *Computational Fluid Dynamics* (1995). An internationally known educator, he has won numerous awards, was elected Vice President of the AIAA for Education, and won the Atwood Award from the ASEE and AIAA.

Computational Fluid Dynamics

John D. Anderson, Jr.

We are in for a little bit of change of pace. I would like for everybody to sit back and relax a little bit because the subject that I have been asked to discuss with you, "Computational Fluid Dynamics," is a little bit different than the other subjects that we have been discussing yesterday and today in the sense that it is kind of a fundamental technical discipline area.

On top of that, it is relatively new. Computational fluid dynamics (CFD) as an identifiable discipline is about thirty years old. So there is not a whole lot of meaningful history here. Therefore, I will be making a few historical remarks. But for the most of my discussion, we are going to be in for a kind of a tutorial on computational fluid dynamics.

First, I would like to give you an idea of just what computational fluid dynamics is and then address how the Air Force used this in the past, has used it in the recent past, and is using it today. We will first of all be discussing some general introductory remarks, where I would like you to think about computational fluid dynamics. Last, I am going to be a little more precise, and go into applications of CFD in the Air Force.

Now, to start with, if you look at anybody who has been educated in the physical sciences and in engineering, generally that education and the way the people carry out their job in the work place in the past has been what I consider to be kind of two-dimensional. We have operated in the world of pure theory, and we have operated in the world of pure experiment. What we have today is a new third dimension in the way we carry out our business, and that is in computational fluid dynamics or in other areas. We will just say general computational mechanics.

Let me elaborate on this for a minute. Historically, the beginnings of the experimental tradition in physical science occurred in France in the middle of the seventeenth century. People like Christian Heigens, who was in the Paris Academy of Sciences, and Marriott, who was also one of the first basic, we will say aerodynamicists, in the seventeenth century, established a tradition of experimentation which is carried on today. At the end of the seventeenth century, with the publishing of the *Principia*, Isaac Newton introduced the world of theory, rational analysis. This two-dimensional world is what we have been operating in for the last couple of centuries. When I was a student, I took courses in pure theory and I went in the laboratory and messed around with pure experiments. This is the way you did your business and the way you carried out your job in this two-dimensional world.

In the last thirty years, a third dimension has been added to this. As far as fluid mechanics is concerned, computational fluid dynamics is not a flash in the pan. Computational fluid dynamics is a major addition, a new third dimension that is going to be with us forevermore, as long as we maintain our society as we know it today. This is a fundamental change, and today in the fields of, say, fluid mechanics and aerodynamics, we operate in this three-dimensional world where all three of these complement each other.

People used to say very enthusiastically that computational fluid dynamics is going to replace experiment and replace theory. Nothing could be further from the truth. These three areas have been and will continue to work together in a synergistic way to provide a means of attacking physical problems. When we talk about computational fluid dynamics, we are talking about a fundamental new third dimension in the way we carry out our business as physical scientists and engineers.

Computational fluid dynamics allows you to calculate flow fields, velocities, and pressures. It gives you lots of numbers and, therefore, information about how flow fields are generated and the consequence of these flow fields, like pressure distributions on the surface. By the way, that pressure distribution is nature's way of grabbing hold of that airplane and exerting a lift on it.

Computational fluid dynamics is used for a lot of other things than just airplanes. It has been used, for example, to show how liquid iron — molten iron — feeds into a the cavity of a mold. This application could give manufacturing engineers an idea of the details of the physical process, which could be useful in terms of designing manufacturing processes.

Computational fluid dynamics was used to calculate the flow of air around a proposed complex of buildings for calculating flow fields over automobiles and trucks. Detroit is discovering computational fluid dynamics in a big way. The Europeans had discovered it about ten years earlier for automobile aerodynamics. All this is just to give you an idea that computational fluid dynamics is quite general. It is just another way for us to grab hold of solutions and get information on fluid flow problems.

Finally, I have to say that computational fluid dynamics is in many respects an inexpensive way to get information about aerodynamic problems. We have seen the cost of a given calculation on a computer go down by a factor of ten every eight years as new computers have appeared. This is why people today, for example at Boeing, can talk about using computational fluid dynamics to design the 777 and save a lot of costly wind tunnel time by using computational fluid dynamics. Part of this is the interplay between theory, experiment, and computations.

290

This was just to give you an idea of what computational fluid dynamics can do, some general introductory comments to try to give you some kind of comfortable feeling for what we are talking about. Now what we have to do is get a little more serious and really ask a more precise question: What is computational fluid dynamics?

Computational fluid dynamics does not come out of thin air. The way that the numbers come out of the computer when you exercise a CFD calculation is that you put something into the computer. What you are putting into the computer are basic physics. In fact, all of aerodynamics is really based on three fundamental principles. I always tell my students aerodynamics is easy. All you have to do is remember three fundamental principles.

One of those is simply that *mass is conserved*. The second one is *Newton's second law*: force is equal to mass times acceleration. The final physical principle is that *energy is conserved*. The first two sets of equations have been known a very long time. The first one, the continuity of the conservation of mass equation comes from Leonhard Euler, about 1753. The second, Newton's second law, called the momentum equations, comes from Stokes, England, and Navier, France, about the middle of the nineteenth century. So, two out of the three sets of these equations are really old. The energy equation comes out of the science of thermodynamics about the middle of the nineteenth century. Again, we have had these things for a long time. We just have not been able to solve them. For a century aerodynamicists have reworked these equations and chopped them up and made all kinds of simplifying assumptions in order to try to solve these.

Today, we cram these equations into a computer. But how do you do that? You cannot just feed a sheet of paper into the computer. What you have to do is to take these equations and convert them into little algebraic expressions. What we do with these algebraic expressions, we will see in a minute.

Here is my definition of CFD. Computational fluid dynamics is the *art* of replacing the integrals or the partial derivatives in the governing equations of motion with discretized algebraic forms. Notice I did not say *science*, that might be arguable. Replace all those derivatives with algebraic forms, basically numbers, which in turn are solved to obtain numbers. That is important.

What comes out of the computer is numbers—not equations. For the flow field values, discrete points in the flow; not everywhere, but at very discrete points, either in time and/or space. The end product of computational fluid dynamics is indeed a collection of numbers, in contrast to a closed-form analytical formula.

That is what computational fluid dynamics is. Now, what does this mean? Well, we take those equations I just mentioned, and we put them in algebraic forms. Lets take a look.

We do this by taking the space of which we are making the calculations, like the space in this room, and divide the space into a series of discrete points.

They are called grid points. We make the calculations of the flow at each one of these little grid points. That is where we get our numbers.

If you keep this picture in mind what happens is that—and again, this is the end of the mathematics—we just take these derivatives which come out of calculus and replace them with little algebraic difference quotients. For example, the velocity of one grid point minus the velocity of an adjacent grid point, divided by, say, twice the distance between those grid points. That is called a *finite difference*. What goes into the computer is a bunch of algebraic equations, set up on some grid. These grids can be pretty exciting and complicated.

We could set a grid up around an F–16 fighter to calculate the flow field over the F–16. It is rather artistic looking and was not easy to do. Until recently, it would take someone about three person-months just to construct a grid for this kind of three-dimensional case.

What happens, in essence, is again that we are just cranking in numbers for the pressures, temperatures, and velocities at each one of these grid points, carrying out a solution and getting numbers back out for the flow field out of the computer. That is what computational fluid dynamics is, in a nutshell.

Now, we have just finished talking about what is CFD. I think I will take this moment to give a little bit of history because what I have just talked about, this matter of replacing the governing partial differential equations with these algebraic equations and then cranking out numbers. That idea goes back a little more than 100 years in the development of numerical solutions of differential equations.

Carl Runge, for example, in Germany, 100 years ago, had set out some of the theories necessary to carry out this kind of calculation. Certainly by the 1930s, the basic mathematical underpinnings of these numerical calculations were in hand. But nobody could do anything with them, because you would spend a couple of years of your life working this out by hand or punching it into the Frieden calculator in the 1940s. Nobody ever did that. It took the advent of the high-speed digital computer to make these mathematical approaches practical. I can remember when I was a graduate student in the late 1950s using an IBM 7090, and then came along the IBM 7094.

These first-generation digital computers allowed us to calculate certain minimal kinds of aerodynamic problems, like boundary-layer calculations. Some of the early aerodynamic and aerothermodynamic work done on intercontinental ballistic missiles were done on these early computers. That work represented the beginnings of computational fluid dynamics, particularly solving the boundary layers and the aerodynamic heating calculations distributions for ICBMs in the 1950s, although they did not call it that in those days.

Finally, in the late 1960s, computational fluid dynamics became kind of an identifiable subject. We started calling it that. In 1969, a major breakthrough was made and a new technique by Bob McCormick, an applied mathematician at NASA Ames, which revolutionized practical CFD.

Since then, the applied mathematicians have grabbed hold of this idea and up through the present day have been working on a continued basis to refine the accuracy of the algorithms and the accuracy of these calculations in general.

So that is really a summary of the history of CFD. That is why I said that there is not much history there. It is too young a science. It is too early to really make some broad-based conclusions. But with that in mind, let me go on to say something about applications of CFD in the Air Force.

The Air Force has reason to be proud of their activities in computational fluid dynamics. I can say this because I am not a member of the Air Force, although I spent three years at Wright Field a long time ago. But the Air Force has reason to be proud of their contributions in CFD.

In 1986 at Wright-Patterson AFB, Ohio, an Air Force engineer named Joe Chang in the Flight Dynamics Laboratory made the first ever computational fluid dynamic calculation of solving equations (called the Navi-Stokes equations) of a complete airplane configuration.

He chose the X–24C lifting body craft for his subject of study. He made the calculation at Mach 5, hypersonic speed. CFD products are color graphics; CFD people love color graphics. As a matter of fact, one of the leading laboratories in the development of CFD is at Mississippi State University. They have a major National Science Foundation-sponsored center in computational fluid dynamics there under Joe Thompson. Joe has about 120 people working for him, including three full-time faculty members from the art faculty at Mississippi State, just to do things like this.

Well, anyway, Joe Chang made a historic calculation, the first-ever flow field over a complete airplane configuration, showing the pressure distribution. On the color graphics, the different colors represent different values of pressure exerted on the surface of the X–24. White areas are real high pressures you get at the leading edges, for example, the triple tail, the canopy, the nose, and so forth. In 1986 this was a very complex calculation to make, a complex flow field. He also calculated heat transfer distributions, same sort of thing, giving you points where the maximum heating is taking place.

I want to emphasize again that, historically, the first such complete configuration calculated by the CFD solution of the Navi-Stokes equations was carried out in the Air Force, basic research by Joe Chang at Wright-Patterson.

Joe Chang's computational fluid dynamics results were pretty good, to within a few percent, especially on the lift-over-drag ratio. That is not bad. So once again I will emphasize how important it is in this area of research that the Air Force made a very important contribution about nine years ago.

I want to point out that computational fluid dynamics is a *tool*. It gives you numbers. How you use those numbers is up to you. You use it as a tool that can be used for carrying out research, to carry out numerical experiments. You can use it for design. It can be a design tool, which will help you design an airplane, for example.

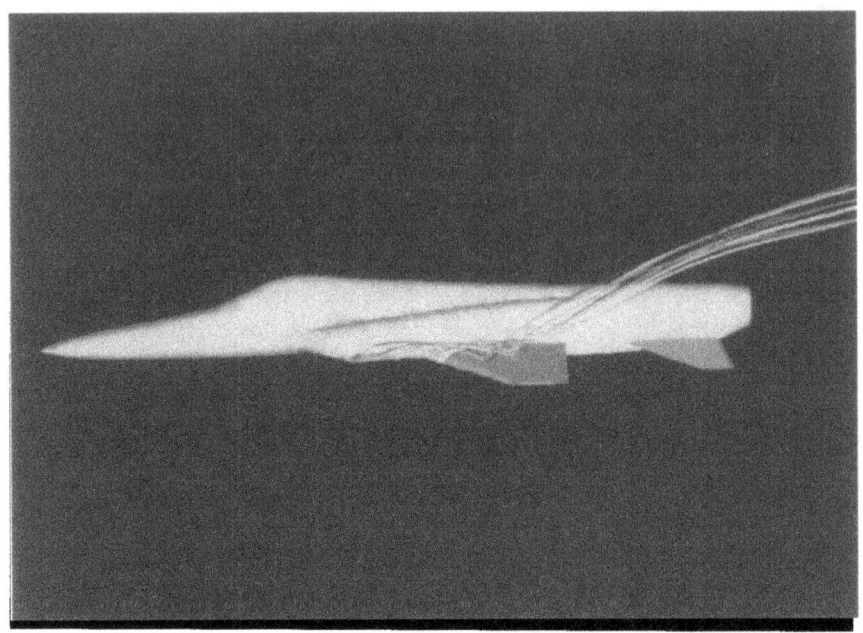

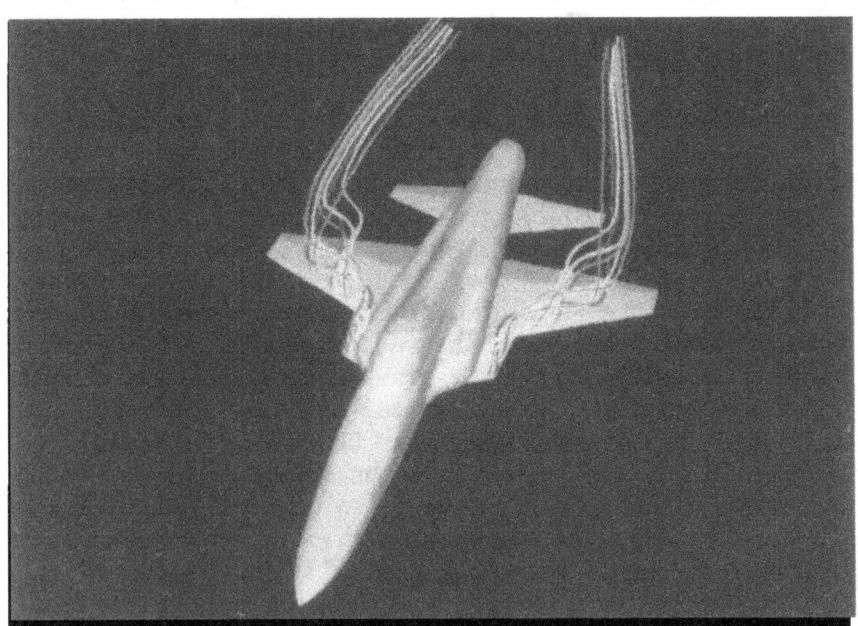

Computed vortex generation and shedding over the wing of a
Northrop F–20. Angle of attack: 25 degrees; Mach number: 0.26.
(Courtesy of Merle Jager, Northrop/Grumman.)

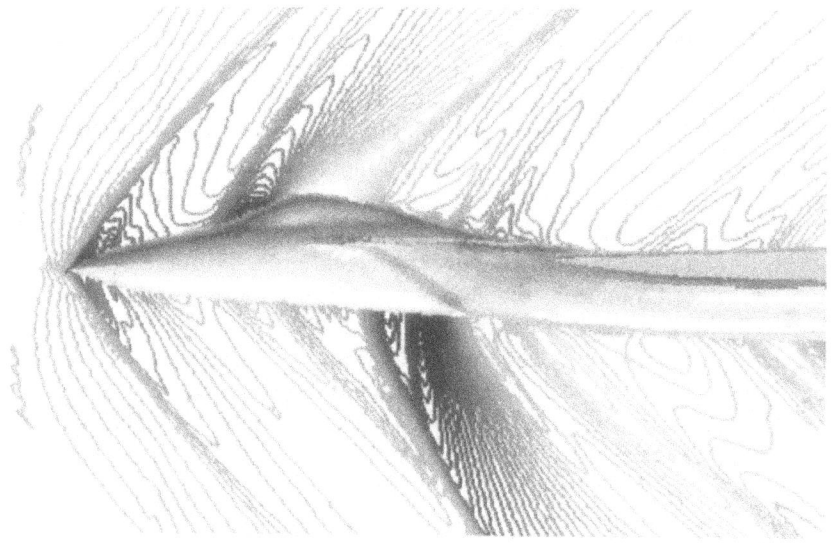

The computed Mach number contours in the flowfield
around a supersonic fighter. (Courtesy of Joe
Thompson, Mississippi State University.)

Comparison of surface streamlines (lines of surface shear stress) over the
Northrop F–20. Computed streamlines, left, compared with experimental
oil streak photos, right. (Courtesy of Merle Jager, Northrop/Grumman.)

Technology and the Air Force

What is interesting about this tool, it is unlike a wind tunnel, which is kind of hard to carry around underneath your arm. It is kind of hard to go over to Wright-Patterson and rip out one of those wind tunnels and carry it over here to Andrews to use it. In contrast, computational fluid dynamics is transportable. In the old days when you had decks of computer cards, you could still carry those decks of cards underneath your arm and carry them wherever you wanted. Then along came the computer terminal and so you could just, in essence, access this on the computer without carrying around the cards.

Today I can transfer a computer program to somebody in Palo Alto, California, by hitting a button on my terminal, so that these numerical tools are infinitely transportable by electronics. In any event, please keep in mind the CFD as a tool to be used for research and to be used for design.

I have a kaleidoscope of Air Force applications of CFD to talk about. For example, a CFD calculation for the F–16 was carried out in 1988, also at the Flight Dynamics Laboratory under Joe Chang. He did the complete flow field, the first time a complete flow field configuration flow field had been calculated for a fighter-type configuration. It was used to analyze some aspects having to do with the little plates added on the surface of the F–16 for structural purposes. CFD was used to assess if these plates would interfere with the aerodynamics, and they did not.

The Flight Dynamics Laboratory also carried out B–1 calculations and calculations of Halon injected through little cavities as a fire suppression activity. This is a two-phase flow where you had not only air, but foreign substances as well. Computational fluid dynamics calculates these kinds of flows.

DFD was used to examine the KC–135 specially designed to be an airborne laser carrier—I was going to say laboratory, in the old days, but it really was for laser weapons applications. But this was a CFD calculation of basically the Mach number distribution over the KC–135 with big splitter plates on the side of the fuselage for generating a more or less uniform flow, through which laser beams are going to pop out, a flow that would not disturb the optical quality of the laser beam. CFD was used by the Air Force to help design this modification to the KC–135.

Tom Julian spoke to you yesterday morning about aerial refueling. It is still a hot topic and CFD is being used to examine the interacting flow fields between the tanker and the receiving airplane downstream to help decide what are optimum locations in this interacting flow field. Again, the calculations were made at Wright-Patterson using CFD.

For the C–17, a study was carried out to try to find out where the air flow goes behind the C–17, because if you have people jumping out of that airplane, they are likely to follow the air flow. So this was a study carried out to try to help assess what is going to happen to paratroopers jumping out of this airplane. This is a really solid, a very important type of application using CFD, again as this kind of tool to study these flows.

Just before lunch we heard a very interesting presentation on stealth. The people who practice CFD are not unaware of these things. A splinter group, a certain section of computational fluid dynamics, has sort of split off into something we call computational electromagnetics, using similar techniques. This is also going on at Wright-Patterson, using their expertise in CFD.

What we have seen so far is really mainline applications of CFD used for research. But it has also been used for research into the flow field inside a supersonic ramjet combuster. It looked at water vapor formation due to the combustion of hydrogen with air. It tells you where the combustion is taking place in this case. The aerodynamicists interpreted this data just like they would wind tunnel data, except it came out of a computer.

This has been carried out at the Air Propulsion Lab at Wright-Patterson and it is an example of how the Air Force is using CFD in the research mode to find out more fundamental information, in this particular case, scramjet engines.

What I have been trying to tell you is that CFD is a research tool and a design tool. It is being used effectively by the Air Force. We have seen this. It is even being enhanced and advanced in the Air Force as new results and new algorithms appear. One of the major fields is computational electromagnetics that is being pioneered out of the Flight Dynamics Laboratory at Wright-Patterson. CFD is here to stay. This is not any flash in the pan, it is something fundamental, a new third dimension in our way of doing business. It is here to stay.

That has been my purpose this afternoon. It is a little bit different than our previous presentations, short on the history, long on what it is that we are talking about. What we need to do is come back twenty or thirty years from now, and maybe there will be some reasonable history of CFD of which we can really make some sense.

Hon. Paul G. Kaminski is the Under Secretary of Defense for Acquisition and Technology. A graduate of the Air Force Academy, he earned an M.S. in aeronautics, astronautics, in electrical engineering from M.I.T and a Ph.D. from Stanford. He was Chairman and CEO of Technology Strategies and Alliances, chaired the Defense Science Board, and served on the Defense Policy Board. Dr. Kaminski had a twenty-year career in the Air Force, including service as Director for Low Observables Technology. He was Special Assistant to the Under Secretary of Defense for Research and Engineering. Earlier he helped develop inertial and terminal guidance components for precision-guided missiles, and spacecraft and payload technology. Dr. Kaminski is a member of the National Academy of Engineering, the American Institute of Aeronautics, the Institute for Electrical and Electronic Engineering, the American Association for the Advancement of Science, Tau Beta Pi Sigma Xi, and Sigma Gamma Tau.

298

Low Observables: the Air Force and Stealth

Paul G. Kaminski

It is a pleasure for me to be with you and share perspectives of a few of the issues that arose as the Air Force, and I might add, DOD as a whole, pursued stealth technology. I will cover in this discussion the period from the mid-1970s through the early phases of the B–2. I would like to identify this morning the key challenges faced in the F–117 program, the decisions made in pursuit of that program, and some of the lessons that acquisition managers and planners might take away from that experience to apply to our current programs.

Let me start with some of the motivation, the why, the reason that the Air Force and the Department of Defense were interested in stealth at all. To do this, go back to the mid-1970s, in about 1974 or 1975. It was a time when the United States and the Soviet Union were engaged in the great worldwide Cold War struggle. It was a time when military advantage and, therefore, political advantage were driven by deploying superior capabilities in a seemingly endless cycle of move and countermove.

The United States was pursuing what I would call the offset strategy. It was a strategy in which we were attempting to exploit technology to develop superior forces to offset the larger numbers of the Warsaw Pact forces. The Air Force's roles were to prevent Soviet tanks from coming across the Fulda Gap using conventional reconnaissance strike forces, as we might think of them today, and to provide deterrents via theater nuclear forces and two legs of the strategic triad at that time.

By the mid-1970s, the Soviets had deployed an enormous internetted radar defense capability in Europe. The system was part of an integrated air defense system that supported numerous complementary radars (early warning, acquisition, and fire control) and surface-to-air missile systems in various forms and models that were netted together. The acronym IADS refers generally to this integrated air defense system.

Many of these systems had been employed with devastating effectiveness in the 1973 Yom Kippur War between Israel and its Arab neighbors. The Israelis, using American-designed aircraft and tactics, lost 109 aircraft in eighteen days against the Soviet-designed system that had been integrated and operated by the Egyptians and Syrians in the integrated air defense system.

299

Technology and the Air Force

At the time, some in the United States projected that U.S. air forces, operating against those kind of defenses in Europe, would be overwhelmed in an engagement that could be as short as seventeen days. I would say that was not a universal projection, but there was a vocal element who had that view at the time. Therefore, from that framework, it was time for a countermove. Many saw stealth technology as a silver bullet in the form of a limited number of aircraft that could blow a hole through those defenses to create penetration corridors for other aircraft.

That whole line of logic would give you a frame of reference that said this was all done by requirements pull. But I was observing pieces of this a little later on, and I know that was not the only side of the equation. There was at least an equal amount of technology push in terms of here was an exciting new opportunity, and there was a good piece of what can we really do with this opportunity.

The technology push side of the story also had its foundations in about 1974, when the then Defense Advanced Research Projects Agency (DARPA), along with the Air Force as a sponsoring element, released a request for proposal (RFP) for a stealth aircraft.

The RFP was released at that time in the open, searching for new ideas to move ahead. There were five fighter aircraft manufacturers who were invited to participate in a design competition. The competition and what happened is a long story, but I will jump to the end.

Lockheed was not one of the original five participants, but they were allowed to come into the program and join the DARPA competition late. In April 1975, a breakthrough occurred at Lockheed that is interesting in a historical sense. A Lockheed radar specialist named Denys Overhalser, with whom I worked for many years, was reading some Soviet literature. He stumbled onto something that was very, very critically important at the time, given our limited ability to do electromagnetic computations. He found an algorithm for accurately calculating the radar cross-section of particular three-dimensional geometric shapes, allowing us to analyze and determine the contributions of those shapes to radar scatter. In fact, it was those fundamental shapes that Lockheed employed in their design.

Oddly enough, in perspective, at the height of the Cold War, the Russians had delivered to us some algorithms fundamental to our construction of this design. The paper, translated by the Air Force's Foreign Technology Division, was called "Method of Edge Waves and the Physical Theory of Diffraction." It had been published nine years earlier in 1976 by Pyotr Ufensev, the chief scientist at Moscow Institute of Radio Engineering.

By April 1976, Lockheed had won both phases of the DARPA design competition. At this point Lockheed was given the go-ahead to build two prototype aircraft (Have Blue) of roughly 10,000 to 12,000 pounds. The purpose of the Have Blue aircraft was to show that we could achieve in flight what we

The angles and flat surfaces of the F–117 contribute to its stealth.

had predicted in our analysis and what we had achieved in scale-model tests on a radar cross-section measurement facility.

In this technology push program, in a sense, we were building the very best *antenna* we could. Every now and then we checked to see if it could fly! That was the thrust that had to be taken at this point in the program. We were trying to push the low observable technologies to the maximum degree possible.

It was an aircraft with very unusual flying characteristics, but it was a key demonstrator, a key predecessor for what became the operational F–117 stealth fighter in 1983. The first Have Blue flight occurred on December 1, 1977, a little over nineteen months from go-ahead in the program. The flight test program ended on the next to the last mission. That is, we had one more mission to go before completing the program, when we lost the second of two aircraft.

These aircraft did not have very friendly handling qualities, in fact, they needed a large flat area in the rear to be able to provide suitable controllability at slow approach speeds. It was something we thought our test pilots could handle, and for the most part, they did. We would never have configured an operational airplane this way. Those difficulties led to the loss of both aircraft, but nearly at the completion of the flight test program. Aside from the crashes, all of the other objectives of the flight test program were met. This program showed us the way ahead.

In August 1976, at about the same time that Lockheed was proceeding with the fabrication of these Have Blue demonstrator aircraft, the Air Force initiated a study of two operational stealth aircraft. The A model had a weight of about 50,000 pounds and a payload of about 5,000 pounds, a fivefold scale-up over the Have Blue airplane. The B model was about a twofold scale-up beyond that, about the size of an FB–111, with a 10,000-pound payload.

Almost one year later, at the end of June 1977, about five months before the first Have Blue flight, I left the Industrial College of the Armed Forces and became special assistant to Bill Perry, who was then in the job that I have now. At the time it was DDR&E, but it soon became the Under Secretary of Defense for Research and Engineering. Within two weeks after I appeared on the job, President Carter announced the cancellation of the B–1 program, another interesting backdrop of the program.

About the time of that cancellation, then Maj. Gen. Bobby Bond set up a new five-man organization within the air staff. He handpicked the five people in the organization, and he located it in an existing organization, sort of as a cover, as a protection for what was going on. He located it in an organization called RDPJ, the office that was the predecessor to today's SAF/AQL organization, which was doing strategic reconnaissance work at the time. They had the U–2, the SR–71, and other programs. It was a convenient place to bury a five-person office that would not be visible.

In October 1977, Lockheed was awarded a one-year concept definition contract for the A model and the B model stealth airplanes. This award was made a little over one month before the first flight of the Have Blue prototype. We were betting on the results to look at missionization concepts, having gotten some confidence in what was happening in Have Blue even before it flew.

For the next six months, a critical objective of the Air Force would be, as you might imagine, how to make up for the cancellation of the B–1A. Of these two aircraft, the A model and the B model, you might guess which one was the favorite of the leadership at the time. There was much more interest in the B model to fill the void left by the cancellation of the B–1A, and the requirements pull was strongly in that direction.

Unfortunately, the technology push and the requirements pull did not line up. There were two bright and very objective colonels in the RDPJ organization at that time who I felt made a very difficult and objective appraisal. They were Joe Ralston, later commander of Air Combat Command, and Ken Staten, who retired as a major general. They looked to see how we were doing with our margins as we were scaling up the designs of these A and the B aircraft.

Pieces of this story have not come out in many of the publications. Only the good news, the positive pieces have come out. What we found was that the B aircraft was not making it; the margins were going away. The facetted design approach simply did not have enough oomph to get us there. I attended a briefing to Secretary Perry in which, with the risks of the program and with our margins disappearing, their recommendation was to defer the B aircraft and focus our attention on the A aircraft, which ultimately became the F–117.

That decision did not go down very well in some circles, particularly at the Strategic Air Command. Gen. Richard Ellis had some strong views about wanting to proceed with the B aircraft, independent of where the technology was at the time. Those views were aired at very high levels. Gen. Lew Allen was personally involved, and I thought the chairman at the time—Gen. David Jones—made a very objective appraisal.

Both the chief and the chairman were in support of proceeding with work on the A model. The chairman's view was that we would probably learn a fair amount from the A model that we would later be able to apply in other arenas. No one had yet really made any real mission commitment at this point in the program; it was still conceptual. In November 1978, a full-scale development contract was signed for seventeen production and five RDT&E A model vehicles, with options for additional production aircraft.

I followed this program very closely as Dr. Perry's only staff assistant on the program, the advisor on the technical piece and on the programmatic piece of the program. Just prior to the time that the administration changed in late 1980, Secretary Perry and I had both decided that this program showed sufficient promise that we were going to be making other very major investments in stealth technology. Recognizing that security constraints had drastically limited the review of this technology, we convened a very special task force of the Defense Science Board to do an independent review. They generally endorsed proceeding with the program and made several recommendations that we followed.

From a personal perspective, there was a big change for me at that time. With the change in the administration, I stayed on as the special assistant to the Under Secretary of Defense until such time as Dick Delauer, the new under secretary, was confirmed. When he was confirmed in March 1981, I was reassigned to be the director of that nonoffice that was responsible for stealth technology in the Air Force, and I joined that still nameless office located in RDPJ.

A few months later, on June 18, 1981, the F–117 made its first flight, but about one year later than planned. If you look at the tail number of the first airplane, it is 780, picked on the basis of the time we first expected the aircraft to fly. This was about thirty months after the FSD award. Lockheed test pilot Hal Farley was at the controls that day, and as we started the early piece of that test program, we discovered some significant problems.

One of the significant problems we discovered—and again, not many of these problems appear in the literature—was directional stability. We missed some wind tunnel data on the aircraft, and we found that the restoring force in yaw due to side-slip was about half of what it should be. The fix for that problem was doubling the area of the vertical stabilizers of the aircraft, something the structure could not support.

One of the interesting aspects of this small, tight-knit group, and the security that went with the program, is that we worked our way through that. We found a path to increase the area of the vertical stabilizers by about 50 percent, a change that the aft structure would bear with reasonable modifications, but we had to put in a roll rate limiter at high angles of attack because we used up all our control authority to deal with the reduced stability. As it turned out, that did not have any significant effect on the mission. With a very small group of people, we were able to deal with the tactical air commander and go through all the trades pro and con.

Had this been a white, highly public, highly visible program, we probably would not have been able to do that. We would have been forced into a several hundred million dollar modification program to restore the original characteristics, which probably would have added at least a year to the development of the program. Operationally, it was a nonsignificant factor, and later in the program, we felt it was a very reasonable design compromise. By 1983, we declared the system operational, and by 1986, thirty-six F–117s had been delivered, with the remaining twenty-six delivered by July 1990.

We overran the development contract on this program by close to 50 percent, but we actually underran the production piece of this program. So the net was very close to being on target. We did not build the aircraft at a very efficient rate. However, it was the first program that I had ever worked on where we built at the rate that we facilitized for. Every other one I worked on, we overfacilitized and did not build to that pace. We built nominally at the rate of one per month, and still, even at that slow production rate, we achieved a unit flyaway cost of just a little over $42 million per aircraft.

Following the termination of the B model, a new program ensued that was another approach at a bomber. A whole new Lockheed design was developed, but at a slower pace, trailing the F–117 program. Well into that program, we received an unsolicited proposal from Northrop, which was also interested in a bomber design. That activity ran for some period of time, with Lockheed having the lead effort. As long as it met the requirements, Lockheed's design was the system that the Air Force was going with, and the Northrop proposal was a fall-back. As the program developed, however, eventually it got to the point where Northrop became a full-fledged competitor, and in fact, there was a large, formal source selection that was concluded early in 1981.

At this point, the program was growing by leaps and bounds. One of the advantages of the security is that we had very little oversight and intrusion, but one disadvantage is that we did not have sufficient review to be sure we were not going to be nipped by some reaction to the technology that we had missed. One of the first things that I did after coming to that office was to set up a formal counterstealth program.

We invested about one percent of our budget, which was substantial at the time, to put together a very aggressive red team that systematically looked at redirecting existing defense systems and new developments tailored specifically to counter the stealth technology. It is interesting today, as I look at the components of the current U.S. counterstealth program, that every single one of the concepts that has been further developed in the concept program came from the red team work that identified promising approaches.

I will spare you the long version of the bomber story. The bottom line was that source selection was eventually completed, but prior to the full completion of that source selection, the Defense Science Board Task Force made their report. They generally endorsed the program, but they felt, as many other of us in the Department of Defense felt, that the requirements that were written for the advanced technology bomber were too narrow. This was an aircraft that would be in the inventory for twenty or thirty years, and at that time, it was being designed with only a high-altitude capability.

After looking at this more carefully, both the board and the department concluded that we ought to modify those requirements and provide a low-altitude penetration capability as well. So that caused a hiccup in the source selection as both contractors went back and modified their designs to be able to fly at low altitude as well as at high altitude.

That delayed the source selection for a bit, but I think you all know the outcome of the source selection. Northrop was selected to proceed with the B-2. That is in itself a very interesting story that I do not have time to go into today because the Northrop B-2 design went far beyond the stated requirements.

The Lockheed approach was a very narrow approach to just barely meet the requirements. The Northrop approach was very much of a technology push, offering substantial new kinds of capabilities in both range and payload. As you know, it was ultimately accepted, and we committed to a buy of 132 B-2 aircraft.

One interesting facet of this Northrop development is that Northrop had committed to a digital design capability by late 1983. They were developing very good analytical tools, tools that we did not have in place on the F-117 program. They discovered something that was judged to be a real problem, but actually it was an enormous benefit. They discovered that, in low-altitude flight, we did not have sufficient control authority in the aircraft to deal with gusts. We had insufficient structural rigidity to be able to alleviate gusts, and we would have either had a damaged structure or we would have had to put a significant limitation on the airplane.

In the past, we would have discovered this type of problem early in our flight test program, but analysis revealed the problems. We committed to a very significant redesign of the aircraft in 1983 that fundamentally changed the nature of the sawtooth shape on the trailing edge and gave the program a far better base.

The B–2 derives its stealth in part from its shape, from the materials used in building it, and by placing the engine intakes and exhausts on top of the wing.

Let me go back now to the F–117 for a moment and just reflect on a few lessons learned. Probably the least understood aspect of the F–117 program was the effort required to marry the stealth technology with the employment doctrine. We were really all over the map about how to best use the stealth technology. A few of our senior operational leaders felt that we ought to be thinking of this as a very small force — a silver bullet — of a few airplanes that you might think of operating in sort of an assassin's role: the ability to go deep, surgically remove a particular target, and not be seen or heard of in any other way. Others thought about leveraging it to give better mileage to the rest of our forces. Technology was clearly driving the operational concepts. Many in the Air Force were not comfortable with new operational concepts like flying only at night, refueling at night, delivering precision weapons at night, or flying a so-called fighter with no missiles or guns.

While there were many in the operational community who were "under-sold" on this program, there were also many who were "oversold." Some of our very senior leaders and planners thought this platform was invincible, that it was totally invisible to all defenses, and we could simply go barrelling our way through anything that we wanted to do.

I had a difficult confrontation at that stage of the program interjecting myself into the operational planning showing the problems that were likely to occur if we used the airplane in that way. It was at that time, however, that we began to make a significant investment, first in modeling and simulation, and ultimately in mission-planning tools, so that we could consider the limitations

306

in stealth technology and route our way around the defenses. We very aggressively tested the F–117 against our most our most advanced representation of the IADS. This was an iterative process in which I insisted that we use our models to predict the results of a test before it was conducted. When we did the test, we could then compare prediction with reality and, as a result, continue to improve our models. Our interest was to validate our modelling and simulation tools, so that the system could be employed in an effective way. Many of the unsung heroes of the F–117 program were involved in conducting these tests and developing these models.

I would observe that, as we look at the operations of the F–117 in Desert Storm, the aircraft behaved exactly as those models predicted. It had no apparent limitations in that environment because it was used as the modelling and simulation directed, and the limitations, therefore, were not apparent.

As I look back on the overall development of stealth technology, several other issues deserve some discussion. Probably the first and foremost is that the F–117 acquisition cycle time was greatly compressed. IOC was achieved within fifty-nine months after program inception. I believe a large part of that accomplishment was due to the decisions made on what to buy and how to go about that decision process.

The A model fighter was clearly the right choice, given the technology at the time. We chose to field what I would describe as a second generation flat-plate solution, rather than attempt to jump to a third or a fourth generation. At that time, the computer-based tools that were needed for design were not really completely in hand. They were being developed then and were applied about five years later as the B–2 was coming into fruition. We chose not to push the envelope on too many technologies at once.

We managed risk by making a conscious decision to rely on as much off-the-shelf hardware as possible. For example, the F–117 borrowed the GE F404 engines from the F–18, the fly-by wire control system computers from the F–16, the navigation system from the B–52, the environmental control system from the C–130, and cockpit gear from other existing aircraft. We had enough troubles developing the new stealth technology. We did not want to be burdened with an unproven baseline system.

The compressed acquisition cycle is also due in part to the program's security classification. It afforded a degree of flexibility and empowerment not possible under less streamlined and more cumbersome procedures that were in place to deal with day-to-day Congressional and public scrutiny. Security also provided a good shelter in another sense. The F–117 was not in visible competition with other Air Force programs. Had it been in competition with other programs, for example, threatening the F–15 or F–16, we might have been much more cautious with the program. We might not have done the program at all. Security also gave us a whole different set of approaches on testing. In this program, it was in our best interest to find all the problems early and fix them.

Technology and the Air Force

If you think about finding all the problems early with a highly visible program, then day after day, all the test problems of the system are on page one of *The Washington Post*. Most program managers are incentivized not to do the aggressive tests until you know that the problems have been solved. Then the tests can be done, and you have successful test results. Our results were quite the opposite. We had all kinds of problems early in the test program, but we fixed them and addressed the issues as the program proceeded.

Because of the program's security, we had great difficulty integrating the system into the war plans. Knowledge of the system was not widespread, and there were no people willing to depend on it. It took a period of twelve to eighteen months to get the system integrated into our war planning so that it would be used.

Security also facilitated open and nonadversarial relationships with the Congress. I was always brutally honest with those in the Congress I dealt with in the program about our problems and our opportunities. That honesty was very well accepted, and the program was very well supported as a result. In fact, it was the Congress that really pushed the Air Force into exercising an option to expand the force beyond the twenty or so aircraft that had been committed to early on. Also, I attribute great benefits to budget stability. The F–117 program was generously funded by both the Carter and the Reagan administrations. As a result, we proceeded to build at the rate for which we had facilitized the program.

As we look to the future, there are several other lessons to take away. One of the things I look back on with interest were the cycles up and down on the early development of stealth. The cycles were very up with the rollout and first flights of Have Blue. They were down a little bit, as you might guess, when we crashed both of the demonstrator aircraft. The cycles were up very much at the early phase of the F–117 program. They were down when we encountered all the test problems early in the program. The cycles were up when we delivered the first production aircraft. They were down when the first production aircraft crashed. We had not had any fundamental problems with the first five RDT&E aircraft, but we misconnected a yaw and a pitch channel in the fly-by wire control system for the first production airplane. It was impossible to fly with that misconnection and crashed immediately after takeoff. We put some procedures in place to deal with that. The next set of problems occurred when the aircraft was thought by some to be invisible, but was not. We had to do model development and integrate the models with the mission planning tools to be able to field the system.

One last big lesson that is very important has to do with the weaponization of the platform. Here we were, happily running along with this major multi-billion dollar investment when two young, bright people from our operational community came to me as the program executive officer and said, "What about a weapon for this airplane?" They were pointing out to me that we did not have

308

a suitable 2,000-pound penetrator weapon. These were two young, bright guys. One of them was Dave McCloud, now a general officer in our XO community; the other was John Casper, now a NASA astronaut.

For the measly sum of less than $10 million, we developed a penetrating front end for a 2,000-pound laser-guided bomb. It turned out that it was the key weapon used in Desert Storm. It was the weapon that made a major difference, but it was an afterthought in this program. It came very late in the development program, just before the airplane became operational. It is another lesson that we need to come back to and look at again and again.

Today, the environment we are operating in is much different. In the past, we were in an environment that was largely unconstrained by cost. We did not think about costs very much in the F–117 program; the drive was to have enough performance to defeat those air defenses. Today, we are in a much different environment. Cost has a place at the table, and we are required to better manage and deal with operational risk. But there is a big risk in the situation of overmanaging operational risk and being overly concerned with the limitations imposed by 90-percent kinds of solutions.

I think we must push very hard today to make sure that this risk management does not inhibit future breakthrough technology along the lines of the F–117. It is one of the reasons that I, personally, and the department are pushing so hard on stressing this idea of advanced concept technology demonstrations where, in some cases, we are using off-the-shelf technology to explore, with very modest investment, new operational concepts that may offer breakthrough capabilities. I do not expect that the majority will, but the investment will be minimal to allow us to be able to look at those kinds of opportunities and to sort them out. Also, we need to continue reforming our acquisition process to remove the barriers to effective and empowered program execution.

I did not talk much about it today, but the program management office for this program was truly empowered. It was a small team of people who met once a month. The price of admission to the meeting was that you could not go home and check with your boss. Decisions were made at the meetings, and the fluff attendees dropped off after about the second meeting. It was a businesslike, decision-oriented approach. Not every decision was correct, but every decision was timely, and those that were not correct were usually picked up in the next monthly meeting. Empowering our program management staffs to operate in that kind of a mode and operating lean is the key principle that I bring with me to my current job.

It is a pleasure to have had the opportunity to discuss just part of this development of low observables with you today. It is a small piece of stealth technology development that I had the personal privilege of observing and playing some part in over a seven-year period. There is much more here for us to bring out and learn from, and I hope to be apart of it as well.

Lt. Gen. Carl G. O'Berry, USAF (Ret.) was Deputy Chief of Staff, Command, Control, Communications, and Computers, Headquarters U.S. Air Force. He was responsible for operational policy and high-level management of approximately $16 billion in C^4I systems, including formulation of Air Force communications and computer doctrine, policies, and plans. Gen. O'Berry advocated and defended Air Force positions and resource requirements to all levels of review, including Joint Chiefs of Staff, Office of the Secretary of Defense, and Congress. The general enlisted in the Air Force in January 1957 as a communications specialist and served tours at Osan Air Base, South Korea; Maxwell AFB, Alabama; and Robins AFB, Georgia. He was commissioned through the Officer Candidate School, Lackland AFB, Texas, in December 1961.

Information Systems and Applications

Carl G. O'Berry

Information Technology, like most modern technology, has a rich basis in history. After all, man began exchanging information for a multitude of purposes almost as soon as his biological forbears stood upright—perhaps even before that, if our studies of primates today are valid indicators of primitive communication.

In a sense, lessons from history are no less interesting and informative with respect to information technology than to any other form of modern endeavor. Certainly, we can look back and say we could have done some things better, given a bit more foresight—just as we can about almost any human activity.

We should be able to apply the lessons of history to our advantage today, and we probably have done so, to some extent; but the current way of looking at information technology, with things changing so quickly, tends to drive us in to reactive, rather than reflective, processes.

I thought I'd try today to establish an information technology baseline in historical terms. Following that, perhaps I can address some of the processes and procedures being applied by the Air Force today, with the goal of delineating where we ought to put our investment and how we ought to deal with the rapid growth and potential of modern information technology.

Looking back, from a military perspective, one need not go too far to be in the age of carrier pigeons and semaphore. But there was a dawning realization, even at the turn of the century, that greater bandwidth and mobility was required to satisfy growing needs for command and control, logistics, and other military functions.

Taking a quick jump to more modern times, say 1945, shows the first use of primitive computers, like those at the Aberdeen Proving Ground. The primary function of those rudimentary machines was to calculate ballistic trajectories, since Aberdeen's principal interest had to do with rounds fired from tanks or heavy artillery. Another short time hop leads to the growth of radar, modern tracking systems—and look at what it all has evolved to: wide band communications, satellite systems, and the age of fiber optics, all within a few decades of the carrier pigeon!

Technology and the Air Force

Dr. David Sarnoff, in 1964, when computers really were not very well known, said, "A computer complex will be at the heart of a total system in which it is possible to achieve . . . effective real time command over any situation or combination of situations anywhere in the world." I will go through that a little bit more in a moment, but this is a very, very astute observation made by a fellow who is known for astute observations.

How did we get from 1945 to where we are now, in historical sense? I decided to go through decade by decade. I witnessed a lot of this personally, because I enlisted in the Air Force in 1957 and was a communications specialist. I started off as a teletype operator and was sent to Korea in 1957, and I am very familiar with that first line of "modern" military systems. About the only transoceanic information transfer capability available back in those days for was high-frequency radio.

We operated high-frequency radio teletype, high-frequency voice systems, and radio facsimile. Not very wide band stuff, but there were no undersea cables in 1957. One dealt with these things in terms of how many voice channels one could get on a radio, and typically it was three or four.

The 1960s to 1970s saw the beginning of ways to expand our military reach through wide band tropospheric scatter, the development of VHF tactical wide band systems, with new, very sophisticated multiplexing and with the increased reliability of space and time diversity and redundancy. The age of line of sight microwave dawned in 1960 or thereabouts. Coaxial cable systems began to show up in 1960 or so, and someone made an observation that the laser was a solution looking for a requirement. The growth of fiber optics and light wave communication over hair-thin glass wires began soon thereafter.

The 1980s and 1990s saw the dawn of what we refer to today as the information age. 1964 saw the birth of a little noticed modern wonder—digital packet radio—in the form of an exploratory system called the ARPANet. How many of us, even those of us who were involved in those early days of digital transmission, could have foreseen the evolution of the ARPANet, consisting of a dozen or so military laboratories and universities, into the all-pervasive system known today as the Internet? Even more to the point, who could have forecast the birth of the microcomputer, the missing link required to spawn that Internet?

It's generally accepted that information technology history prior to the year 1990 has been overshadowed by the awesome growth of the global Internet. Its millions of connected computers and a server structure that has seen exponential growth from a few hundred thousand users in 1990 to forty million today are continuing to expand at a rate of 180% per year on a global scale!

That leads us to today. From my Air Force perspective, I see information technology as enabling three grids: a terrestrial cable grid, which connects enterprises and facilities through coaxial, copper, or fiber optic cables; an earth-based, primarily earth-coverage radio system, which we call the terrestrial RF grid; and the extra-terrestrial satellite grid.

312

What is wrong with this in my view is that, because of some oversights as these grids developed, they are not integrated. Each one of the grids provides a different set of services and capabilities; but they do not interact as effectively or as efficiently as they should.

Have we entered the Information Age? You had better believe it. Between 1984 and 1992, the pattern of spending switched, in terms of capital investment in Industrial Age tools and facilities versus Information Age capital spending—that is, computers and the things that pertain thereto—a very, very dramatic shift in the way capital spending.

What does all that mean from an Air Force perspective? Well, it means that as capital spending in information technology continues to increase, we must develop a better understanding of the leverage and synergy made possible by such investment—not to mention the increasing rate at which new technology is being introduced.

We've tended historically to bring on new information technology on a catch as catch can basis. We bought thousands of computers, without thinking that the time would come when those boxes might be used to form interactive networks of computers. Then we created computer networks without thinking about optimizing the interfaces between networks of computers. And, worst of all, perhaps, we did all that without thinking too much about the requirements of the individuals using the computers and networks.

Only after the numbers of computers and networks grew very large did we begin to think about matching functionality with properly designed and architected information technology tools—both hardware and software—to address operational requirements in an integrated fashion. The result, of course, is a legacy of suboptimized networks that cannot operate together effectively.

As Dick Hallion pointed out earlier, the Air Force is engaged in a lot of investing, about $16 billion over the current program. I'm sure that you've heard this several times in the last couple of days. That's one of the interesting things about being the last guy to gain the stage in a forum like this—just about everybody who's been up here before has been talking about my business in one fashion or another.

What we need to have, in the final analysis, is the ability to transfer information at a very rapid rate between what I have tended to refer to as information appliances. But even that does not help much if the appliances are not properly designed to interoperate, and if the functions to be performed by the appliances are not interrelated through a carefully thought out architecture. That's the principal issue facing us today: we've gone about trying to define the application of information technology, the information transfer medium, and the functional requirements of today's military environment as though they were separate and distinct entities with separable objectives.

Desert Shield and Desert Storm demonstrated just how wasteful we can be with the currently limited bandwidth and services associated with information

313

technology. There was not a whole lot of cable in the desert, and we found that we simply could not keep up with the demand, despite the addition of substantial commercial satellite capacity and the movement of a Defense communications satellite into an Indian Ocean orbit from elsewhere. Desert Warriors said we were not providing them with sufficient bandwidth. One was tempted to respond: "The bandwidth you do have is being wasted. You simply do not have the wherewithal to take advantage of it."

What do I mean by that? Well, I mean that it is our mentality that determines how we state requirements and provide information services. What is the solution of choice for any communication requirement? It is a circuit from point A to point B. That is the way we were raised, and that is the way we think about communications: a circuit from me to [Dr.] George Abrahamson [former Air Force Chief Scientist] is required because I need to talk to George, or I need to send him a fax.

We put in a circuit to connect the chairman of the Joint Chiefs of Staff with the commander in chief in the field. Now, this is not a big bandwidth requirement; it took maybe fifty-six kilobytes because it was an encrypted voice circuit. It sat there all the time during Desert Shield and Desert Storm, being used about .05 percent of the time. It was hard-wired and the bandwidth could not be used for anything else. It subtracted fifty-six kilobytes of bandwidth from being applied to other requirements—not a very smart way of doing business.

But that is the way we think about communications today. That's the way the providers think about it; that's the way the world has a tendency to think about it. It does not make any difference whether it is 56 kilobytes, or a T1 circuit, or OC48, or whatever it is. It always goes from point A to point B and you pay for it, twenty-four hours a day, seven days a week, whether you are using it or not.

The typical T1 circuit, that is, 1.544 megabytes of capacity, gets charged to you twenty-four hours a day, seven days a week, because that is what you require occasionally, when you really stress the communications process. You use it typically, even in a high-density environment like the Persian Gulf, about 30 percent of the time, not a very efficient way of doing business.

I don't care whether your information appliance is a JSTARS airplane, or a U–2, or what it is. The appliance and the requirement may change form, it may be combined with others, it may evolve—a lot of things may happen. But we must overcome the current inclination to connect everything to everything else with fixed bandwidth circuitry.

No circuit should exist until it is required, and when it is required, it should be created virtually, with sufficient bandwidth for the purpose at hand, which is usually a short-term requirement. Bandwidth can be reusable! Bandwidth resources can be made to serve one purpose, then be put back into a pool of temporarily idle bandwidth resources to be applied against another requirement. The user need not be aware that the network is operating thus; in fact, to the user, it can be made to look as though he is continuously connected. We have attempted to do this, as I alluded to a little bit earlier in talking about modeling and taking the battlefield environment and looking at it in terms of how one ought to invest in this.

Suffice it to say I started out trying to characterize the Air Force not as a command and control entity or as a war-fighting entity. Other people are doing that all the time, so the information guy did not need to do so. Instead, I tried to characterize the enterprise in terms of its information flow requirements. That is, the mission and support information transfer needs of the Air Force.

I have alluded to this many times in a way analogous to a bunch of doctors of different specialties looking at Gray's Anatomy. Depending on the specialty involved, a doc opens Gray's Anatomy, and sees a mark one, mod one human-being (well, a couple of varieties) differently than his or her peers. If you are a cardiologist, obviously you see the things that cardiologists are interested in. If you are a neurologist, you see dendrites and neurons. If you are a bone guy, you see muscles and joints and so forth, but we are all talking about basically the same model. It is just that we have to be able to characterize it in different ways if we are going to treat different aspects of it in some fashion that makes sense.

That's what we are trying to do with the view of the entity related to information: define the Air Force in terms of its information flows. What does the Air Force enterprise look like as a series of information nodes and information flows across those nodes?

We found that it was not as difficult as we feared at first. I was interested, from the Headquarters, United States Air Force perspective, in doing that with a manageable number of nodes and links. I did not need to define them in detail, because remember, I was trying to get away from the circuit mentality. I was trying to think in terms of who needs to talk to whom. Who needs this from here to there, and how do you characterize that flow? We found we could do that fairly easily. We found we could characterize the Air Force in these terms with about 137 major nodes! Now, obviously, beneath that, if one picks the skin off and looks more closely, it decomposes to a lot more detail as one works down through the major command to the operations center to the combat unit to the weapon system to the janitor. But the overarching architecture needed to bring order out of the legacy of chaos can be, and has been, defined. So, too, has a small set of minimal constraints upon the development of nodes and links below that top level that will ensure that the whole entity will work.

We wanted to create an environment where an individual can operate day for day with a set of procedures and equipment and, when "the balloon goes up" he reaches under his desk, pulls out a packing case, puts his machine in there, climbs on an airplane, and goes to war. When he gets to the other end, he takes the appliance out of the packing case out, plugs it in, and continues to operate. It's the same person, the same procedures, and the same equipment, operating in the same way as he did back home. Because it does not make sense to operate one way in peacetime, then deploy and have to do things entirely differently. What we are talking about here is something that is extendable all the way down to the flight line, to the munitions loader, to the cook, to the medic, to everybody else.

That kind of approach to things reveals that every one of those nodes can be characterized, through a process called intersection analysis, to establish the information flow requirements across each node—in the same way as a college professor uses nodal analysis to define electric current flows. It's basically the same, whether it's a voice, data, or video transfer; it's all zeros and ones today.

What I'm interested in is making sure that the required zeros and ones get through the node and arrive where they're needed, on time, every time. When this architecting job is done, the result is an interoperability table, or a table of attributes, associated with each of the nodes. So, we have built what we call the Air Force Horizon on this basis, to steer the investment leading toward an efficient information technology model and, for once, to get out in front of the technology horse race.

The Horizon model is readily extendable from the Air Force to other domains. All that's required is to start with a basic model and add the Air Force mission areas, combat operations, intelligence support, mission support, mobility operations, and so on. If I wanted to make it an Army model, I could change the operational elements to fire support and maneuver units, that sort of thing. The process is still viable; it works exactly the same way, but then has an Army flavor. The model holds for the Joint world, too, and could be extended to the whole Department of Defense.

What is the significance of all that? Why are we going to so much trouble to do this? We started out with a number of interesting notions here, but if one thinks of information technology capability on the vertical axis and time on the horizontal axis, the inclination has been to look at capability as growing in some sort of linear fashion. The reason for that, I profess, is that we do not live very long, and we have a tendency to see a fairly small part of the spectrum. But interestingly enough, information technology is changing so fast that we can see things happening in a much more rapid sense than we did before.

The reality, in my view, is for capability growth to progress along a more or less linear line until some sort of "critical mass" is attained, at which point, capability improvement suddenly becomes exponential. A good example is the Internet, to which I alluded earlier. In 1964, ARPA created something referred to initially as the ARPANet. Its purpose was to demonstrate the efficacy of digital packet transmission and switching. That early network had twelve nodes, located at universities and government laboratory installations. What became of it probably surprised its originators as much as everyone else. From the early to the mid-1960s until 1991, the population of ARPANet users grew gradually as more and more hosts—largely from universities again—were added to this net.

It was found that the efficiency of intercomputer exchanges of information was growing in ways that were disproportionate to the links associated with the network. Now, that did not come as any surprise to those who designed it, because that was what it was all about to begin with.

In 1991, however, something began to happen that astounded everyone. At that time, the population on the Internet had grown to perhaps 375,000 steady users. In 1994, it is reported that about 30 million people were using the Internet routinely. With all these nodes coming on with ".com" instead of ".edu" the commercial world is taking it over, along with the government—you have ".gov" there, too, on a lot of home pages. The growth of use on the Internet became almost an instantaneous and very steep exponential curve.

More to the point, perhaps, from a technology perspective, is the notion again that we think about capability growth in linear terms. Moore's law is named for the guy who started the Internet. He projected in 1975 or 1980 that the power of computer chips would double about every eighteen months. If one looks at the back at the 4004 CPU, moving all up through the Pentium, that is kind of the way it happened. Now, it has shortened up a little bit between the Pentium and the P6.

It causes me to wonder if we have not observed the linear growth period, and with the advent of the P6, which is starting to take on some characteristics of massive parallelism in its operation, we may have reached the critical mass point. There were some early projections that a wall was being reached in terms of a limit in the number of devices that a silicon chip could support, and of course the more devices the more power and speed on chips. It was sort of like running a marathon; at nineteen miles, you are supposed to hit the wall, right? So you can't really get much faster or much more energy out of the system once you reach that wall.

But the wall in silicon has proven to be a myth. The P6 has a .35-micron space between devices, a fraction of a P5, the Pentium. My guess is that we are about to see an exponential jump in terms of CPU capabilities, with interdevice spacing in the range of 0.1 microns or less. There is some evidence to that effect on the street now.

We ought to be taking lessons from history and deciding how to deal more effectively with all these info-tech things. Let me bring it back out of the military domain and talk about it in just you-and-me terms.

It seems to me that what we need to do is make the network very smart, in terms of providing information transfer, that we ought to have an information transfer utility, a global one that is not unlike the electrical power system today. The analogy is not perfect, but it is close enough. Out there is a huge power grid that you, as individuals, unless you happen to be in the business or happen to be an engineer who is interested in this sort of thing, do not have a clue how the heck it operates, how the electrons get from wherever they originated to that wall plug right there. But you do know that the electrical power system does not do the work of electricity. It provides the capacity, and right there is a Mark I model on human interface to the power grid. Interesting, is it not?

I am one of those people who drives down the road and looks for the transposition points in the power lines just because I am an electrical engineer. An electrical engineer is supposed to do wacky stuff like that, but you do not know whether the potential at that plug is derived from Niagara Falls or Three Mile Island. It is a complex system, and it requires a lot of smarts because it does not cost the same to generate power hydroelectrically as it does through a nuclear power plant or a fossil fuel plant.

There are a lot of things that have to take place here in terms of translating to a fair rate the charge to the consumer for use of electrical power. Furthermore, there are thousands of companies involved in this process. Some of them are power generation companies, some are transmission companies, and some are distribution companies. But the fact is the grid works very well. You plug into it at the power substation, which may be a 200-amp service entry into your house, or it may be a very much larger one, and you consume electrons by plugging appliances into it. The appliances are what do the work of electricity, taking advantage of electrical power capacity to generate useful products and work for you and me.

You are charged only for what you consume. So you do not have to pay for 200 amps twenty-four hours a day, seven days a week, just because you have the capacity, like you do for a 1.544-megabyte T–1 circuit. You are only charged for what you consume.

Now, why cannot the information grid work the same way? Whatever level you need, it should be there. It could be brought down to the enterprise level, the user's level, through some sort of a substation; meter the packets going in or coming out, or both, I do not care how you do it. Instead of having a 200-amp service entry, you might have a gigabyte service entry, a common interface level, and appliances could use that information capacity to produce useful work. That is where we need to be, on a global scale, to make the picture I talked about earlier become reality. Maybe what we end up with is what I have referred to as a fourth grid: but what I really mean is a single integrated environment that is no longer necessarily a geocentric grid. You could put high earth orbit satellites up there, two of which could cover the entire earth and could cover the entire plane of the solar system. It might be worth thinking about doing it that way. This, then, could really be what the Internet is all about.

In December 1945, Dr. Theodore von Kármán, in his letter accompanying *Toward New Horizons*, wrote, "The men in charge of the future Air Forces should always remember that problems never have final or universal solutions, and only a constant inquisitive attitude toward science and a ceaseless and swift adaptation to new developments can maintain the security of this nation through world air supremacy." I think we can apply a lot of this to meet the challenges of the future in terms of information technology. My concern is that if we do not move out smartly with some of these things, get out in front of the power curve in terms of the capability associated with information technology, we continue to be reactive and driven by it, as opposed to taking advantage of it.

I think von Kármán's words apply just as well today as they ever have. I would like to see us be a bit more circumspect in our stewardship of the resources made available to the United States Air Force.

This is what I derived from my historical look at information technology. I will be glad to answer any questions you might have, and I hope that this has been useful.

Summation

Richard P. Hallion

It is difficult to sum up what we have had over the last two days. If we think about it, we have had a very rich dish. We have had a number of presentations in a variety of areas about science and technology, what they mean to the Air Force, and some of the lessons that we have learned. Hopefully, some of the lessons we will continue to call to heart.

I will just throw out a few that we might want to keep in mind in the future. I think the one thing that has come through very strongly in all these papers is that planning must be constant. It must be historically rooted, and it must be tied to defined military needs. When you do not have that, you tend to set yourself up for some real trouble. Tied to this is the appropriateness of technological choices. We saw in the fields of jet engines and supersonic flight tremendous opportunities and tremendous challenges, and sometimes the choices that people made were not necessarily the most appropriate ones.

Another point that I think has come through very strongly is that successful technologies have generally advanced far more rapidly than their adherents would have thought possible or would have claimed possible at the time that they were being developed. C^4I is a good example of this. Other examples include fly-by-wire flight control technology, precision weapons technology and, for that matter, jet aircraft. Thinking of the capabilities of the modern jet airlift and where we came from illustrates something we need to keep in mind.

The danger of unwarranted technological optimism is another point well worth contemplating. This goes back, I think, to the notion of appropriate planning and the use of appropriate planning methodologies and tying planning to requirements. We heard about some programs that were afflicted by technological optimism. Again, supersonic flight is one of those. More noticeably, or perhaps more significantly, the story of nuclear-powered flight clearly indicates this. The field of hypersonics has also been afflicted with this. We did not hear about this in the symposium, but we will have the opportunity to see it in the written proceedings of this conference.

Another factor that comes through, if we listened to the subtext coming through the presentations and the papers, is the critical dependency on an appropriate laboratory structure. With the tremendous drawdown governmentwide today, not merely militarywide, but governmentwide, the future of the U.S. laboratory structure causes some concern. Obviously, to keep the kind of technological edge that we as a service feel is very necessary, and looking at the track record of Air Force science and technology historically, it is very important that the Air Force very much keep its laboratory structure intact.

Another approach, or idea, that comes through the presentations is the importance of thinking of new technology and its incorporation in military systems and military operations and the importance of thinking in terms of the total system. An important example of this that came out of Paul Kaminski's talk is the idea that if you develop a stealth airplane, it is nice to have something to drop from it. A stealth airplane going to war with the weapons that it might otherwise have had obviously would not have been nearly as effective as one going to war with hardened, penetrating, precision-guided munitions.

Another aspect we need to keep in mind is that there are a number of ideas, usually of longstanding origin, that have profound implications, and while they may be quite simple, they may also be challenging to pursue. We heard about several of these simple ideas with profound implications that really gave the Air Force tremendous leverage: air refueling, smart munitions, GPS, AWACS, and JSTARS.

We have to recognize that, when those technologies prove themselves, we have to be very concerned that, by their success, they might immediately generate counters. I think Bill Holley alluded to this in his luncheon talk the other day. For example, one lesson a likely opponent of ours might gain from the Gulf War is to deny us the medium-altitude attack environment. If we can be forced to make our attacks from a greater standoff distance, it is going to have a serious impact on the leverage that we get with precision munitions.

The AWACS and JSTARS are two very, very valuable platforms to us. At the same time, simply because they are so valuable and have demonstrated such capabilities, they are also now highlighted as very serious threats that must be countered by an opponent. We need to think about what we want to do to render them less vulnerable or, perhaps, to render them less vulnerable by going to a different apparatus for conducting such operations.

I think another point that is very important, and really the last point I want to make, is that it is even more important now than it has been not to miss the next significant revolutions. Because if we look at the lengthy development times that take place in science and technology today, it is very, very important to enter the arena and exploit these ideas early enough that we do not find ourselves outclassed by our competitors. We simply do not have the time any longer to play catch-up. The price is simply too great. Take, for example, the history of low observables. We know that low observable theory was at least available in the mid-1960s, and that it could have been used by our potential opponent to possibly give them a leverage to use against us by the early 1970s. It really is a very cautionary tale, if you think about it.

This proves that we do not have to radically transform the nature of the threat that you confront. If we simply improve our performance on the margins that are at a critical point, just that marginal effectiveness or that marginal improvement may be enough to do the job. Certainly, that was a relatively close-run experience for us.

Today we have information dominance or information warfare. The concern is whether these are mere shibboleths or buzzwords that people are coming up with. But if we take a look at the state of information exchange in society and the implications that it has for national security — as General O'Berry has just pointed out — it is really profound.

What does this all mean? I think, if we look back on this century, that we are going to remember that this has really been the century of three dimensionality, both commercially and in military operations. We move now as a three dimensional people. Take a look at the revolution in military affairs effected in this century by the submarine and the airplane and then take a look at the impact the revolution in commercial aviation has had on transportation.

To put this in a way we can relate to, the steamship came about in the middle of the nineteenth century. In 1958, after only a very few years of trans-Atlantic air travel, and really only one year of trans-Atlantic jet travel, the steamship was put out of business as the major carrier on the North Atlantic run. Consider the railroad, once again a mid-nineteenth century system. By mid-1950, again after relatively few years of transcontinental air travel, more people were travelling across the United States by plane than by train. Technologies tend to revolutionize things very, very quickly.

Those of you who were fortunate enough to hear Fred Frostic last night would certainly agree that there is indeed a new calculus in military affairs, or a newer calculus, as he terms it. It is no longer possible to say with the certainty that people once did that surface warfare forces are the primary means whereby a nation secures victory in war.

We have not been yet as successful publicly as we could be in getting that message across to the outside world. Indeed, we have not been that successful in getting that message across to senior leadership, even within our own service. For example, a few weeks ago, a senior retired Air Force officer was on *Meet the Press*. He stated in response to a question on air power in Bosnia, "Well, you know, air power has some serious limitations. It cannot seize or hold ground."

Well, that is true. But neither can artillery. But both are the gatekeepers and control what moves on the surface. Air power controls who has access to the battlefield. It controls who can move across that surface and their degree of mobility. And we certainly saw this in the Gulf War and subsequently in Bosnia.

Changes may occur because of location, terrain, or weather, but they do not affect the ultimate result. They simply change the time scale of the application.

Technology and the Air Force

Ours, I think, is the most technologically dependent of all the services. We are a service that really defines the expression high technology, and for that reason, we cannot afford to pursue outdated technology.

Our technology choices have to be appropriate. We cannot afford the luxury that we have sometimes seen in other services in which their technological choices have been more traditional and, indeed, archaic. For example, the continued reliance by the United States Navy on the large carrier battle group, despite some of its obvious limitations, and the continued reliance of the United States Army on mechanized vehicles at a time when the future of the mechanized vehicle is in doubt as an arbiter of strength on the battlefield.

We in the Air Force cannot afford that. We cannot be a service that is simply trapped by the notion of using the airplane or the notion of using the missile. We must look for new and creative ways to use this technology. Coming out of the symposium that we have had over the last two days, one overriding lesson we see is that we must be as bold in our thinking as our profession has been over the years.

Thank you all very much for taking the time to come, and I think I speak for all of us when I say the last two days have really been quite remarkable.

Index

*Numbers in **bold** indicate illustrations.*

Index

Index

Index

Index

www.ingramcontent.com/pod-product-compliance
Lightning Source LLC
Chambersburg PA
CBHW080758180526
45168CB00006B/2249